D1524395

Solar Neutrinos and Neutrino Astronomy
(Homestake, 1984)

AIP Conference Proceedings
Series Editor: Hugh C. Wolfe
Number 126

Solar Neutrinos and Neutrino Astronomy

(Homestake, 1984)

Edited by
M. L. Cherry and K. Lande
University of Pennsylvania
and
W. A. Fowler
California Institute of Technology

American Institute of Physics
New York 1985

L.C. Catalog Card No. 84-63143
ISBN 0-88318-325-0
DOE CONF- 8408132

PREFACE

Twenty years ago, in the fall of 1964, Brookhaven National Laboratory and the Homestake Mining Company agreed to create a deep underground neutrino laboratory, one devoted to investigating the energy-generating mechanisms that operate in the interior of the sun. Here Raymond Davis built an innovative detector, utilizing the neutrino-induced chlorine-to-argon transition. The observations that he carried out for the past two decades have profoundly changed not only our understanding of the energy generation in stars, but also our approach to investigating the hitherto invisible neutrino world around us.

The Homestake detector utilized an amazing neutrino detection technique, the conversion of a liquid (the chlorine in perchloroethylene) into a gas (argon) by neutrinos. This technique was first suggested by Pontecorvo in 1946 and then independently proposed by Alvarez in 1949. The recognition by John Bahcall of the existence of an excited state in argon that is the analog of the chlorine ground state, and the development by Ray Davis of the extraction and counting techniques necessary to detect a few atoms of argon in hundreds of tons of perchloroethylene, completed the steps necessary to establish the solar neutrino detector.

But without a doubt, the crucial element in carrying out the investigation of neutrino generation in the Sun was Ray Davis. It was Ray's quiet, forceful determination to succeed, his devotion and love of science, his careful and thorough investigation of all possible sources of error and of backgrounds, his careful design of the detector and meticulous execution of the experiment, that produced such a beautiful and convincing measurement.

During the past two decades Ray has had a number of excellent and devoted associates: Don Harmer and John Galvin, who were associated with the detector in its early days, John Evans, who joined Ray in the early 1970's, and Bruce Cleveland and Keith Rowley, who are working with Ray today. It is most unusual in today's world of very large collaborations to find such a monumental experiment carried out by one or two scientists. It is equally unusual to find another example of a single observation that has had as much impact on science as has the Homestake solar neutrino program.

The solar neutrino program could not have been carried out without the continuous, enthusiastic, and generous support of the Homestake Mining Company. This support existed at all levels in the company from the chief executives, Donald McLaughlin, Paul Henshaw and Harry Conger, to the Lead operations management, Don Delicate, Joel Waterland, Al Winters, Jim Dunn and Al Gilles.

As we set out to prepare this program and edit these proceedings, we are impressed by the large number of theoretical and experimental ideas on solar energy and neutrino generation and detection that are being developed. The future looks very impressive. No doubt the results of the next two decades will answer the questions we pose today and in turn will leave us with a deeper set of questions about the inner workings of stars and their role in the evolution of the universe. Neutrino astronomy has not only a stimulating present, but a most exciting future.

One of us (K.L.) would like to take this opportunity to thank Ray Davis for his kindness, understanding, help and guidance over the past dozen years and express his strong wishes that Ray Davis' future work at Homestake will be as much fun as the past has been.

This conference could never have been held without the tremendous help and support of the towns of Lead and Deadwood, South Dakota, and the generous support of the staff of the Homestake Mining Company. In particular, we would like to thank Bill Ausmann, the Principal of the Lead High School, for providing us with an excellent conference site and attending to all the details necessary to make us feel at home, Sam Grover, Norm Bakke and Joe O'Dell for conducting the tours of the underground laboratory, Jim Dunn and Al Gilles for helping in every step of the planning and conference arrangements. C.K. Lee, Steve Corbato, Dave Kieda and Dennis Hothem, as usual, did all the unusual required to have a smoothly flowing and effortless conference. Elizabeth Fulton and Lisa Granata did a magnificent job in assembling these proceedings. Finally, we would like to thank the "Homestake" wives and families who tolerate our activities and who provide the real and crucial support.

Table of Contents

IV) OTHER SOLAR NEUTRINO DETECTORS

THE CHLORINE SOLAR NEUTRINO EXPERIMENT

J. K. Rowley, B. T. Cleveland, and R. Davis, Jr.
Brookhaven National Laboratory, Upton, New York 11973

ABSTRACT

The chlorine solar neutrino experiment in the Homestake Gold Mine is described and the results obtained with the chlorine detector over the last fourteen years are summarized and discussed. Background processes producing ^{37}Ar and the question of the constancy of the production rate of ^{37}Ar are given special emphasis.

INTRODUCTION

The first underground experiment in the Homestake Gold Mine was the chlorine solar neutrino experiment, which was initiated nearly 20 years ago. The so-called solar neutrino problem is the discrepancy between the results of this experiment and the result predicted by solar model calculations using the best available input physics, i.e., by the standard solar model. The neutrino capture rate in the chlorine detector calculated using the standard solar model has changed with time as new data have become available. However, since 1969, in spite of great effort producing many new and improved measurements of nuclear reaction cross-sections, new opacity calculations etc., the capture rate predicted by the standard solar model has not changed in a major way. The chlorine detector has been operating regularly since 1970 with 61 experimental runs completed at the present time. The purpose of this paper is to give a brief description of the chlorine solar neutrino experiment and to discuss at greater length the results of these experimental runs.

The standard solar model is based upon the set of nuclear reactions shown in Table I.

Table I The proton-proton reaction chains

	Reaction	Neutrino Energy in MeV
	$H + H \rightarrow D + e^{+} + \nu$ (99.75%)	0-0.42 spectrum
	or	
PPI	$H + H + e^{-} \rightarrow D + \nu$ (0.25%)	1.44 line
	$D + H \rightarrow {}^{3}He + \gamma$	
	${}^{3}He + {}^{3}He \rightarrow 2H + {}^{4}He$ (87%)	
	${}^{3}He + {}^{4}He \rightarrow {}^{7}Be + \gamma$ (13%)	
PPII	${}^{7}Be + e^{-} \rightarrow {}^{7}Li + \nu$	0.861 (90%) line 0.383 (10%) line
	${}^{7}Li + H \rightarrow \gamma + {}^{8}Be \rightarrow 2\,{}^{4}He$	
	${}^{7}Be + H \rightarrow {}^{8}B + \gamma$ (0.017%)	
PPIII	${}^{8}B \rightarrow {}^{8}Be^{*} + e^{+} + \nu$	0-14.1

In addition to these reactions the reactions of the CNO cycles make a small contribution to energy generation. If the sun operated solely on the CNO cycles, the predicted neutrino capture rate would be very high, 35 SNU, and readily calculated. Using the PP reaction chains shown in Table I and the CNO reaction cycles, a neutrino spectrum of the standard solar can be calculated. The fluxes of neutrinos from various sources in the sun as predicted recently[1,2] from standard solar calculations are listed in Table II together with the resulting ^{37}Ar production rates in the chlorine detector.

Table II Neutrino fluxes and ^{37}Ar production rates predicted by the standard solar model

Neutrino type	Flux at earth, 10^{10} cm^{-2} sec^{-1}*	Production rate of ^{37}Ar In SNU	Production rate of ^{37}Ar In atoms day $^{-1}$ in chlorine detection
pp	6.1	0	0
pep	0.015	0.23	0.04
^{7}Be	0.42	1.0	0.19
^{8}B	0.00046	5.0	0.93
^{13}N	0.05	0.08	0.01
^{15}O	0.04	0.26	0.05
Total	6.63	6.6	1.22

1 SNU is defined as one capture per sec in 10^{36} target atoms.
* Fluxes and capture cross-sections are from references 1 and 2.

Because the energy threshold for the reaction: $^{37}Cl + \nu \rightarrow {}^{37}Ar + e^{-}$ is 0.814 MeV and because of the dominant contribution of the transition to the bound isobaric analog state of ^{37}Ar, the capture rate in the chlorine detector is very sensitive to the internal temperatures in the sun. Therefore, the calculated rate depends critically on the assumed reaction cross-sections, opacities, internal composition and dynamic processes in the sun. The fact that the chlorine experiment is very sensitive to these factors was a strong argument that was used to obtain support to build the experiment in 1965.

DESCRIPTION OF THE EXPERIMENT

Since a detailed description of the experimental facility and procedures has been given elsewhere,[3] we will only outline them here and mention recent improvements and tests.

Outline of Procedure. The neutrino target is 2.2×10^{30} atoms (133 tons) of $^{37}C\ell$ in the form of 3.8×10^{5} liters of liquid perchoroethylene, $C_2C\ell_4$. The tank containing this material is located at the 4850 foot level underground in the Homestake Gold Mine. The ^{37}Ar produced in the tank is separated at intervals averaging about 80 days, purified, and counted. Experimental runs were carried out on a rather irregular schedule during the period 1969-1975. During 1976 an attempt was made to perform measurements more frequently, every 35 to 50 days. Since that time measurements have been made on a nearly regular basis of six runs per year. In 1982 the schedule was modified because of a strike at the mine. Occasional special runs have been performed to look for increased neutrino fluxes from various astronomical and solar flare events -- the Lande event in 1974 (Run 32), solar flares (Runs 27, 35 and 54), novae (Runs 38 and 46), and the recent large x-ray flare on May 25, 1984 (Runs 84 - still counting).

The steps in the experimental procedures are:

1. After the end of a run approximately 0.2 standard cc of either ^{36}Ar or ^{38}Ar is added (alternately) to serve as a carrier for the ensuing run.

2. The tank is exposed for the desired length of time.

3. After exposure, the argon in the tank is removed by circulating about 4×10^5 liters (1 tank volume) of He through the gas and liquid phases in the tank, then through a condensor at -32 °C and a molecular sieve at room temperature, and finally through a charcoal trap cooled to the temperature of liquid nitrogen where the argon is absorbed. Gas circulation is accomplished and the tank is stirred by two large pumps, each connected to an educator system. About 95 percent of the argon is removed (and retained on the charcoal trap) by circulating this volume of helium.

4. The argon is transferred from the heated large charcoal trap to a line where it is purified, its volume measured, and it is loaded into a small (0.3 - 0.5 cc) proportional counter along with tritium-free methane which serves as a counting gas. The counters containing experimental samples are calibrated approximately every two months.

5. The sample is counted for approximately 8 months and often longer.

6. The carrier yield is determined by mass spectroscopy.

In order to show that ^{37}Ar, produced in the form of a recoiling ion by the low energy neutrino capture process becomes a neutral Ar atom, a special test was performed.[4,5] Tetrachloroethylene labelled with $^{36}C\ell$ decays by beta-minus emission to ^{36}Ar:

$$C_2C\ell_3{}^{36}C\ell \rightarrow C_2C\ell_3{}^{+}(\text{fragments}) + {}^{36}Ar^{+} + e^{-} + \bar{\nu},$$

the dynamics of this process are nearly equivalent to the inverse beta process $^{37}C\ell(\nu,e^{-})^{37}Ar$. The Ar^{+} ion produced is believed to become a neutral atom[6,7] which should be recovered with the carrier ^{38}Ar. After standing a definite length of time the quantity of neutral ^{36}Ar produced was determined by activation analysis after separation using a helium purge. In this experiment (performed by Evans, Vera Ruiz, and Davis) the ^{36}Ar was recovered with 100 ± 3% yield. A similar experiment was performed in the Soviet Union with the same result.[8]

The ultimate test that the ^{37}Ar is quantitatively recovered would be to place a neutrino source of known neutrino intensity and neutrino energy in or adjacent to the chlorine detector. A 1.0 megacurie source of ^{65}Zn would be required to obtain an ^{37}Ar production rate of 4 ^{37}Ar atoms day^{-1}, a rate ten times higher that presently observed in the detector (presumably from solar neutrinos). In 1981-1982 a trial irradiation was performed in the HFIR reactor at Oak Ridge to measure the ^{65}Zn production in zinc (^{64}Zn is the 48.6% abundant). This test irradiation indicated that only 0.5 megacurie could be prepared at HFIR and further tests at HFIR were abandoned. A measurement of the neutrino capture cross-section using the spectrum of neutrinos from μ^+ decay could test the cross-section calculations to all excited states.[2] An experiment was proposed[9] to carry out this measurement at the Los Alamos Meson Physics Facility and the fast neutron background at the experimental site was measured. This experimental test is now entirely feasible and we believe this is the best way to verify the calculations of the neutrino capture cross-sections for $^{37}C\ell$.

Counting. The pulses in a proportional counter originating from the low-energy Auger electrons resulting from decay of ^{37}Ar differ from most electrons from β decay by having a short pulse rise-time. Consequently, both energy and pulse rise-time can be used profitably to characterize ^{37}Ar decays. Figure 1 shows a plot of pulse rise

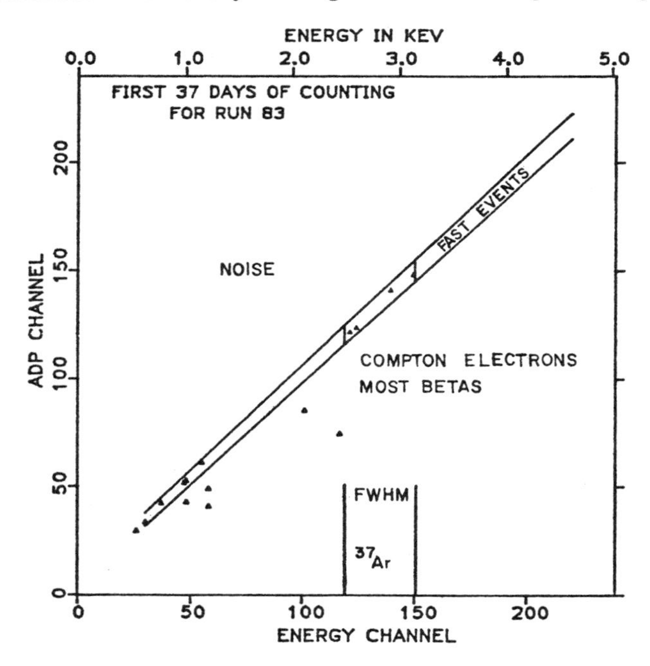

Figure 1. Plot of pulse rise-time versus energy for the first counting period of Run 83.

time (called ADP for amplitude of the differentiated pulse) versus pulse height for the first counting period of a very recent run (Run 83). The areas where pulses with different origins appear are labelled in this figure. An area within which 68% of ^{37}Ar decay pulses are located is defined by a calibration procedure with an ^{55}Fe source. This area is shown in the figure as the box containing four events. Although the counter background is greatly reduced by defining this area, some of the contained counts are background counts.

Over the course of this experiment many improvements in the counting have been made. Reductions in background have been achieved by using low radioactivity material in the counters and other counting components, by using a large well-type sodium iodide crystal as an anti-coincidence shield, and by moving the counting apparatus to the underground laboratory at Homestake.

Analysis of the Counting Data. The counts obtained in a given run within the proper pulse-height range and the proper ADP range are analyzed by the method of maximum likelihood.[3,10] For each run the counts are assumed to arise from a constant background and a constant production rate yielding a single decaying component. Except when it is desired to determine the half-life of the decaying component, this half-life is assumed to be equal to 35 days, that of ^{37}Ar. From the recorded times of all counts with the proper pulse-height and proper pulse rise-time a likelihood function is calculated. This likelihood function includes fluctuations in ^{37}Ar production, fluctuations during extraction and processing, and fluctuations during counting. The most likely values of the production rate and of the background rate are those that maximize the likelihood function. Results from many runs are combined by multiplying individual likelihood functions to give a combined likelihood function.

This method for treating the data has been used since 1977. It has been extensively tested by Monte Carlo simulations with varying input production rates and background counting rates. The results of these simulations show that the combined most likely values of the ^{37}Ar production rate and of the counter background state agree well with the input values.[10]

BACKGROUND PROCESSES

Once the total production rate of ^{37}Ar in the detector has been calculated using the maximum likelihood method, there remains the problem of subtracting any ^{37}Ar production from known non-solar sources. Four principal known background sources underground are known. These are: 1) cosmic ray muons and products of their interaction ($\pi^{\pm}$, energetic protons and neutrons, and evaporation protons), 2) fast neutrons from the rock wall by way of (α,n) reactions and from spontaneous fission of ^{238}U, 3) alpha particle interactions from uranium and thorium in the perchlorethylene, and 4) cosmic ray neutrinos. The background ^{37}Ar production rates from alpha particles[3] and from cosmic ray neutrinos[11,12] are both estimated to be quite small. We will discuss the other two background sources at greater length.

Underground muons, produced by cosmic rays in the upper atmosphere, penetrate deeply underground and produce cascades containing energetic pions, protons and neutrons together with evaporation protons which ultimately produce ^{37}Ar by the reaction ^{37}Cl (p,n) ^{37}Ar. The magnitude of this background process for the chlorine detector has been estimated[12,13] by exposing three 600 gallon tanks of perchloroethylene at higher levels in the mine, measuring the ^{37}Ar production rate at each of these levels and extrapolating these production rates to the depth of the chlorine detector. In making this extrapolation it is necessary to know the effective depth of each exposure site. Recently we have carefully calculated again the effective flat surface depth, in hg cm^{-2} of standard rock, of each location where these tanks were exposed and of the chamber containing the chlorine detector.

These calculations were performed in the following manner. We first read from topographic maps the surface elevations at evenly spaced points on the circumferences of concentric circles centered on a point directly above the site of interest. These readings were carried out to a horizontal distance from the center of the circles which was large enough that the largest circle included more than 90% of the total muons passing through the site of interest. Rock samples were collected at various levels near the vertical Yates shaft and their densities were measured. The average densities of the rocks in a given rock formation agreed well with those tabulated by the Homestake Mining Company. Weighted for rock formation thickness, the resulting overall average density above the neutrino chamber is 2.84 g cm^{-3}. Using the overall average rock density for each site together with the known elevations and zenith angles and the dependence on depth and zenith angle of the underground muon intensity, we performed a numerical integration to obtain the total muon flux at the site. Working backwards we could then obtain the equivalent flat-surface depth (in hg cm^{-2} of standard rock) of the site. For the neutrino chamber an additional correction was made for the average Z, $\overline{Z}$, of the overhead rocks. Based upon some determinations by the Homestake Mining Company of typical rock compositions in different rock formations, $\overline{Z}$ is equal to 10. Consequently, a multiplying correction factor[14] of 0.97 was used to correct for the fact that $\overline{Z}$ was not equal to 11, that of standard rock.

The resulting effective depth of the chlorine detector is $4.0 \times 10^{-3} \pm 200$ hg cm^{-2} of standard rock. The largest source of error is uncertainties in the rock densities. This uncertainty in effective depth translates into an uncertainty in the cosmic ray induced background of about 30%. However, since the vertical flux of muons in the chamber containing the chlorine detector has been measured (see later discussion), this is probably not the best way to estimate this error.

The results of all of the new effective depth estimations are shown in Table III along with the old depth estimations and the measured ^{37}Ar production rates in C_2Cl_4 scaled to the volume of the chlorine detector.

Table III Newly calculated depths of sites where cosmic ray induced ^{37}Ar production rates were measured

Mine level (location)	Depth in hg cm^{-2} of standard rock old	new	^{37}Ar production rate in 10^5 gallons of C_2Cl_4
300 foot (600 feet from Kirk portal)	290	254	320 ± 50
300 foot (1100 feet from Kirk portal)	405	327	178 ± 25
1100 foot (adjacent to Yates shaft)	1080	819	24 ± 13
4850 foot (neutrino chamber)	4400	4000	0.07 ± 0.03*

* extrapolated value

The result of the new extrapolation of the measurements at the other locations listed in Table IV to the depth of the neutrino chamber, 0.07 ± 0.3 ^{37}Ar atom day^{-1}, is essentially the same as the older estimate.

Another way to calculate the muon induced ^{37}Ar production rate is to utilize a measured value[15] of the vertical muon flux in the neutrino chamber (equal to 5×10^{-9} cm^{-2} sec^{-1} $sterad^{-1}$) to calculate the total muon flux which must then be multiplied by the calculated ^{37}Ar yield per muon appropriate to the average muon energy in the neutrino chamber. This average yield per muon may be estimated by interpolation from the results of calculations of Zatsepin, Kopylov, and Shirakova.[16] The final result of such a calculation is a muon induced background ^{37}Ar production rate equal to 0.09 day^{-1}, in good agreement with the value given in Table IV. We adopt the value, 0.08 ± 0.03 day^{-1} (0.4 ± 0.16 SNU) for this production rate.

A more extensive discussion of cosmic-ray induced background is presented in this volume in the paper by Fireman et al.

Fast Neutrons. Another background source of ^{37}Ar is fast neutrons which can produce ^{37}Ar by the reactions, $^{35}Cl(n,p)^{35}S$ followed by $^{37}Cl(p,n)^{37}Ar$. Since the threshold in neutron energy of this reaction sequence is about 1 MeV, the magnitude of this background process is made neglible by a water shield. In the absence of a water shield the magnitude of this background process for the chlorine detector has been estimated to be equal to 0.04 ^{37}Ar atoms/day by Davis[7] on the basis of measurements of the neutron flux in the chamber containing the chlorine detector combined with a yield of 7×10^{-7} ^{37}Ar per neutron absorbed. Barabanov et al.[17] used calculated (α,n) yields, a calculated neutron attentuation factor, and granitic rock composition to estimate a ^{37}Ar production rate from fast neutron in the chlorine detector equal to 0.08 atoms per day. Since 12 runs (Runs 18, 19,20, 38, 39, 76, 77, 78, 79. 80, 81, and 82) have been performed without a water shield and 49 runs with a water shield, it is interesting to compare the combined production rates of these two sets of runs. The results are:

49 runs with water shield	0.45 ± 0.05 ^{37}Ar day^{-1}
12 runs without water shield	0.51 ± 0.09 ^{37}Ar day^{-1}

Although the results without a water shield are about 12% higher than those with a water shield, the error in measurement is large enough that the difference cannot be considered statistically significant. We conclude that, since most of the runs were performed with a water shield, the effect of the runs without a water shield on the combined production rate of all runs is small and well within the quoted error. No correction for a fast neutron background has been applied to the ^{37}Ar production rates given in the following section.

RESULTS

In this section we report and discuss the results of 61 completed runs covering the period from 1970 to 1984. In these 61 runs the total counts having the correct energy and correct rise-time are 774. These counts were divided by the maximum likelihood procedure described earlier into about 435 background counts and 339 counts resulting from ^{37}Ar decay (these numbers depend on whether the runs are treated individually or are combined). Before discussing the production rate of ^{37}Ar, we will discuss counter backgrounds briefly.

Counter Background Rates derived from the maximum likelihood treatment have been variable. The average counter background rate over 61 runs in the two dimensionsal ^{37}Ar region is 0.033 day^{-1}. The range is from 0 to 0.137 day^{-1}. Over the last twelve runs the average background counting rate is 0.010 day^{-1}, i.e., 3.6 counts per year.

Table IV shows the results of the 61 experimental runs. Listed in the table for each run are the time at the start of exposure, the time at the end of exposure, and the mean time of exposure, defined as

$$t_{mean} = t_{start} + (1/\lambda)\,\ell n[\frac{1}{2} + \frac{1}{2}\exp(\lambda\, t_{end} - \lambda\, t_{start})]$$

Table IV Exposure times and ^{37}Ar production rates from individual runs using the chlorine detector

Run Number	Exposure times, years: Start	End	Mean	Atoms per day: ^{37}Ar Production Rate	Upper Error Limit	Lower Error Limit
18	70.279	70.874	70.780	0.214	0.0	0.498
19	70.874	71.180	71.098	0.490	0.150	0.830
20	71.180	71.462	71.383	0.349	0.067	0.630
21	71.462	71.755	71.675	0.0	0.0	0.555
22	71.755	71.951	71.885	0.289	0.0	0.779
24	72.168	72.380	72.311	0.497	0.226	0.768
27	72.517	72.848	72.765	1.226	0.820	1.633
28	72.848	73.073	73.002	0.0	0.0	1.165
29	73.073	73.287	73.218	0.608	0.211	1.006
30	73.287	73.668	73.581	0.147	0.0	0.365
31	73.668	73.952	73.873	0.505	0.0	1.080
32	73.952	74.070	74.023	0.277	0.0	0.928
33	74.070	74.487	74.398	0.302	0.066	0.539
35	74.500	74.591	74.553	0.0	0.0	0.509
36	74.591	75.121	75.028	0.671	0.355	0.987
37	75.121	75.454	75.370	0.877	0.455	1.298
38	75.454	75.733	75.654	0.279	0.0	0.755
39	75.733	76.062	75.978	0.580	0.252	0.909
40	76.065	76.180	76.134	0.419	0.078	0.760
41	76.180	76.270	76.232	0.569	0.152	0.987
42	76.270	76.386	76.340	0.605	0.0	1.534
43	76.386	76.542	76.485	0.058	0.0	0.260
44	76.542	76.676	76.625	0.047	0.0	0.371
45	76.676	76.772	76.732	0.337	0.025	0.648
46	76.772	76.924	76.868	0.491	0.137	0.844
47	76.924	77.076	77.020	0.979	0.583	1.376
48	77.076	77.290	77.221	0.407	0.138	0.676
49	77.290	77.385	77.345	1.075	0.593	1.558
50	77.385	77.594	77.526	0.910	0.588	1.232
51	77.594	77.824	77.754	0.849	0.525	1.173
52	77.824	78.054	77.982	0.588	0.308	0.867
53	78.054	78.361	78.279	0.200	0.0	0.404
54	78.361	78.595	78.523	0.626	0.365	0.887
55	78.595	78.827	78.755	0.468	0.162	0.774
56	78.827	79.051	78.980	0.853	0.546	1.160
57	79.051	79.150	79.110	0.215	0.0	0.668

Table IV Exposure times and ^{37}Ar production rates from individual runs using the chlorine detector (continued)

Run Number	Exposure times, years			Atoms per day		
	Start	End	Mean	^{37}Ar Production Rate	Upper Error Limit	Lower Error Limit
58	79.150	79.375	79.304	0.853	0.295	1.410
59	79.375	79.586	79.517	0.237	0.034	0.439
60	79.586	79.818	79.746	0.0	0.0	0.158
61	79.818	80.065	79.991	0.090	0.0	0.387
62	80.065	80.281	80,211	0.023	0.0	0.254
63	80.281	80.451	80.391	0.0	0.0	0.325
64	80.451	80.604	80.548	0.488	0.222	0.754
65	80.604	80.739	80.687	0.224	0.0	0.649
66	80.739	80.892	80.836	0.361	0.0	1.614
67	80.890	81.059	80.999	0.319	0.051	0.588
68	81.059	81.290	81.218	0.359	0.175	0.544
69	81.290	81.519	81.448	0.477	0.166	0.788
70	81.519	81.673	81.616	0.081	0.0	0.301
71	81.673	81.826	81.770	1.209	0.844	1.574
72	81.826	81.966	81.913	0.636	0.337	0.935
73	81.966	82.210	82.136	0.077	0.0	0.228
74	82.210	82.361	82.305	0.478	0.237	0.720
75	82.361	82.810	82.719	0.503	0.176	0.830
76	82.810	83.040	82.968	0.475	0.144	0.806
77	83.040	83.194	83.137	0.461	0.237	0.684
78	83.194	83.366	83.305	0.752	0.465	1.040
79	83.366	83.531	83.471	0.604	0.332	0.875
80	83.531	83.654	83.606	0.824	0.299	1.348
81	83.654	83.884	83.812	0.330	0.089	0.571
82	83.884	84.095	84.026	0.545	0.257	0.832
All sixty one runs combined				0.462	0.421	0.502

together with the production rate and the upper and lower 1 - σ error limits on this production rate derived by the maximum likelihood method from the time sequences of counts. We show in Figure 2 these same results along with the production rates resulting from combining the runs for each year. The vertical error bars show 1 - σ errors in both halves of the figure. The combined production rate of 61 runs show at the right of the upper part of the figure, is equal to 0.46 ± 0.04 ^{37}Ar atoms day^{-1}.

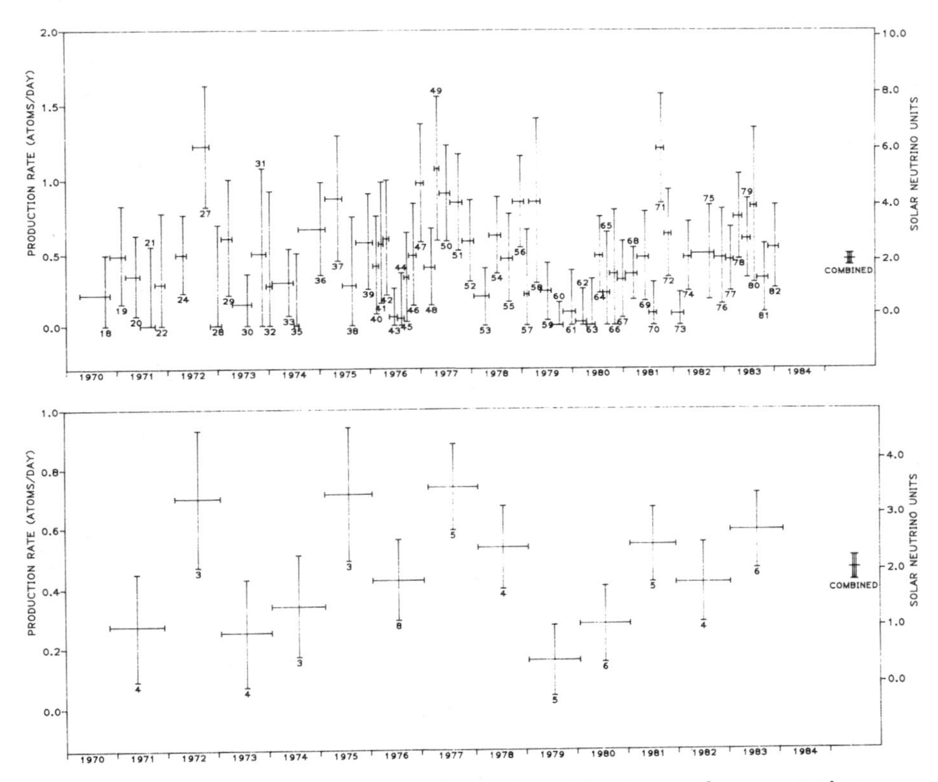

Figure 2. Experimental results from the chlorine solar neutrino experiment: upper -- individual measurements; lower -- yearly averages. The number beneath each point in the lower part shows the number of runs represented by that point.

In Figure 3 we present as a function of time the combined ^{37}Ar production rates. Each point represents a combined production rate for all runs starting with Run 18 and ending at the time where the point is plotted. Run 18 in 1970 was the first run for which pulse rise-time was used to characterize ^{37}Ar decay events. The 1 - σ error bars decrease with time as more runs are included in succeeding points until the present 10% error is reached. The pulse-rise time measurement has improved enormously the ability to distinguish ^{37}Ar decays from background.

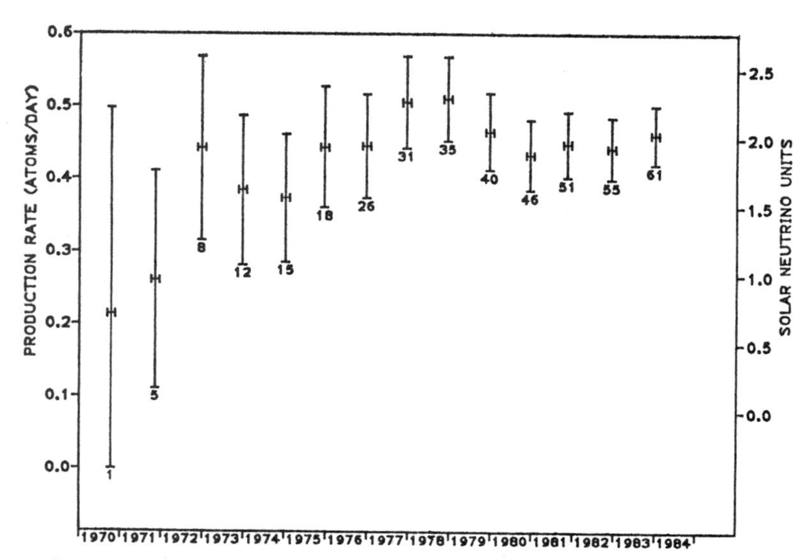

Figure 3. Experimental combined production rate beginning with Run 18 as a function of time. The number beneath each point shows the number of runs combined to give the production rate represented by that point.

In an independent maximum likelihood treatment of the counting data from all 61 runs, <u>both</u> the production rate and the half-life of the decaying component were assumed to be variables whose most likely values were to be derived from the treatment. The resulting values are a half-life of 37 ± 6 days and a production rate of 0.45 ± 0.05 ^{37}Ar atoms day^{-1}. That this half-life is in agreement with that of ^{37}Ar (35 days) offers convincing evidence that the decaying component is indeed ^{37}Ar.

If we subtract the cosmic ray induced background rate for ^{37}Ar from the combined production rate, we obtain:

Combined Production Rate =	0.46 ± 0.04 ^{37}Ar atom day^{-1}
Known Background Production Rate =	0.08 ± 0.03 ^{37}Ar atom day^{-1}
Net Production Rate =	0.38 ± 0.05 ^{37}Ar atom day^{-1}

This net production rate is clearly above that caused by known background processes. If we attribute this net production rate to solar neutrinos and translate it into SNU units (1 SNU ≡ 1 neutrino capture per second per 10^{36} target atoms), the result is: 2.0 ± 0.3 SNU, which may be compared with three values recently predicted using the standard solar model:

Bahcall (1984)[1]	6.6 SNU
Filippone et al. (1983)[18]	5.6 SNU
Fowler (1982)[19]	6.9 SNU

There is clearly a discrepancy between the experimentally derived number and all of the predicted values.

MONTE CARLO SIMULATIONS

To form some feeling for the results shown in Figure 3 it is helpful to compare these results with those from a maximum likelihood treatment of Monte Carlo simulations assuming a constant production rate. The results of two such simulations are shown in Figures 4 and 5. Both of these simulations use the sequence of counter background rates that was observed, i.e., corresponding to the same actual run. The upper figure shows the results when a constant ^{37}Ar production rate of 7.6 SNU is assumed. The lower figure shows the results when a constant ^{37}Ar production rate of 1.8 SNU is assumed. Upon comparison with the real results in Figure 3 two conclusions are evident:

1. The results of the chlorine experiment are clearly different from the results of the simulation with a production rate 3 - 4 times higher.

2. The results of the chlorine experiment resemble closely the results of the simulation with a production rate of 1.8 SNU. The value of the standard deviation derived from the actual runs is equal to 0.32 day^{-1} and that derived from the simulated runs shown in Figure 5 is also equal to 0.32 day^{-1}. Note the occurrence of runs with a production rate of zero per day in both the actual runs (5 times in 61 runs) and the simulation (5 times in 47 runs). Note also the occurrence of high runs in both figures. Although a quick comparison of this kind indicates no reason to conclude that the real results are inconsistent with the constant production rate, this question will be considered in more detail in the last section of this paper.

The results of treatment by the method of maximum likelihood of other Monte Carlo simulations of the production, separation, processing and counting of ^{37}Ar can be summarized by the following expression for the fractional 1 - σ error in the production rate (in this treatment the half-life of ^{37}Ar was assumed to be a known parameter):

$$E^2 = \frac{0.027}{\eta \; \varepsilon SP} \left(1 + \frac{3.7\ B}{\varepsilon SP}\right)$$

where E is the fractional 1 - σ error in the production rate, η is the number of runs, P is the production rate in day^{-1}, B is the background counting rate in day^{-1}, ε is the fractional counting efficiency, S is the saturation factor defined as $1 - \exp(-\lambda t_e)$, λ is the decay constant for ^{37}Ar, and t_e is the exposure time. This expression is applicable for 300 days of counting time for each run and for an extraction efficiency of 95% for ^{37}Ar; these are both typical values. For two months exposure (t_e = 60 days), P = 0.47 day^{-1}, and ε = 0.40 the term 3.7 B/εSP may be evaluated for different values of B. When B = 0.01 day^{-1}, this term is equal to 0.28 and only 22% of the error comes from the term containing the

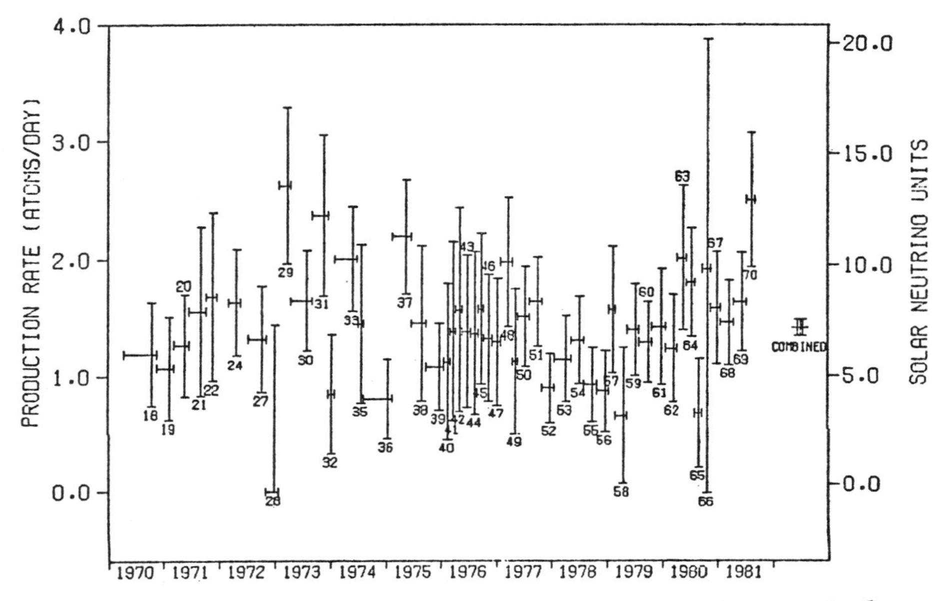

Figure 4. Results of maximum likelihood treatment of Monte Carlo simulations with constant production rate of 7.6 SNU.

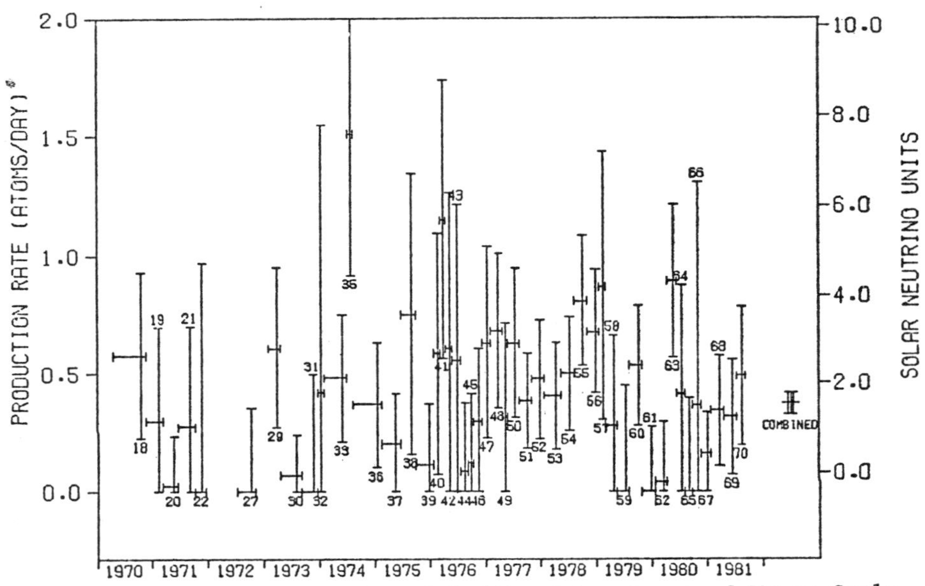

Figure 5. Results of maximum likelihood treatment of Monte Carlo simulations with constant production rate of 1.8 SNU.

background rate. From this viewpoint no large improvements in reducing the error can be achieved by further decreases in the counter background rate. If, however, both the production rate and the half-life are considered as free parameters to be derived from the counting data, there is still much to be gained by reducing counter background rates. In any case, it is a desirable goal to continue trying to reduce counter background rates.

SOLAR ACTIVITY AND THE CHLORINE SOLAR NEUTRINO EXPERIMENT

One of the major aims of continuing operation of the chlorine experiment during the last thirteen years has been to improve the statistical accuracy of the measurement of the ^{37}Ar production rate and to assure that ^{37}Ar is indeed being observed.[3] To achieve these goals we have continually improved counter backgrounds by using better electronic circuitry, a NaI anti-coincidence detector and a thick Pb-Hg-Boron-plastic shield and by carrying out the counting measurements underground. The present results of the chlorine experiment clearly demonstrate that ^{37}Ar is being produced in the detector, but that its production rate is inconsistent with that predicted by the standard solar model. Since the problem of explaining this discrepancy will be discussed by others at this conference, we will not discuss it here. Instead we would like to consider further the question of the constancy of the ^{37}Ar production rate.

We now have a long record of data that can be used to determine if the sun's neutrino output is constant during a period comparable to an eleven-year solar cycle and to search for pulses of neutrinos. We would like to discuss some interesting correlations that appear in the data that could be of great importance to our understanding of the sun and cosmic rays.

Because of the sun's slow response to changes, the neutrino luminosity of the core of the sun is generally believed to be constant for periods in the range of 10^4 - 10^6 years. Processes that could perturb the rate of the thermal fusion reactions would not be expected to be observed over a period of a solar cycle.[20] However, in view of the fact that the interior of the sun is not very well understood it is important to look carefully at the results for changes in the ^{37}Ar production rate with time. One must bear in mind that observational neutrino astronomy is a new subject and there may be surprises. Because the chlorine detector is the only detector with a high neutrino (ν_e) sensitivity and a long observational record, a number of investigators have examined the results from the chlorine experiment for variations.

Solar Cycle Variations. In this report we will not review the literature on this subject, but will give some references and an indication of the analysis applied (see ref. 21). A number of investigators have pointed out that the ^{37}Ar production rate anticorrelates with solar activity as measured by sunspot cycles. This correlation is shown in Figure 6 in which the yearly average

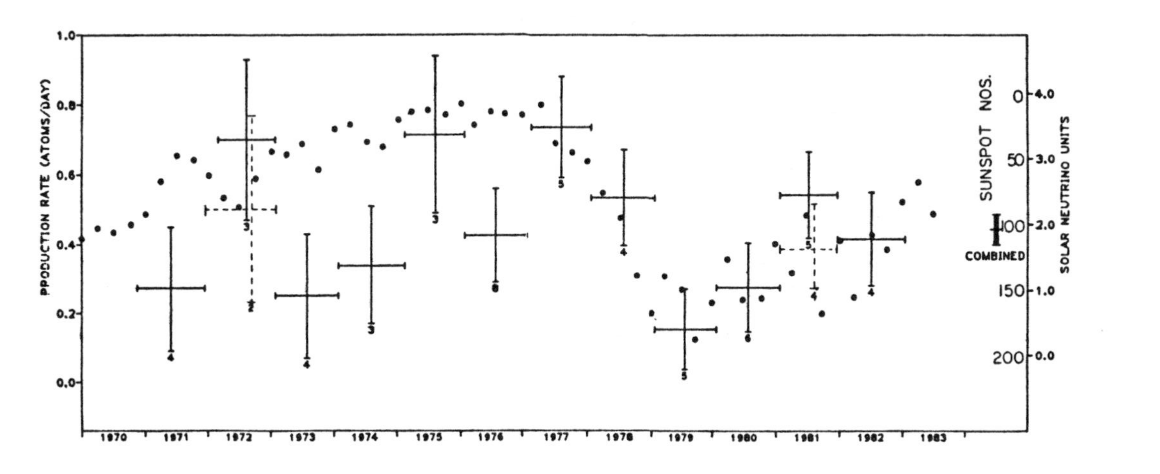

Fig. 6 (above). Yearly average results for the chlorine experiment superposed on monthly mean sunspot numbers (plotted inverted). The number of experimental runs included in the average is indicated for each year. The averages for 1972 and 1981 shown with dotted lines are without Runs 27 and 71, the runs that appear to correlate with the large solar flares of Aug. 4-8, 1971 and of Oct. 12, 1981.

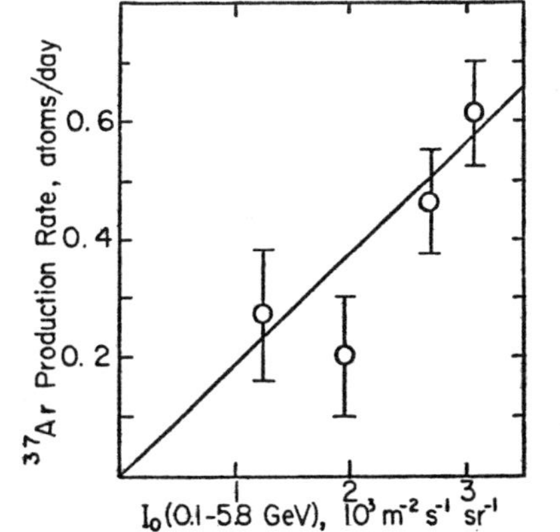

Fig. 7 (at right). The ^{37}Ar production rate divided into four bins according to the intensity of cosmic ray protons plotted against the intensity of cosmic ray protons in the energy range, 0.1-5.8 GeV, (from Bazilevskaya et al.[19]).

^{37}Ar production rate is compared to the sunspot numbers. In this plot the sequence of points appear to anticorrelate with sunspot numbers during the rise and fall of solar activity in cycle 21. It is well known, and understood, that the galactic cosmic ray intensities of protons and alpha particles with energies less than a few Gev also anticorrelates with solar activity. In Figure 7 we show, following the example of Bazilevskaya et. al.[22,23] a plot of the cosmic ray intensity in the energy range 0.1 to 5.8 Gev against the ^{37}Ar production rate. Although the errors in the ^{37}Ar production rates are large, there does appear to be a correlation of the rate with the cosmic ray intensity. This effect could not arise directly from low energy cosmic ray particles because they would not reach the neutrino detector at a depth of 4000 hg/cm^2 ($\bar{E}_\mu$ = 340 Gev). It is conceivable that neutrinos from μ^+ decay that are associated with low energy cosmic rays could be responsible for the signal observed. The $^{37}C\ell$ capture cross section for electron-type neutrinos from μ^+ decay is 1.1×10^{-40} cm^2, a value one-hundred times greater than the cross-section of neutrinos from ^8B decay. Present estimates of the flux of low energy neutrinos from cosmic ray interactions in the atmosphere are too low to account for a flux of low energy neutrinos (20 to 100 Mev) as high as 10^4 cm^2/sec, as explained by Stanev at this conference.[24] However, it has been reported that the Kamioka proton-decay experiment has observed neutrino-like events in the energy range of 35 to 100 Mev. Large scale water Cherenkov detectors in the future could achieve the required sensitivity to observe low energy neutrinos.[25]

Resolution of the question of the validity of this apparent correlation is of great importance to our present interpretation of the observations of the chlorine neutrino detector as a flux of neutrinos from the core of the sun. Although it is unlikely that the neutrino luminosity of the sun varies, this possibility should not be absolutely ruled out. The chlorine detector responds primarily to the ^8B and ^7Be neutrino fluxes and the production of these neutrino emitters is a very sensitive function of the temperatures in the solar interior (the ^8B neutrino flux is proportional to the twenty-fifth power of the central temperature of the sun). Therefore, the chlorine detector is the best of the radiochemical detectors to search for possible variations. A gallium detector would be far less responsive. The 3000-ton chlorine detector planned in the USSR[26] would, of course, be a great improvement over the Homestake detector.

Solar Flare Effects. It was pointed out ten years ago that run 27, the highest experimental run, corresponded in time with the great solar flare of August 4-7, 1972.[27] If this flare was responsible for the increase, we missed observing its full magnitude because the ^{37}Ar decayed by a factor of six before the sample was removed from the tank. At that time, and also at the present time, it is believed that measurable fluxes of neutrinos could not be produced in a solar flare[28] or by solar flare particles interacting in the earth's atmosphere.[29] Shortly after the time of occurrence of the large solar flare of October 7-12, 1981 and before any results of run 71 were known, Stozhkov predicted that this flare might be detected by the chlorine experiment.[22,23] Our second highest run resulted. Since that time, this Soviet group noted that there are, in all, two flares that occurred on the visible disc and one on the invisible disc of the sun that produced in the earth's atmosphre a large flux of protons with energy above 150 Mev and that a measurable increase in the ^{37}Ar production might be detected.[23] These events are listed below, in Table V, along with the decay period before collecting the sample and the ^{37}Ar production rate (calculated in the usual way assuming a uniform flux).

Table V Solar flares with large observed proton fluxes and the measured ^{37}Ar production rates for the same periods

Run No.	Flare Date	Solar Coordinates	Proton Intensity (> 150 Mw) cm^{-2} sec^{-1} Sr^{-1}	Decay Period in Days Before Counting	Observed ^{37}Ar Production Atoms/Day
27	8/ 4/72	14N; 8E	~ 90	94	1.23 ± 0.41
	8/ 7/72	14N; 36W	~ 3	91	
51	9/24/77	10N; 120W	~ 3.5	35	0.85 ± 0.32
71	10/ 7/81	19S; 88E	~ 0.5	21	1.21 ± 0.37
	10/12/81	22S; 35E	~ 4	16	

If these three flare events were responsible for the increase in rate above the average rate (0.40 ± 0.04 atoms/day), the approximate numbers of ^{37}Ar atoms produced in the detector in runs 27, 51 and 71 were 250 ± 130, 20 ± 15, and 56 ± 30 respectively.

As mentioned earlier Monte Carlo simulations of our data indicate that experimental runs as high as these are expected, approximately 1 - 2 for 60 experiments. It is unlikely that high values would correlate in time with a solar flare event, but to be convinced one would like to observe a flare event as large as the 1972 event immediately after its occurrence. It is interesting to note that the great flare of 1956 was over one-hundred times larger than the 1972 flare.[30] A special run was performed in early May 1984 to see if the April 25, 1984 x-ray flare was observed (run 84). The Soviet group detected little or no increase in proton flux at the earth from this flare. There was not an enhanced ^{37}Ar production in run 84.

There is no ready explanation for a measurable neutrino flux to be developed by a solar flare, or by the interaction of flare protons with solar matter,[28] or with the earth's atmosphere.[29] However, a large water Cherenkov detector could observe neutrinos if their energies were over 30 Mev and the flux were as high as indicated by our measurements. Using the threshold and sensitivities given in reference 13 for the Kamioka detector, the production of one ^{37}Ar atom in the chlorine detector would correspond to 0.6 events in a Kamioka detector. On this basis neutrinos from a large solar flare should be readily measured if the chlorine detector observes another large flare.

THE FUTURE OF THE CHLORINE SOLAR NEUTRINO EXPERIMENT

The chlorine detector is the only neutrino detector with a well-known neutrino (ν_e) energy response, low threshold, and high sensitivity which has a long observational record. Because of its unique character and importance to astronomy, we believe observations using this detector should continue.

ACKNOWLEDGMENT

This research was carried out at Brookhaven National Laboratory under contract DE-AC02-76CH00016 with the U.S. Department of Energy and supported by its Office of High Energy and Nuclear Physics. This experiment has involved the help of many people. Co-workers are J. C. Evans, J. P. Galvin, D. Harmer, K. Hoffman, L. Rogers and V. Radeka. We would like to express our gratitude to many people from the Homestake Mining Company. Without their cooperation and vital help this experiment could not have been performed. We are especially grateful to A. Gilles and J. Dunn who have helped in so many ways at Homestake. We are also especially grateful for the continuous support and encouragement from W. A. Fowler and J. N. Bahcall.

REFERENCES

1. J. N. Bahcall, paper presented at the conference and private communication.
2. J. N. Bahcall. Rev. Mod. Phys. 50, 881 (1978); Rev. Mod. Phys. 54, 767 (1982).
3. R. Davis, Jr., Proc. Informal Conf. on Status and Future of Solar Neutrino Research, Brookhaven National Laboratory, G. Friedlander, Ed., BNL-50879, 1, 1 (1978).

4. R. Davis, Jr., Proc. Solar Neutrino Conference-Irvine, February 1972, V. Trimble and F. Reines, Eds.
5. J. N. Bahcall and R. Davis, Jr., Science 191, 264 (1976).
6. J. J. Leventhal and L. Friedman, Phys. Rev. D. 6, 3338 (1972).
7. S. J. Buelow, D. R. Worsnop and D. R. Herschbach, Proc. Nat. Acad. Sci., U.S.A. 78, 7250 (1981).
8. V. V. Gromov, A. V. Kopylov, G. Ya. Novikova and I. V. Orekhov, Proc. Int. Conf. on Neutrino Physics and Astrophysics (Neutrino-77), Baksan Valley 1, 73 (1977).
9. R. Davis, Jr., E. C. Fowler, S. L. Meyer and J. C. Evans, Jr., Research Proposal No. 53, Los Alamos Meson Physics Facility, June, 1973.
10. B. T. Cleveland, Nucl. Instr. Meth. 214, 451 (1983).
11. G. V. Domogatskii and R. A. Eramzhyan, Izv. Akad. Nauk 41, 1969 (1977).
12. A. W. Wolfendale, E. C. M. Young and R. Davis, Jr., Nature (Phys. Sci.) 238, 130 (1972).
13. G. L. Cassiday, Proc. 13th Int. Conf. Cosmic Rays, Denver 13, 1958 (1973).
14. A. G. Wright, Proc. 13th Int. Conf. Cosmic Rays, Denver 13, 1704 (1973).
15. E. J Fenyves, M. Cherry, M. Deakyne, K. Lande, C. K. Lee, R. I. Steinberg and J. M. Supplee, Proc. 17th Int. Conf. Cosmic Rays, Paris, 10, 317 (1981).
16. G. T. Zatsepin, A. V. Kopylov and E. K. Shirokova, Sov. J. Nucl. Phys. 33, 200 (1981).
17. I. R. Barabanov, V. N. Gavrin, G. T. Zatsepin, I. V. Orekhov and L. P. Prokop'eva, Sov. Atomic Energy 54, 158 (1983).
18. B. W. Filippone, A. J. Elwyn, C. N. Davids and D. D. Koetke, Phys. Rev. Lett. 50, 412 (1983).
19. W. A. Fowler, Amer. Inst. Phys. Conf. Proc. 96, 80 (1982).
20. R. Ulrich, Science 190, 619 (1975).
21. Twenty-five month variation, K. Sakurai, Nature 269, 401 (1979), Solar Physics 74, 35 (1981), Proc. 18th Int. Cosmic Ray Conf., Bangalore, India 4, 210 (1983) V. N. Gavrin, Yu. S. Kopysov and N. T. Makeev, JETP Lett. 35, 608 (1982); superposed epoch analysis, R. Ehlich, Phys. Rev. D25, 2282 (1982); power spectrum analysis, H. J. Haubold and E. Gerth, Astron. Nacht. 304, 299 (1983), and paper in this conference; observations inconsistent with a constant flux, A. Subramanian, Current Science 52, 342 (1983), A. Subramanian and S. Lal preprint TIFR-BC-83-12, July 1984, D. Basu, Solar Physics 81, 363 (1982), P. Raychaudhuri, Proc. 18th Int. Cosmic Ray Conf. 4, 108 (1983), Solar Physics (in press) 1984, preprint August 1984, G. A. Basilevskaya, S. I. Nikolskii, Yu. I. Stozhkov, T. N. Charakhch'yan and A. M-A. Mukhamedzhanov JETP Lett. 35, 343 (1982), Proc. 18th Int. Cosmic Ray Conf. 4, 218 (1983), preprint submitted to J. Nucl. Phys (USSR) Dec. 1983; observations consistent with a constant flux, L. J. Lanzerotti and R. S. Raghavan, Nature 293, 122 (1981), A. V. Kopylov and V. N. Gavrin Pis'ma Astron. Zh. 10, 154 (1984).

22. G. A. Bazilevskaya, Yu. I. Stozhkov and T. N. Charakhch'yan, JETP Lett. 35, 341 (1982).
23. G. A. Bazilevskaya, A. M.-A. Mukhamedzhanov, S. N. Nikolskii, Yu. I. Stozhkov and T. N. Charakhch'yan preprint, submitted to J. Nuclear Phys. (USSR) December 1983.
24. T. Stanev, paper presented at this conference.
25. H. Chen, paper presented at this conference and Internal Report, UCI Neutrino No. 120, July, 1984. A. Mann, Conf. on Interactions Between Particle and Nuclear Physics, Steamboat Springs, CO, May 23-30, 1984 (AIP Conf. Report).
26. I. R. Barabanov, et. al., these conference proceedings (paper presented at conference by V. N. Gavrin and A. V. Kopylov).
27. R. Davis, Jr. and J. C. Evans, Jr., Proc. 13th Int. Cosmic Ray Conf., Denver 3, 2001 (1973). R. Davis, Jr. and J. C. Evans, Jr., Proc. Int. Conf. Particle Acceleration and Nuclear Reactions in Space, Leningrad, p. 91 (1974).
28. I. N. Erofeeva, S. I. Lyutov, V. S. Murzin, E. V. Kolomeets, J. Albers and P. Kotzer, Proc. 18th Int. Cosmic Ray Conf. 7, 104 (1983). A. Dar and S. P. Rosen preprint, September 1984, R. Lingenfelter, this conference.
29. K. O'Brien and A. de la Zerda, Trans Am. Nucl. Soc. 46, 641 (1984).
30. M. A. Pomerance and S. P. Duggal, Rev. Geophys. and Space Phys. 12, 343 (1974).

COSMIC-RAY DEPTH STUDIES AT THE HOMESTATE MINE WITH $^{39}K \rightarrow {}^{37}Ar$ DETECTORS

E. L. Fireman
Smithsonian Astrophysical Observatory
and
B. T. Cleveland, R. Davis, Jr., and J. K. Rowley
Department of Chemistry, Brookhaven National Laboratory, Upton, NY 11973

ABSTRACT

The argon-37 production rates in potassium detectors are measured at depths down to 4850 ft (4000 m.w.e.), the location of the chlorine solar neutrino detector in the Homestake Mine. The depth variation of the ^{37}Ar production rates in K agrees with that expected from cosmic ray muon interactions. At shallow depths, the ^{37}Ar production rates in K parallel those in Cl, with the ^{37}Ar rate per ton K equal to the ^{37}Ar rate per 615 tons of C_2Cl_4. At 4850 ft depth, the measurements give 0.07 ± 0.07 ^{37}Ar day^{-1} per ton of K, which is to be compared to the value of 0.46 ± 0.04 ^{37}Ar day^{-1} per 615 tons of C_2Cl_4 in the solar neutrino detector. The comparison indicates that there is an excess of ^{37}Ar over the cosmic-ray production in chlorine at 4850 ft depth; however, the error in the K measurements at 4850 ft depth is presently too large to determine accurately the cosmic ray background of the solar neutrino experiment. This error will be reduced by additional measurements.

INTRODUCTION

The primary reason for measuring argon-37 in potassium as a function of depth is to determine the cosmic-ray background for the chlorine solar neutrino experiment. The hydrogen fusion reactions, believed to furnish the sun's energy, generate electron neutrinos with energies up to 14.1 Mev.[1] The chlorine solar neutrino experiment is based on the inverse beta reaction:

$$\nu_e + {}^{37}Cl \rightarrow \bar{e} + {}^{37}Ar \;, \quad E_{TH} = 0.82 \text{ Mev} \;. \tag{1}$$

The ^{37}Ar production rate measured in 615 tons of C_2Cl_4 at 4850 ft depth in the Homestake Mine (4000 m.w.e.) is 0.46 ± 0.04 ^{37}Ar atoms day^{-1}.[2] This rate is larger than the calculated cosmic-ray background, 0.07 ± 0.03 day^{-1} [3,4] but smaller than the calculated solar neutrino production rate 1.5 ± 0.3 day^{-1}.[1] Most discussions of the solar neutrino experiment are concerned with the fact that the measured ^{37}Ar rate is a factor of 3 smaller than that theoretically calculated. Our experiment concerns itself with a question rarely discussed. Is the ^{37}Ar background rate equal to the calculated cosmic-ray value? The argon-37 in potassium experiment duplicates many of the features of the chlorine solar neutrino experiment except for the fundamental one of solar neutrinos playing a role. It is energetically impossible for solar neutrinos with energies of less than 14.1 Mev to produce ^{37}Ar from potassium.

Neutrinos can produce ^{37}Ar from ^{39}K only by the following neutral
urrent reactions:

$$\nu + {}^{39}K \rightarrow \nu + {}^{37}Ar + N + P \quad , \quad E_{TH} = 25 \text{ Mev}$$

$$\nu + {}^{39}K \rightarrow \nu + {}^{37}K + 2N, \quad {}^{37}K \xrightarrow{\beta^+} {}^{37}Ar \quad , \quad E_{TH} = 31 \text{ Mev} \qquad (2)$$

n addition to the high threshold energy, the cross-sections for the
eutral current reactions are exceedingly small. The neutrinos created by
osmic ray interactions in the earth's atmosphere produce only 1.8×10^{-5}
^{37}Ar atoms day^{-1} ton^{-1} K.[5]

At the depths of interest, swift muons are responsible for the cosmic
ay ^{37}Ar production both in chlorine and in potassium. The nuclear event
ates caused by swift muons in chlorine and in potassium have the same
epth dependence, even though the ^{37}Ar event rate in chlorine is
pproximately a hundred times smaller than in potassium. A detailed
iscussion of the ^{37}Ar rates and depth dependence is given in later
ections. The high cosmic-ray muon production rate of ^{37}Ar in potassium
nables one to study the cosmic-ray background radiochemically at large
epths underground with relatively small potassium detectors.

EXPERIMENTAL PROGRAM

This study is an expanded, improved KOH version of a $^{39}K \rightarrow {}^{37}Ar$
tudy[6,7] by Fireman that was done with 0.68 ton of K in the form of
$C_2H_3O_2$ powder. The expansion increased the amount of K from 0.68 ton to
.3 tons. The improvement consisted of the use of ^{37}Ar counting systems
hich are much more sensitive than the counting systems used by
ireman.[6,7] The use of KOH permits additional expansions with no loss
n sensitivity by adding units similar to those in operation.

Fireman obtained accurate ^{37}Ar values with his detector down to
850 ft depth; however, at 4850 ft depth, the location of the solar
eutrino detector, only a limit value of approximately 1.0 ^{37}Ar day^{-1}
on^{-1} K was obtained. This limit is nearly 20 times the value expected
rom cosmic rays. It was clear that the 0.68 ton potassium experiment had
o be expanded and improved to achieve the goal of an accurate ^{37}Ar value
n potassium at 4850 ft depth.

Within the past 2 years, four KOH detectors containing a total of
.3 tons of potassium were put into operation at the mine. Two are
ationary tanks, with 1.15 tons of potassium each, located next to the
lar neutrino detector at 4850 ft depth. Two are mobile tanks, with
.0 ton of potassium each, that can be moved over the mine rail system to
y level. Also put into operation at the mine, was an ultra-low-level
Ar counting system which is similar to the one used in the solar
utrino experiment. The expanded, improved system has been in operation
ly since November 1983, but has already yielded the results given in
ble I and plotted as the dark circles in Fig. 1. In the first run, one
bile KOH tank was at 1100 ft depth and one at 1850 ft depth and the
gon samples extracted from the two stationary tanks at 4850 ft depth
re combined. The results compared well with those Fireman obtained from

many runs during several years of operating the 0.68 ton detector. At 1100 ft depth, a KOH detector gave 20.0 ± 1.5 ^{37}Ar day^{-1} ton^{-1} K compared to 20.7 ± 1.5 ^{37}Ar day^{-1} ton^{-1} K in the old detector. At 1850 ft depth, a KOH detector gave 3.4 ± 1.1 ^{37}Ar day^{-1} ton^{-1} K compared to 5.7 ± 0.7 ^{37}Ar day^{-1} ton^{-1} K. At 4850 ft depth, the KOH tanks gave 0.18 ± 0.26 ^{37}Ar day^{-1} ton^{-1} K compared to the previous limit value of $\leq$ 1.0 ^{37}Ar day^{-1} ton^{-1} K. The KOH and the old detector results at 1100 and 1850 ft depths give confidence in the results that will be obtained at 4850 ft depth. We expect to obtain an ^{37}Ar value with 25% accuracy at 4850 ft depth within 3 years of operation with the KOH detectors. The error in each run is reduced by a factor of 2 by using the four KOH detectors at 4850 ft depth instead of the two used in the first run. The counting error is reduced by the factor, $n^{-1/2}$, where n is the number of runs. We plan to make a run every 2 months and to count each extracted argon sample for 8 months. The ultra-low-level system can accommodate four counters.

The argon is extracted from the KOH by a helium purge during a 5-day visit to the mine approximately every 2 months. We have completed four extractions. The ^{37}Ar counting is in progress for the last three. These data are summarized in Table 1 and plotted in Fig. 1.

The overburdens (m.w.e. = 100 g cm^{-2}) that correspond to the depths in the Homestake Mine have been recently recalculated. These overburden values are approximately 300 m.w.e. smaller than the values used in previous publications.[3,4,6,7] These overburdens are the thicknesses of "standard" rock (ρ = 3.05 g cm^{-3}, $\overline{Z}$ = 11) which correspond to the depths. The ground surface at the mine is not flat and changes in rock densities occur with depth. These factors were taken into account by numerical integrations. A series of concentric circles were drawn around the point on the surface directly above the detector location, the circumference of each circle was divided into 24 equidistant points, and the average thickness experienced by the muons that entered each circle and passed through the detector was calculated. With the expression for the muon angular dependence, an integration to obtain the "effective" overburden was then performed. The values for the "effective" overburden at 1100, 1850, and 4850 ft depths are 820, 1415, and 4000 m.w.e., approximately 300 m.w.e. lower than the values of 1080, 1775, and 4400 m.w.e. used in the previous publications.

The data presented as the crossed points in Fig. 1 were mainly taken from Fireman;[7] the data presented as the open circles were taken from Wolfendale _et al_.,[3] and Davis _et al_.;[2] the dark circles were obtained from Table I. The solid curve in Fig. 1, which is discussed in detail in the next section, is the fast muon ^{37}Ar production rate versus depth. The dashed curve in Fig. 1, also discussed in the next section, is the stopped muon ^{37}Ar production rate versus depth. The crossed points, dark circles, and open circles at all depths except 4000 m.w.e. are in excellent accord with the sum of the solid and dashed calculated curves. At 4000 m.w.e. depth the dark circle is on the calculated curve but the open circle is above the calculated curve. It is vitally important to reduce the error associated with the potassium measurement at 4000 m.w.e. to obtain the cosmic-ray background for the chlorine experiment more accurately. The error bars are 1σ values.

Table I Exposure depths, counting data, and ^{37}Ar production rates derived from the KOH experiment.

Run	Tank	Depth (ft)	Exposure begins (date)	Exposure ends (date)	Total counts in energy-rise time window (counts)	Counting (days)	^{37}Ar (day^{-1} ton^{-1} K)	Overburden (m.w.e)
1	2 Stationary	4850	11/21/83	02/02/84	6	123	0.18 ± 0.26	4000
	1 Mobile	1850	11/28/83	02/01/84	46	233	3.40 ± 1.1	1415
	1 Mobile	1100	11/28/83	02/01/84	172	62	20.00 ± 1.5	820
2	2 Stationary	4850	02/02/84	03/30/84	15	171	~ 0.20	4000
	2 Mobile	4850	02/06/84	03/31/84	12	136	~ 0.20	4000
3	2 Stationary +2 Mobile	4850	03/31/84	06/14/84	4	67	0.12 ± 0.12	4000
4	2 Stationary +2 Mobile	4850	06/14/84	08/29/84	1	35	---	4000
1 + 2 + 3 + 4		4850	--	--	38	532	0.07 ± 0.07	4000

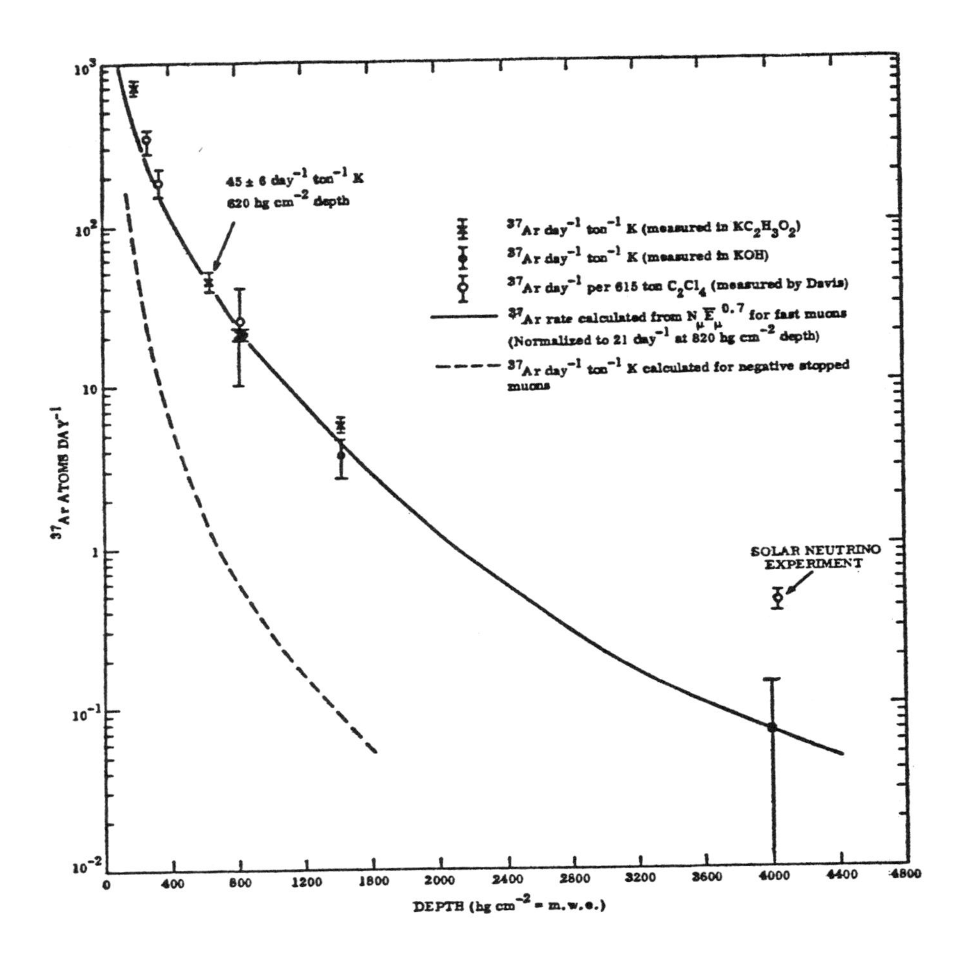

Fig. 1. Calculated and measured ^{37}Ar production rates.

COSMIC-RAY ^{37}Ar PRODUCTION RATES UNDERGROUND

Calculations of the ^{37}Ar production rate in Cl by cosmic-ray muons underground (Wolfendale et al.,[3] Cassiday,[8] Zatsepin *et al.*[4]) have been made. These authors state that any nuclear event rate produced by fast cosmic-ray muons should be proportional to the quantity $N(h)\ \overline{E}_{\mu}^{0.7}$, where N(h) is the muon intensity and $\overline{E}_{\mu}$ is the average muon energy at depth h. Since N(h) and $\overline{E}_{\mu}$ are known from muon measurements, it is only necessary to normalize the nuclear event rate to obtain the event rates at all depths. The ^{37}Ar production rates in three 600 gallon C_2Cl_4 detectors were measured at shallow depths in the Homestake Mine.[3] The measured rates scaled from 1800 gallons up to 10^5 gallons (615 tons), at 254, 327 and 820 m.w.e. depths are given in Table II and shown by the open circles in Fig. 1. Wolfendale *et al.*[3] normalized $N(h)\overline{E}_{\mu}^{0.7}$ to fit these measurements and obtained ~ 0.05 ^{37}Ar day^{-1} per 615 tons C_2Cl_4 from the normalized $N(h)\ \overline{E}_{\mu}^{0.7}$ curve at 4850 ft depth (estimated to be 4400 m.w.e.). Zatsepin *et al.*[4] used the same procedure with improved estimates for N(h) and $\overline{E}_{\mu}$ to obtain 0.070 ^{37}Ar day^{-2} per 615 tons C_2Cl_4 for the cosmic-ray background of the solar neutrino experiment, which is the background value most commonly used at present. The solid curve in Fig. 1 is the normalized $N(h)\overline{E}_{\mu}^{0.7}$ curve in units of ^{37}Ar atoms day^{-1} per 615 tons of C_2Cl_4.

The ^{37}Ar production rates in potassium are also nuclear events caused by fast muons. The $N(h)\overline{E}_{\mu}^{0.7}$ proportionality should apply equally well to the calculation of its depth dependence. We normalized $N(h)E_{\mu}^{0.7}$ to fit the ^{37}Ar measurements in K at 820 m.w.e. depth and obtained the same solid curve (Fig. 1) in units of ^{37}Ar atoms day^{-1} per ton of K.

The first column of Table II gives the mass overburden (newly calculated values) at mine depths where ^{37}Ar measurements have been done or are contemplated. The second and third columns give the muon intensities, N(h), and average energies, $\overline{E}_{\mu}$, obtained from Miyake[9] and Zatsepin *et al.*[4] The fourth column gives $\overline{E}_{\mu}^{0.7}$ N(h) normalized to 21.0 ^{37}Ar day^{-1} at 820 m.w.e., which is plotted as the solid curve in Fig. 1. The dashed curve in Fig. 1 and the fifth column in Table II give the ^{37}Ar production rates in K for negatively charged stopped muons. These stopped muon rates were calculated with the data presented in Table III. The total ^{37}Ar production rate is the sum of the fast muon and stopped muon production rates. Stopped muons contribute 20% of the total at 177 m.w.e., 2% of the total at 1415 m.w.e., and a smaller fraction at deeper locations. The sixth column of Table II gives the measured ^{37}Ar production rates in K. The data for KOH and $KC_2H_3O_2$ are combined in Table II but are shown separately by the dark circles and crossed points in Fig. 1. The measurements are in good accord with the calculated values. The error in the ^{37}Ar production rate at 4000 m.w.e. is too large to have its maximum scientific impact.

We shall now examine the fast muon results in terms of cross-sections. The ^{37}Ar data from depths shallower than 4000 m.w.e. indicate that the ratio of the ^{37}Ar production cross-section in ^{39}K to that in ^{37}Cl, $\sigma(^{39}K)/\sigma(^{37}Cl)$, equals 155, which by coincidence is equal to

Table II Muon parameters and calculated and measured ^{37}Ar production rates.

Depth (m.w.e.)	Muon intensity* N(h) (cm^{-2} sec^{-1})	Muon Av. Energy* $\bar{E}_\mu$ (Gev)	Calculated[a] fast muon production rate (^{37}Ar day^{-1} ton^{-1} K)	Calculated[b] stop. neg. muon production rate (^{37}Ar day^{-1} ton^{-1} K)	Measured production rate (^{37}Ar day^{-1} ton^{-1} K)	Measured production rate $\left(\frac{^{37}Ar\ day^{-1}}{615\ ton\ C_2Cl_4}\right)$
177	2.1×10^{-4}	40	520.	100.	(710 ± 70)**	---
254	1.0×10^{-4}	50	291.	23.	---	320 ± 50
327	5.0×10^{-5}	60	166.	10.	---	178 ± 25
473	2.3×10^{-5}	80	93.	4.3	---	---
620	1.0×10^{-5}	96	46.	1.5	(45 ± 6)**	---
820	4.0×10^{-6}	116	21.0	0.55	(20.5 ± 1.1)†	24 ± 16
1415	6.6×10^{-7}	177	4.65	0.090	(4.9 ± 0.7)†	---
2000	1.6×10^{-7}	226	1.26	0.021	---	---
4000	6.0×10^{-9}	340	0.065	0.0009	(0.07 ± 0.07)‡	0.46 ± 0.04

a Calculated from $E_\mu^{0.70}N(h)$ normalized to 21.0 ^{37}Ar day^{-1} ton^{-1} K (estimated error 15%).

b Calculations for stopped negative muons given in Table 3 (estimated error 25%).

* Average of values obtained from Zatsepin et al.[4] and Miyake.[9] K. Lande et al. (priv. comm. 1984) measured muon intensity at 4000 hg cm^{-2} in the Homestake Mine to be 5.0×10^{-9} cm^{-2} sec^{-1}. Use of Lande's intensity value would reduce the calculated rate at 4000 m.w.e. by 20%.

**Measured in $KC_2H_3O_2$; †Measured in both $KC_2H_3O_2$ and KOH (average values given); ‡Measured in KOH.

Table III Stopped negative muon ^{37}Ar production rates in potassium

(m.w.e.)	Stopped Muon ton^{-1} / N(h) (ratio = R)[a]	Neg. Muon Stopped (day^{-1} ton^{-1})[b]	^{37}Ar (day^{-1} ton^{-1} K)[c]
177	150	1200	100
254	750	286	23
327	65	127	10
473	60	52.6	4.3
620	50	19.0	1.5
820	45	6.8	0.55
1415	45	1.13	0.090
2000	45	0.026	0.021
4000	50	0.011	0.0009

a Ratios from Grupen, Wolfendale, and Young.[11]

b Negative stopped muons = (R) x [N(h) day^{-1} cm^{-2}](1/2.27); positive to negative muon ratio = 1.27.

c From the data given in Spannagel and Fireman[12] who used $KC_2H_3O_2$ targets in a negative muon cyclotron beam, we calculate 0.032 ^{37}Ar atoms produced per stopped muon. We assume this yield also applies for a 50% KOH solution.

the ratio of ^{37}Cl atoms in 615 tons of C_2Cl_4 to ^{39}K atoms in one ton of K. The cross-sections, $\sigma(^{39}K)$ and $\sigma(^{37}Cl)$ are 4.2 and 0.027 mb at 820 m.w.e. and vary with depths as $\overline{E}_\mu^{0.7}$.

Fast muons interact electromagnetically with the material in the detectors and in the overhead rock producing protons, neutrons, and pions. These muon interactions are described by the equation,

$$\mu^{\pm} + M \rightarrow \mu^{\pm} + M^{*} + (P, N, \pi^{\pm}), \tag{3}$$

where M represents nuclei in the detector and overhead rock and M* represents the product nuclei.

Protons from the muon reactions (3) produce 90% of the ^{37}Ar from ^{37}Cl.[4]

$$P + {}^{37}Cl \rightarrow N + {}^{37}Ar\ . \tag{4}$$

Two muon processes contribute to ^{37}Ar from ^{39}K. The first is

$$\mu^{\pm} + {}^{39}K \begin{cases} \nearrow \mu^{\pm} + N + P + {}^{37}Ar \\ \searrow \mu^{\pm} + 2N + {}^{37}K \end{cases} , \tag{5}$$

which produces ~ 25% of the ^{37}Ar from ^{39}K on the basis of Bergamasco and Cini's[5] calculated 1.2 mb cross-section for (5) with muons of ~ 150 Gev and the total muon cross-sections of 4.2 mb and 5.9 mb, obtained from our data at 820 m.w.e. and 1415 m.w.e. (where the average muon energies are 116 Gev and 177 Gev, respectively). The remaining ~ 75% of the ^{37}Ar from ^{39}K arises from secondary neutrons, protons, and pions. Of the secondaries, the neutrons are by far the most effective. The reaction of most importance therefore is

$$N + {}^{39}K \left\langle \begin{array}{l} \rightarrow 2N + P + {}^{37}Ar \\ \rightarrow 3N + {}^{37}K \end{array} \right\} \quad . \qquad (6)$$

If the cross-section ratio $\sigma({}^{39}K)/\sigma({}^{37}Cl)$ of 155 is to be checked by ^{37}Ar measurements in K and Cl targets in an accelerator muon beam, the targets must be faced by ~ 500 g cm^{-2} of rock material (several fast neutron interaction length thicknesses). In spite of the simplicity of the normalized $\overline{E}_\mu^{0.7}N(h)$ procedure for calculating the depth variation of ^{37}Ar for K and Cl detectors, the cross-section ratio, $\sigma({}^{39}K)/\sigma^{37}Cl)$ measured to be 155 in the mine, are not calculated because of the complexity of the neutron, proton, and pion cascades.

Reactions (4), (5), and (6) should all follow the $\overline{E}_\mu^{0.7}N(h)$ depth dependence according to Wolfendale *et al*.[3] and Zatsepin *et al*.[4] These authors do not explicitly discuss the effect of a change in detector size on the $\overline{E}_\mu^{0.7}N(h)$ rule; however, since they normalize $\overline{E}_\mu^{0.7}N(h)$ with ^{37}Ar data for 600 gallon C_2Cl_4 detectors and apply the result to the 10^5 gallon C_2Cl_4 detector, it is obvious that they consider the detector size to be of no importance.

SUMMARY

The depth variation of the ^{37}Ar production rate in K is in good accord with the $\overline{E}_\mu^{0.7}N(h)$ proportionality expected for fast muon interactions down to 4000 m.w.e. depth. The measurements at 4000 m.w.e. depth give 0.07 ± 0.07 ^{37}Ar day^{-1} ton^{-1} K, which is consistent with the value of 0.065 ± 0.010 ^{37}Ar day^{-1} ton^{-1} K expected by extrapolating the shallower depth results according to the $\overline{E}_\mu^{0.7}N(h)$ rule. The error in the ^{37}Ar production rate in K at 4000 m.w.e. depth should be reduced by additional measurements.

ACKNOWLEDGEMENT

Research carried out at Brookhaven National Laboratory was performed under contract DE-AC02-76CH00016 with the U.S. Department of Energy and supported by its Office of High Energy and Nuclear Physics.

REFERENCES

1. J. N. Bahcall, W. F. Huebner, S. H. Lubow, P. D. Parker, and R. K. Ulrich, Rev. Mod. Phys. <u>54</u>, 767-799 (1982).

2. R. Davis, Jr., B. T. Cleveland, and J. K. Rowley, Conf. on the Intersection between Particle and Nuclear Physics, Steamboat Springs, CO, May 23-30, 1984, to be published in A.I.P. Conf. Proc. (1984).
3. A. W. Wolfendale, E. C. M. Young, and R. Davis, Jr., Nature (Phys. Sci.) 238, 130-131 (1972).
4. G. T. Zatsepin, A. V. Kopylov, and E. K. Shirokova, Sov. J. Nucl. Phys. 33(2), 200-205 (1981).
5. L. Bergamasco and G. Cini, Il Nuovo Cimento, 1, CN4, 293-298 (1978).
6. E. L. Fireman, Neutrino '77, Baksan Valley 1, 53-59 (1978).
7. E. L. Fireman, 16th International Cosmic Ray Conf. (Kyoto, Japan) 13, 389-393 (1979).
8. G. L. Cassiday, 13th International Cosmic Ray Conference, (Denver, CO) 3, 1968 (1973).
9. S. Miyake, 13th International Cosmic Ray Conf. (Denver, CO) Rapporteur Paper, pp. 3638-3655 (1974).
10. C. Grupen, A. W. Wolfendale, and E. C. M. Young, Il Nuovo Cimento 10B, 144-154 (1972).
11. S. Spannagel and E. L. Fireman, J. Geophys. Res. 77, 5351-5359 (1972).

THE HOMESTAKE LARGE AREA SCINTILLATION DETECTOR AND COSMIC RAY TELESCOPE

M.L. Cherry, S. Corbato, D. Kieda, K. Lande, and C.K. Lee
Dept. of Physics, Univ. of Pennsylvania, Philadelphia, PA 19104

R.I. Steinberg
Dept. of Physics, Drexel University, Philadelphia, PA 19104

The Homestake Large Area Scintillation Detector consists of 140 tons of liquid scintillator in a hollow 8 m x 8 m x 16 m box surrounding the Brookhaven ^{37}Cl solar neutrino detector. The experiment is located at a depth of 4850 ft. (4200 m.w.e.) in the Homestake Gold Mine. Half of the detector is currently running; the full detector will be taking data early in 1985. An extensive air shower array is also currently under construction on the earth's surface above the underground chamber, consisting of 100 scintillators, each 3 m^2, covering approximately 0.8 km^2; the first portion of the surface array will also be providing data in early 1985. Together, the new Homestake detectors (Fig. 1) will be used to search for slow, massive magnetic monopoles; study the zenith angle distribution of neutrino-induced muons; search for neutrino bursts from the gravitational collapse of massive stars; measure the multiplicity and transverse momentum distributions of cosmic ray muons; and study the composition of the primary cosmic rays. In this paper, we present a progress report on the new detectors. In Sec. I we describe the underground device and its capabilities as a monopole detector; in Sec. II we describe the surface array and the cosmic ray studies; the neutrino measurements have been discussed elsewhere[1].

I. Monopoles and the Underground Detector

Arguments based on the Parker limit[2] suggest that the mean flux of monopoles in the galaxy can be no more than about 10^{-15} $cm^{-2}sec^{-1}sr^{-1}$, and other arguments[3] suggest that even this value may be high. Experimentally[4], the limits set by induction experiments are now down below

0094-243X/85/1260032-18 $3.00 Copyright 1985 American Institute of Physics

4×10^{-12} cm^{-2}sec^{-1}sr^{-1}. The expected velocities of GUT monopoles caught in the solar or galactic gravitational fields are in the range 10^{-4}-10^{-3} c. The Homestake Large Area Scintillation Detector has therefore been built with as large an area and as low a velocity threshold as possible: the detector has an aperture for isotropic monopoles of 1200 m^2 sr, sufficient to detect one event in 3 years at a flux level of 9×10^{-16} cm^{-2} sec^{-1}sr^{-1}, and can detect signals as low as 0.1 times minimum ionizing

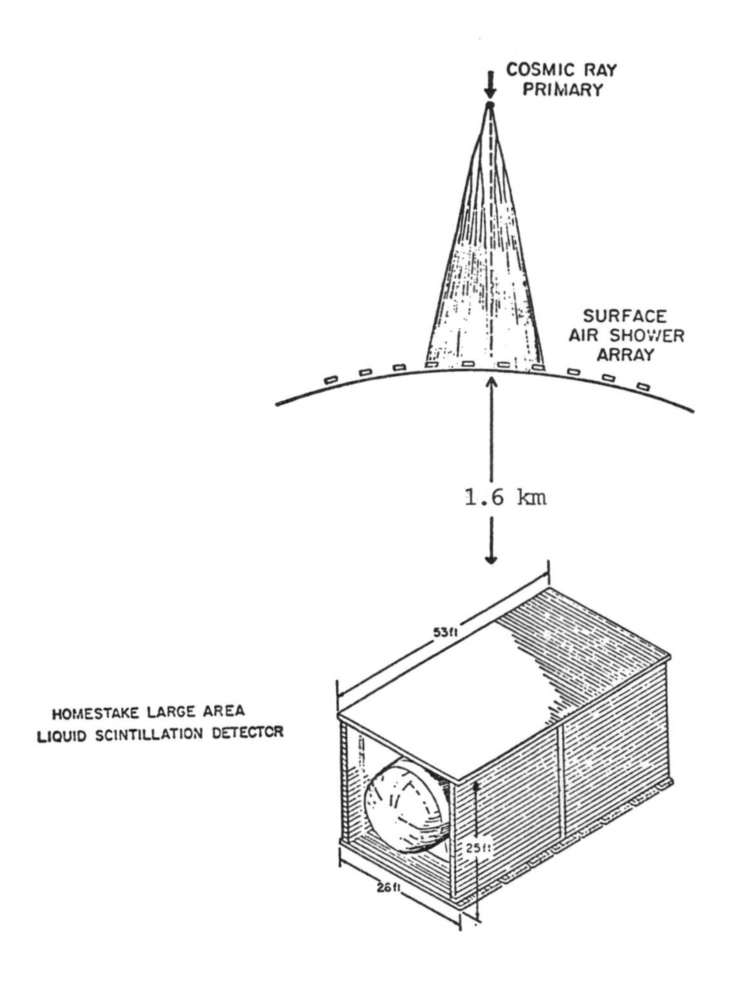

Fig. 1. Surface and underground detectors.

(corresponding to velocities 1.5 - 6 x 10^{-4}c).

At such low velocities, where the maximum energy transfer from a massive monopole to an atomic electron is less than the typical ionization energy, the calculation of expected scintillation yield is somewhat uncertain. Ahlen and Tarlé[5] suggest that the yield from a monopole at 10^{-3}c is approximately 6 times minimum ionizing, and that excitation cuts off for velocities below 6 x 10^{-4}c. Earlier calculations[6] have been more optimistic. Given the uncertainty in the calculation, we have designed the Homestake scintillator to be sensitive to as low a signal as possible. With a scintillator thickness of 25 g cm^{-2}, a signal of 0.1 times minimum ionizing corresponds to a 5 MeV pulse height, well above the level of typical energy deposits due to background radioactivity.

The Large Area Scintillation Detector consists of a hollow 8m x 8m x 16m box of 30cm x 30cm x 8m liquid scintillation detectors surrounding the existing ^{37}Cl solar neutrino tank (Fig. 2). Each of the 200 scintillator elements is a PVC box, lined with teflon (for total internal reflection), containing a low-cost mineral oil-based liquid scintillator developed to have excellent light collection and transmission characteristics, a light attenuation length greater than 8 m (Fig. 3), long-term stability, a high flash point, and low toxicity. Each detector element is viewed by two 5-inch photomultiplier tubes in coincidence, one at each end. Fast muons passing through the midpoint of one of the modules produce an average of 350 photoelectrons at each photomultiplier. A particle ionizing even at 0.01 times minimum would thus produce 3-4 photoelectrons at each photomultiplier and still be visible. The low energy threshold is therefore set not by the scintillator light output, but rather by the background produced by the ambient radioactivity (primarily MeV gamma rays) from the rock walls. The individual detector elements have $\pm$ 1.3ns time resolution, spatial resolution of $\pm$ 15 cm, and a very low muon background flux (1100 m^{-2} yr^{-1}, a factor of 10^7 below the surface flux; see Fig. 4). Cosmic ray muons and neutrino-induced muons will typically produce two pairs of coincident photomultiplier tube pulses, one pair as the muon enters the detector and one

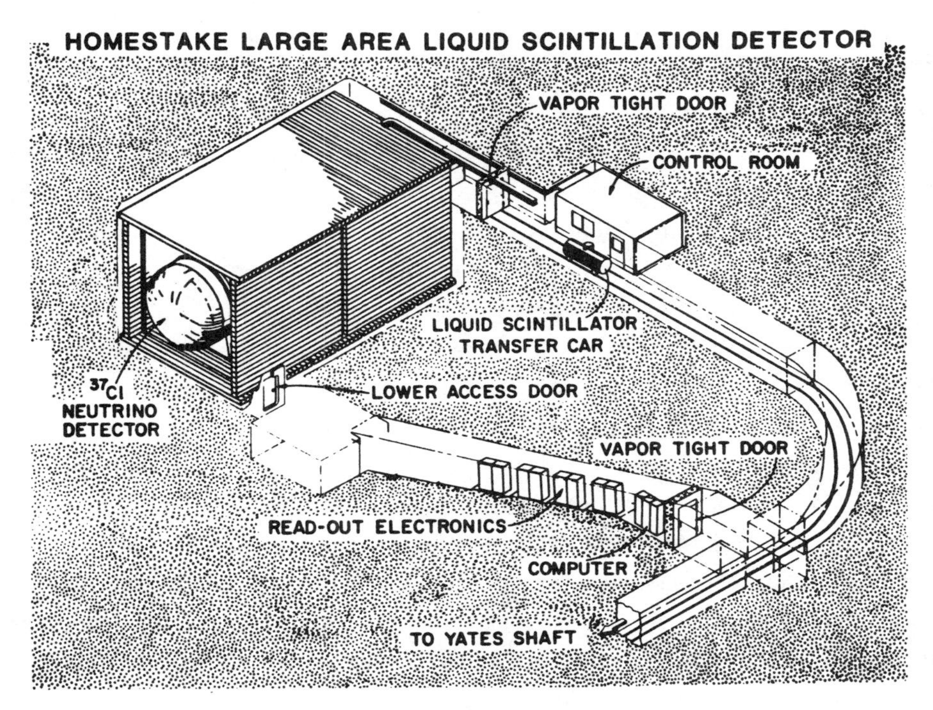

Fig. 2

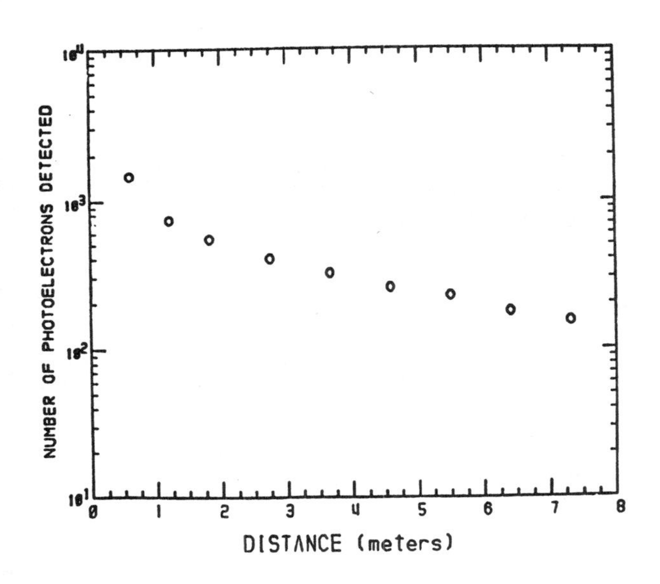

Fig. 3 Measured photoelectrons per muon vs. distance from photomultiplier tube in an 8m x 30cm x 30cm detector module. Muon energy loss is 54 MeV.

delayed pair as the muon leaves the detector. The delay between the entering and exiting pulses will be about 25 ns. From the time difference between the pulses at each end of a given module, and the corresponding ratio of pulse heights, we can locate the position of a muon to $\pm$15 cm. We can also recognize multiple muons passing through a given module by the increased pulse height and the mismatch between time differences and pulse height ratios. From the location of the entering and exiting points, the muon direction can be determined to $\pm$ 3°.

Examples of observed muon events are shown in Fig. 5. In both cases, the southern half of the detector is shown viewed from the east. The clock counts are shown for each photomultiplier tube that views an event, in units of 2.5 ns. The clocks count backwards, so that a high clock reading means that the tube saw the event early. On the left-hand side of Fig. 5, the muon enters the top counter slightly to the western (far) side of the midpoint. It then passes through the C1 tank and reappears moving downward through the five lowest boxes on the far (west) side. In the right half of Fig. 5, a neutrino-induced muon enters the bottom close to the west phototube, and then moves up the west wall, moving from north to south until it exits into the rock.

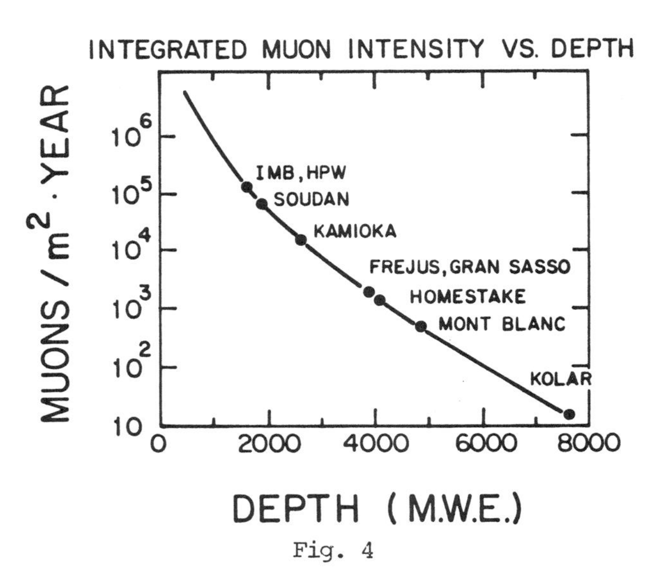

Fig. 4

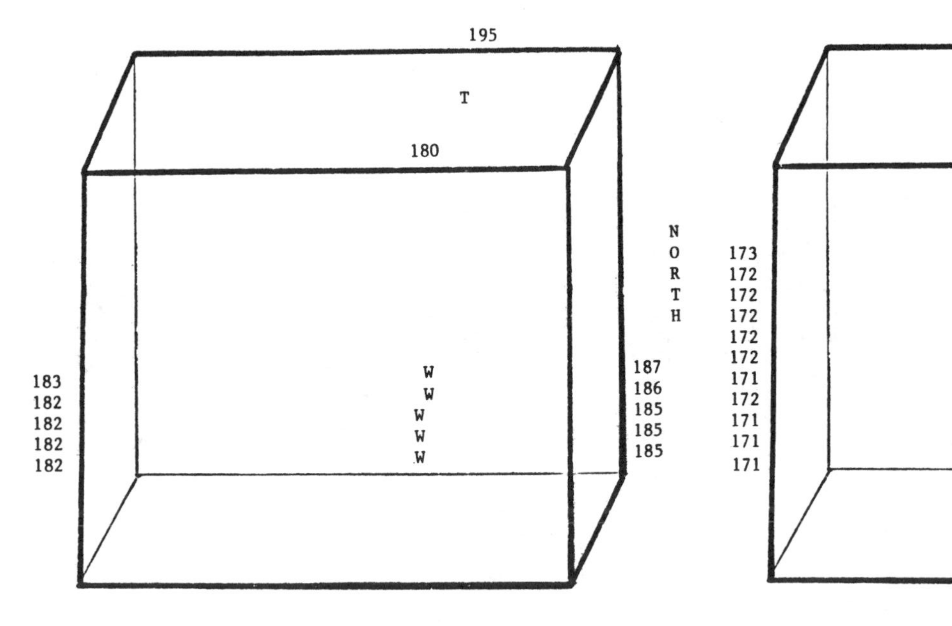

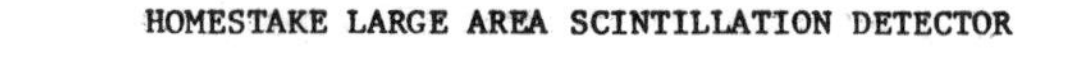

Fig. 5. Observed muon events in the LASD. The left-hand figure shows a muon moving down through the detector; the right-hand side shows an upward-moving event.

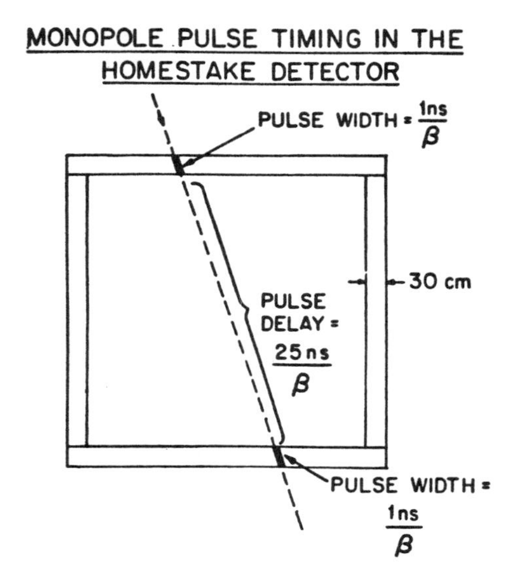
MONOPOLE PULSE TIMING IN THE
HOMESTAKE DETECTOR
PULSE WIDTH = 1ns/β
30 cm
PULSE
DELAY =
25ns/β
PULSE WIDTH =
1ns/β

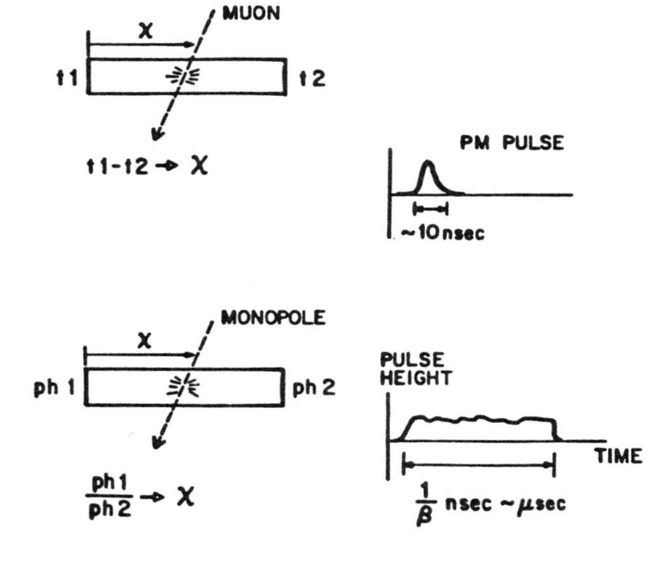
MUON & MONOPOLE SCINTILLATOR SIGNALS
MUON
X
t1
t2
t1-t2 → X
PM PULSE
~10nsec
MONOPOLE
X
ph 1
ph 2
ph1/ph2 → X
PULSE
HEIGHT
TIME
1/β nsec ~μsec

Fig. 6

For a monopole, we expect a pair of slow pulses with width 1 ns/β as the monopole enters the detector and, after a delay of 25 ns/β , a second pair of slow pulses as the monopole leaves (Fig. 6). For the velocities of interest, the delay time between the two entering and exiting pulses will be 25-250 μs. Such long delays can only be correlated in a very low background environment such as that available in a deep mine. The most severe background will be due to two independent, traversing muons, each of which is detected in only one of the two counters through which it passes. The accidental coincidence rate for independent muons mimicking a slow particle of velocity β is $4\mathrm{x}10^{-5}/\beta$ yr^{-1}. At $\beta = 10^{-4}$, this corresponds to 0.4 accidental coincidences per year. This rate will be further reduced by requiring that there be no fast (25 ns) coincidence and that the four individual photomultiplier pulses be wide (1 ns/β). The monopole position in each box will be determined from the ratio of pulse heights at each end of the box; the individual pulse heights are then corrected for the position, and the monopole pulse height is determined.

The detector array provides an aperture of 1200 m^2 sr, corresponding to one monopole event in 3 years at a flux level of 9×10^{-16} $cm^{-2}sec^{-1}$ sr^{-1}. The Homestake sensitivity is shown in Fig. 7 as a solid line for $\beta > 6 \times 10^{-4}$ and a dashed line down to $\beta = 1.5 \times 10^{-4}$, reflecting the uncertainties in the estimates of energy loss at low velocities. The limits obtained with other detectors can be put into two classes: First are the induction experiments, shown as a global limit of 4×10^{-12} cm^{-2} $sec^{-1}sr^{-1}$ representing the combined Stanford, IBM, and Chicago-FNAL-Michigan results[4]. The remaining results are obtained from ionization counters. Of these, the Homestake LASD, with three years of running, will provide the greatest sensitivity.

The electronics are designed to permit us to look at both fast muons and slow monopoles. The muon circuitry is currently running on the southern half of the detector. An amplifier/discriminator is mounted outside the detector module at each photomultiplier tube. A fast discriminator on this unit gives the fast muon timing pulse. A coincidence of discriminator pulses from two ends of a single box provides the trigger. The pulse height from each tube is then measured in a flash analog-to-digital converter. The range of interesting times is from 1 ns to 250 μs. The system is therefore equipped with a fast clock (2.5 ns resolution) which covers the first 500 ns and a slow UTC clock (0.1 μs resolution) to cover the time span thereafter. A 16 word deep memory buffer is associated with each photomultiplier so that multiple pulses for each event can be recorded.

The monopole circuitry is presently being built. It provides the functions of a transient recorder for each photomultiplier. The flash ADC will run continuously, sampling the pulse height every 5-10 ns and filling a deep memory associated with each photomultiplier tube. In parallel with the ADC, the analog pulse is integrated and fed to a slow

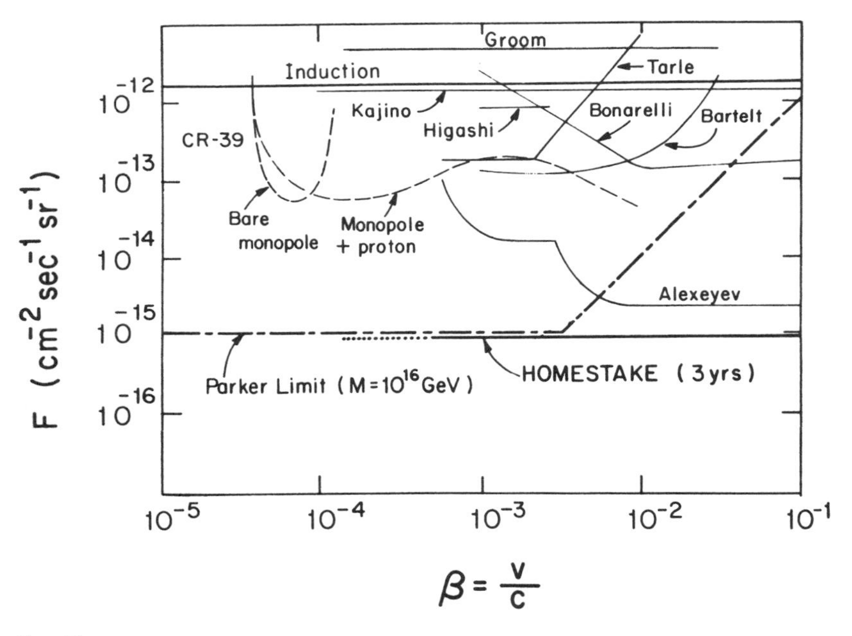

Fig. 7. Slow monopole flux limits. The Parker limit is that given for monopoles of $M = 10^{16}$ GeV by Turner et al.[2] The induction limit is the global limit suggested by Stone[4]. The CR-39 limit is the Price result[7]. The Homestake line is the result of running for three years with the detector described here. Other limits are referenced in ref. 7.

discriminator. When a coincidence of either fast or slow discriminator pulses is seen, the memories are read out for every phototube which fires, giving a record of pulse heights in intervals of 5-10 ns over an interval of up to 250 μs. A muon pulse then appears as a rapidly-rising (10 ns) pulse, whereas a monopole pulse rises more slowly ($1/\beta$ ns).

The mechanical work on the underground detector is essentially finished. The southern half of the detector has been filled with liquid and turned on. We are currently testing and calibrating this portion of the detector while installing the north-side electronics and filling the remaining detector modules with oil. The detector is expected to be fully operational early in 1985.

II. The Homestake Cosmic Ray Telescope

Astrophysically, the region of cosmic ray energies near 10^{15} eV seems to be very interesting: there appears to be a discontinuity in the spectrum,[8] and the Maryland data suggest a heavy composition near this region[9] (Fig. 8a); data on the depth of shower maximum likewise suggest a heavy composition[10] (Fig. 8b); and the anisotropy observed in air showers begins to increase noticeably near this region.[8] Experimentally, however, this is a difficult energy regime in which to work: it is too high for direct balloon and satellite measurements, and too low for most air shower detectors. In addition, the unknown particle physics has been a major cause of uncertainty. Very shortly, though, the particle physics should be known reasonably well with the advent of collider data up to 2 TeV in the center of mass (corresponding to cosmic ray energies of 2×10^{15} eV); and with the results of the Chicago composition experiment aboard Spacelab,[11] the composition should be well known up to energies of 2 TeV/nucleon (10^{14} eV total energy for iron).

By combining a large (0.8 km^2) air shower array on the surface with the deep underground detector, we can measure the cosmic ray composition between 10^{14} and 10^{16} eV. Measuring the total electron number N_e on the surface (i.e., the total energy/nucleus) and the multiplicity of high

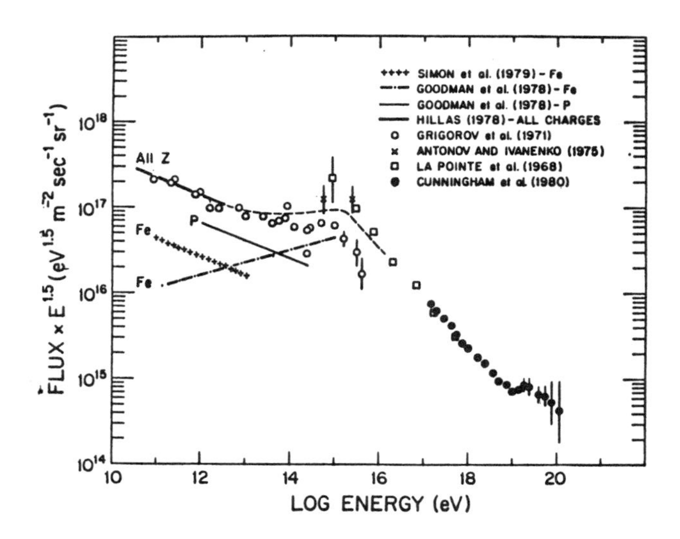

Fig. 8a. Summary of measured cosmic ray energy spectra.

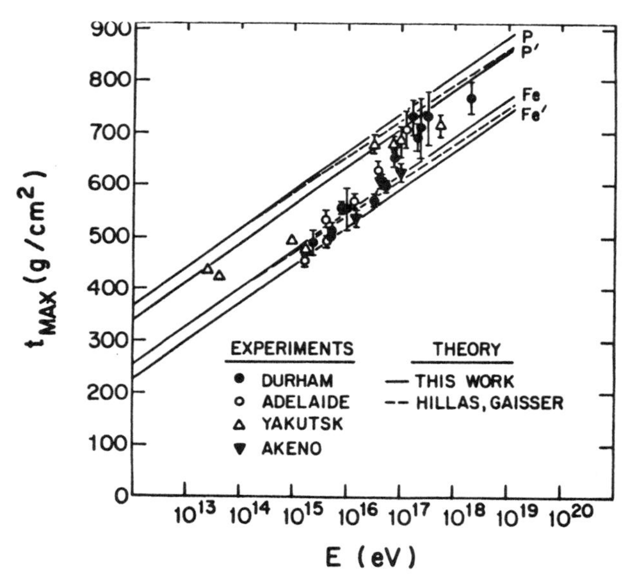

Fig. 8b. Comparison between experimental results and theoretical predictions of the mean depth of shower maximum t_{max} as a function of primary energy[10]. The calculations are made for the case of primary protons and iron.

energy ($E_\mu \gtrsim 2.7$ TeV) muons underground (i.e., the energy/nucleon) permits discrimination between primary species in a way that depends essentially on energetics. In order to reach our depth, muons must have roughly 2.7 TeV at the surface of the earth. Such muons can be produced by proton primaries with energies in excess of 10^{13} eV or, for example, by iron primaries with energies above a few times 10^{14} eV. A proton generally gives rise to a single high energy muon while an iron, consisting of a superposition of 56 separate nucleons, has a large probability of multiple muon production, particularly above 10^{15} eV (Fig. 9). Our data will thus consist of muon multiplicity and separations underground, and shower size at the surface. For small showers ($E \lesssim 10^{15}$ eV) we expect to observe single muons primarily from cosmic ray protons, while for large showers ($E \gtrsim 10^{15}$ eV) we expect a mix of single and multiple

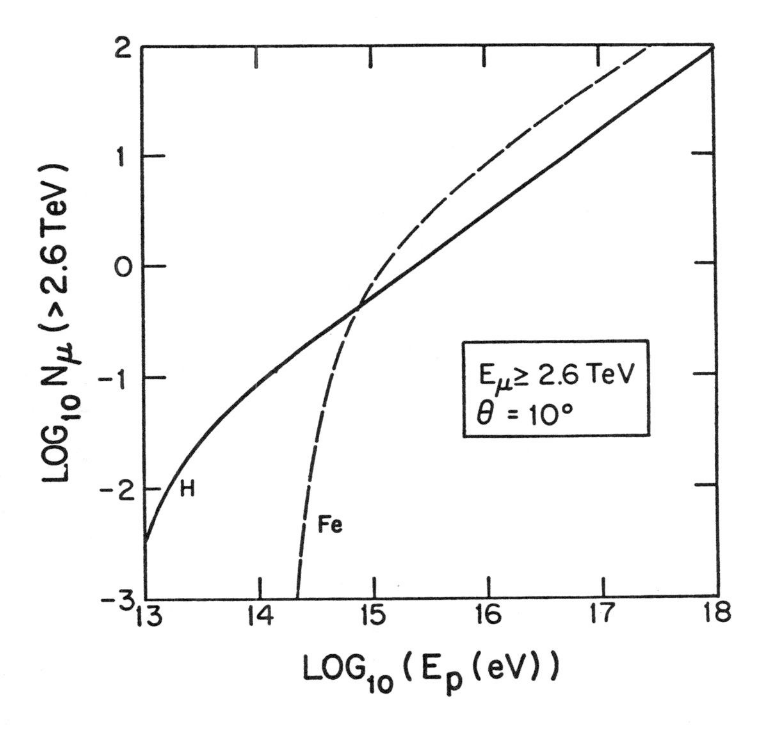

Fig. 9. Expected number of underground muons from a single proton or iron primary of total energy E_p.

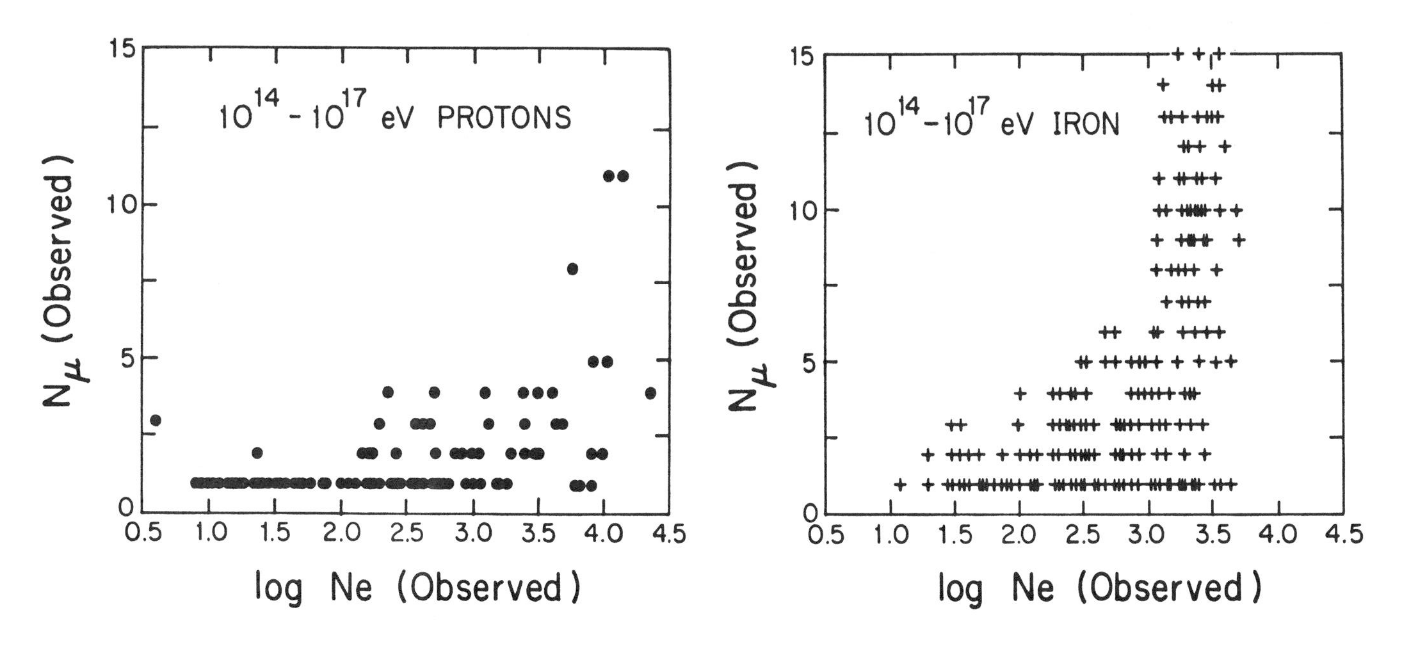

Fig. 10a. Distribution of observed underground muon number N_μ vs. observed electron number N_e, based on a Monte Carlo simulation of one year of operation, wiht a spectrum of 10^{14} - 10^{17} eV protons described in ref. 12.

Fig. 10b. Distribution of $N\mu$ vs N_e for iron

muons from protons and heavy (nominally iron) primaries. In Fig. 10 we demonstrate the expected capabilities of the detector system by showing distributions of observed underground muon multiplicity N_μ vs. observed shower size N_e, based on Monte Carlo simulations of one year's worth of proton and iron events in the full surface-underground telescope.

The only existing or contemplated underground detectors sufficiently deep to make such a composition measurement at 10^{15} eV are at Frejus, Mt. Blanc, Gran Sasso, Kolar, and Homestake. All other present or planned detectors are at depths significantly less than 4200 m.w.e. and consequently vertical muon thresholds less than 1 TeV[1].

Another consideration in choosing the depth has to do with the lateral detector dimensions required to contain events. Accurate measurement of muon multiplicities requires an underground detector with surface area sufficiently large to contain the showers. From the CERN $p\bar{p}$ Collider data and the underground muon results at Utah and Homestake,[13] it appears that typical transverse momenta at 10^{14} - 10^{15} eV rise to $p_t \gtrsim 500$ MeV/c. In order to detect all the secondary muons, a detector should be capable of containing p_t up to 1 - 1.5 GeV/c. The linear dimensions necessary for the underground detector can be related to the minimum pion energy E_π required so that the decay muon can penetrate underground:

$$\ell \sim \frac{p_t}{E_\pi} h.$$

Here h is the typical altitude of the primary interaction ($h \sim 19$ km) and p_t is the typical transverse momentum of the pion in the production process. For the Homestake detector, with $E_\pi \sim 3.6$ TeV and $p_t \sim 500$ MeV/c, we find $\ell \sim 2.6$m, well within the present 8m x 16m dimensions. The typical existing 10m dimensions permit measurements up to $p_t \sim 2$ GeV/c for vertical showers. In Fig. 11, we plot the depth vs. lateral dimensions required to contain showers with p_t = 400, 800, and 1500 MeV/c. Multiple scattering in the rock causes the curves to converge at 2.5 m. The experiment should therefore be sufficiently deep to provide a high muon threshold, yet not so deep that the transverse momentum resolution is

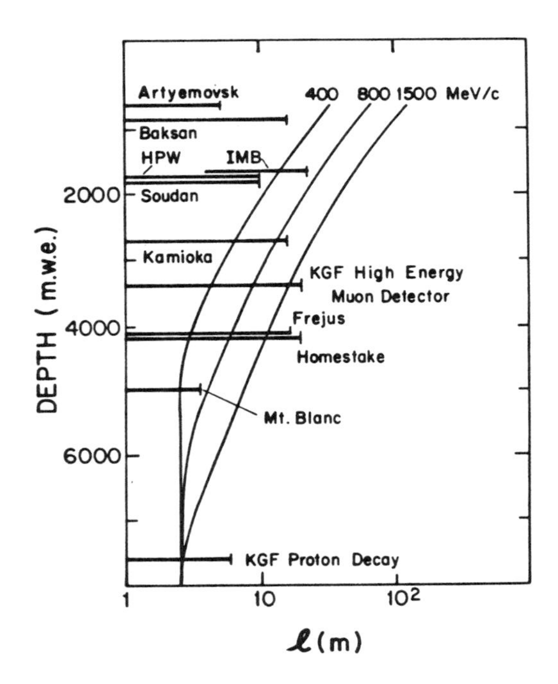

Fig. 11. Depth vs. lateral dimension required to contain showers with p_t = 400, 800, and 1500 MeV/c, showing existing and proposed deep underground detectors.

degraded by scattering effects.

The surface array will consist of approximately 100 scintillation detectors, spaced by 25 - 200 m and deployed over an area of about 0.8 km^2 over the underground chamber. The individual detector elements consist of reinforced concrete boxes 4 ft x 8 ft x 2 ft high with 3" thick side walls, covered on top by a 24 gauge galvanized tin cover plate. The inside of the box is lined with styrofoam insulation and an aluminum light reflector. The active detector is 4" of liquid scintillator, designed to have a high flash point and to remain clear at low temperatures. The scintillator is viewed by two 5" photomultiplier tubes operating in coincidence. Laboratory tests give a variation in response over the face of the counter of less than $\pm 20\%$.

For detectors satisfying the two-fold coincidence between the phototubes, we read out the sum of the pulse heights and the relative time at which the detector fires. A temperature-compensated discriminator and two-stage amplifier provide a dynamic range of 10^3. High voltage and low voltage regulation, pulse amplification, shaping, discrimination and coincidence circuitry are housed in the upper part of the detector box near the photomultiplier tubes. Pulse height, timing, and monitoring data are transmitted to a central trailer via a coaxial and twisted pair cable link. Pulse arrival times from individual detector boxes are measured with a resolution of 2.5 ns, and the pulse heights are individually digitized in 64-channel analog-to-digital flash converters. A system trigger is generated either by N ($N \gtrsim 3$) surface detectors firing simultaneously or by a pulse from the underground detector. The trigger transfers the data, labelled with a time readout from a WWV-correlated UTC clock, to a DEC PDP 11/23 minicomputer. The relative timing of pulses from individual detectors will be known to 2.5 ns, and the absolute time to 100 μs.

The telescope can operate either in a "prompt" mode in which only the surface elements fire, or in a "delayed" mode in which a trigger pulse from the underground detector arrives approximately 14 μs after the surface array signals. The 14 μs delay is the result of the 5 μs muon flight time from the surface to the underground detector plus the 9 μs signal propagation time along the cable connecting the underground detector and the surface array.

The location of the shower core will be calculated from the locations, pulse heights, and arrival times seen by those detectors firing in the shower. Based both on the experience with the Haverah Park array and Monte Carlo calculations, we expect to locate shower cores to within 4 m in the central section of the array and 10 - 20 m in the outer part. Underground, the liquid scintillator elements make it possible to resolve tracks separated by 2 ft; however, the final underground position uncertainty of 2.5 m is determined primarily by scattering in the rock. The angular resolution of the combined surface-underground telescope is then 3 - 10 mrad. The expected surface-underground coincidence rate will be a few hundred per year.

The first twenty-seven boxes have now been installed, filled with oil, and turned on. They will be operated over the Winter, and the remaining boxes will be placed in operation in 1985.

III. Summary

The southern half of the underground detector is presently running and collecting muon data. The full detector is expected to be filled with scintillator and running, together with the first twenty-seven surface array boxes, this Winter.

The Homestake scintillator experiments are a collaborative effort involving the University of Pennsylvania, the University of Leeds (A. Watson, R. Reid, J. Lloyd-Evans), Brookhaven National Laboratory (B. Cleveland), the University of Texas at Dallas (E. Fenyves), and the South Dakota School of Mining and Technology (T. Ashworth). Funding is provided by the U.S. Department of Energy and the National Science Foundation. The assistance and generous cooperation of the Homestake Mining Company are deeply appreciated, as are the advice and assistance of R. Davis, I. Davidson, T. Daily, K. Brown, and E. Marshall.

References

1) M.L. Cherry, I. Davidson, K. Lande, C.K. Lee, E. Marshall, and R.I. Steinberg, Proc. Workshop on Science Underground, Los Alamos, ed. by M.M. Nieto et al., 248 (1982).
2) See M.S. Turner et al., Phys. Rev. D26, 1296 (1982) for a recent discussion.
3) See J. Preskill, Proc. Inner Space/Outer Space, FNAL (1984), and R.A. Carrigan and W.P. Trower, Nature 305, 673 (1983) for reviews.
4) J. Stone, Proc. Inner Space/Outer Space, FNAL (1984).
5) S. Ahlen and G. Tarlè, Phys. Rev. D27, 688 (1983).
6) D.M. Ritson, SLAC Report No. SLAC-Pub-2950 (1982); cf. also R.A. Carrigan and W.P. Trower, eds., Magnetic Monopoles, Plenum, N.Y. (1983), and ref. 3.
7) P.B. Price, CERN preprint EP/84-28 (1984); D.E. Groom et al., Phys. Rev. Lett. 50, 573 (1983); F. Kajino et al., 18th Intl. Cosmic Ray Conf., Bangalore, 5, 56 (1983); S. Higashi et al., 18th ICRC, Bangalore, 5, 69 (1983); J. Bartelt et al., Phys. Rev. Lett. 50, 655 (1983); R. Bonarelli et al., Phys. Lett. 126B, 137 (1983); G. Tarle et al., Phys. Rev. Lett. 52, 90 (1984); E.N. Alexeyev, 18th ICRC, Bangalore, 5, 52 (1983).
8) J. Linsley, 18th International Cosmic Ray Conference, Bangalore 12, 135 (1983).
9) J.A. Goodman et al., Phys. Rev. D26, 1043 (1982).
10) J. Linsley and A.A. Watson, Phys. Rev. Lett. 46, 459 (1981); M.L. Cherry, A. Dar, and S.A. Bludman, preprint (1984).
11) D. Müller, J. L'Heureux, and S. Swordy, Cosmic Ray and High Energy Gamma Ray Experiments for the Space Station Era, Baton Rouge, ed. by W.V. Jones and J.P. Wefel (1984).
12) M.L. Cherry, I. Davidson, K. Lande, C.K. Lee, E. Marshall, and R.I. Steinberg, Workshop on Very High Energy Cosmic Ray Interactions, Philadelphia, ed. by M.L. Cherry, K. Lande, and R.I. Steinberg, 356 (1982).
13) M.L. Cherry, M. Deakyne, K. Lande, C.K. Lee, R.I. Steinberg, B. Cleveland, and E.J. Fenyves, Phys. Rev. D27, 1444 (1983).

THE ^{76}Ge DOUBLE BETA DECAY EXPERIMENT AT HOMESTAKE

R.L. Brodzinski, D.P. Brown, J.C. Evans, Jr.,
W.K. Hensley, J.H. Reeves, and N.A. Wogman
Pacific Northwest Laboratory, Richland, Washington 99352

F.T. Avignone, III and H.S. Miley
University of South Carolina, Columbia, S. C. 29208

ABSTRACT

An ultralow background intrinsic Ge detector has been developed over several generations of experiments. The radioactive background from construction materials has been reduced by more than two orders of magnitude. The sources of background in a standard commercial cryostat have been identified and eliminated. Data taken with this 135 cm^3 prototype installed in the Homestake Gold Mine are presented. A large (1440 cm^3) detector and data acquisition system now under construction are also described.

INTRODUCTION

There are clearly many applications for a γ ray detector having more than two orders of magnitude more sensitivity than a standard, low background, Compton suppressed commercial intrinsic Ge spectrometer shielded by lead. This paper describes a series of experiments performed with a 135 cm^3, intrinsic Ge detector in three different environments. The baseline experiment involved a typical low level counting facility. Sources of background were identified, and a baseline sensitivity was measured as a function of energy. An extensive program, carried out over several years, identified clean sources of material and evaluated their radiopurity. A cooperative effort by Pacific Northwest Laboratory (PNL) the University of South Carolina (USC) and Princeton Gamma Tech Inc. (PGT), was formed and a new cryostat was built from materials identified from our radiopurity measurements. The detector was then tested above ground in a bulk and partial cosmic-ray anticoincidence shield. It was then moved to a site within the Homestake gold mine having 1438 meters of rock overburden which has a water-equivalent for cosmic ray attenuation of 4084 meters. The measurements at this depth showed that the radioactive background was reduced by between two and three orders of magnitude.

A 1440 cm^3 mosaic Ge detector system is presently under construction. While both detectors will have a large number of applications in low background γ ray counting, the sensitivity of each is discussed in the context of an experiment to search for both no-neutrino and two-neutrino double beta decay of ^{76}Ge. This fundamentally important experiment will place limits on or detect the effects of a Majorana mass of the electron neutrino or of lepton nonconservation.

THE NEUTRINO MASS

In an early paper Fermi[1] concluded that beta decay spectral shapes implied that the neutrino rest mass is very small relative to the mass of the electron. Unlike the case of the photon, there is no fundamental reason why the neutrino must be massless. Intrinsic properties of neutrinos hold important keys to understanding very basic questions ranging from cosmological to those bearing on the details of Grand Unified Theories.

The present experimental limits on the mass of the electron neutrino come from several experiments. In 1980 Lubimov and his co-workers[2] reported the analysis of 18 experimental data sets from a study of the shape of the beta spectrum in tritium decay near the end-point. They placed limits of $16\text{eV} \leqq m_\nu \leqq 46\text{eV}$. The same group repeated the experiments after reducing the background by an order of magnitude and improving the energy resolution by a factor of two[3]. Their model independent limit was $m_\nu > 20\text{eV}$ when worst case scenarios concerning the corrections for energy resolution and the atomic and molecular excitations are assumed. When all corrections are applied, the mean value is $m_\nu = 33.1 \pm 1.1$ eV. These limits from direct searches do not depend on the charge conjugation properties of the neutrino.

DOUBLE BETA DECAY

Double beta decay is a second order weak process and is the only possible mode of decay between even-even nuclei whose binding energies are increased by pairing forces sufficiently to render first order beta decay energetically forbidden. Two possible decay modes are:

$$(Z,A) \rightarrow (Z+2,A) + 2\beta^- + 2\bar{\nu}_e \text{ and } (Z,A) \rightarrow (Z+2,A) + 2\beta^- . \quad (1)$$

The first decay is an ordinary $\Delta\ell=0$, second-order weak decay, while the second is lepton nonconserving, $\Delta\ell=2$, and can only occur if the neutrino is a Majorana particle. Studies of this exotic decay mode yield specific information about the properties of the neutrino under charge conjugation[4]. The most important issues are associated with the $\Delta\ell=2$ decay; however, experimental determinations of half-lives of $\Delta\ell=0$ decays are very important for testing the theoretical nuclear matrix elements.

The $\Delta\ell=2$, $\beta^-\beta^-$-decay involves the leptonic current,

$$J^\ell_\mu = \psi^+_e \gamma_4 \gamma_\mu \{(1+\gamma_5) + \eta(1-\gamma_5)\} \, [\psi_{\bar{\nu}}(x)+\psi_\nu(x)]. \quad (2)$$

The γ_5-invariance can be broken by neutrino mass or explicitly by $\eta(1-\gamma_5)$. Detailed nuclear structure calculations yield an expression for the decay constant λ of the form,

$$\lambda \cong \{a\xi^2 + b\eta^2 + c\eta\xi\}, \tag{3}$$

where $\xi \equiv m_\nu/m_e$ and where higher order terms in η and ξ are neglected. In principle then, at least limits on m_ν and η can be determined from the same experiment in cases for which the constants a, b, and c can be reliably calculated. The most elaborate theoretical treatment of the nuclear aspects of double beta decay are the shell model calculations of Haxton, Stephenson and Strottman[5], who treated the cases of ^{76}Ge, ^{82}Se, ^{128}Te and ^{130}Te. The result for the two-neutrino ($\Delta\ell$=0), $\beta^-\beta^-$-decay of ^{76}Ge is $T_{1/2}$ = 4.15 x 10^{20}y while for the no-neutrino, ($\Delta\ell$=2) decay, the expression corresponding to eq. (3) is

$$\lambda = 3.42\times10^{-21}\ \text{sec}^{-1}\{\eta^2+1.54\xi^2 - 0.595\eta\xi\}. \tag{4}$$

THE PNL-USC EXPERIMENTS

The goals of our program were to establish a baseline with a commercially available detector, to reduce the background by testing and carefully selecting construction materials, to construct and test a low background prototype and finally, to design and construct a 1440 cm^3 mosaic Ge detector comprised of two 720 cm^3 separate modules.

The apparatus used in the establishment of the original baseline was an intrinsic Ge detector in a commercially available low-background cryostat, with a NaI(Tl) anticoincidence shield discussed in an earlier work[6]. The effective volume was 125 cm^3 after accounting for the surface escape of β^- particles. Data were accumulated for a total of 4054 hrs. in periods of several days each. The effective energy resolution from the sum of all spectra is 3.4 keV in the region of the decay Q-value which is 2040.71 ± 0.52 keV[7]. The data from the different counting periods were combined after correcting for slight gain shifts.

In our earlier work, we invested significant effort in reproducing the background γ rays, with the correct relative intensities, using radioactive sources of ^{228,232}Th, ^{226}Ra, ^{234m}Pa, ^{60}Co, ^{137}Cs, ^{40}K and a PuBe neutron source. The sources were located at a variety of points with various energy degraders until the proper line shapes and relative intensities were achieved. The reference spectra collected in this manner were normalized to the background peaks in the data and subtracted leaving a smooth continuum due to cosmic rays. This was fit to a function $y = ax^b$ where x is proportional to energy and a and b are constants. Accordingly the sources of background and the approximate contribution from cosmic rays were determined. The results of the identification and the percent contribution of each source are given in Table I.

Table I: Experimentally determined background levels in the no-neutrino and a portion of the two-neutrino energy region.

Source	% total (2041 keV)	% total (1000-1100 keV)
^{226}Ra	44%	24%
^{228,232}Th	22%	13%
neutrons	13%	4%
^{234m}Pa	-	24%
^{40}K	-	16%
^{60}Co	-	2%
Cosmic Ray Continuum	21%	17%

Several steps were taken to reduce the level of the background in the original experiment. First, the NaI(Tl) system was eliminated which significantly reduced the background due to ^{226}Ra and its daughters. The source of the remaining background was then traced step by step. The aluminum end cap, diode cup, support hardware, and electronic parts closest to the detector proved to be the primary sources of primordial and man-made radioactivity. The isotopes ^{228}Ac, ^{212}Pb, and ^{208}Tl from the ^{232}Th chain, ^{234m}Pa from the ^{238}U chain, ^{235}U, and ^{40}K were the significant primordial contributors, with ^{137}Cs and ^{60}Co being the only discernible man-made radionuclides. Prospective construction materials, in quantities of ten to one hundred times that actually used in the detector, were assayed for radionuclide contamination using two 30 cm diameter by 20 cm thick NaI(Tl) detectors operating in coincidence with each other and in anticoincidence with a large plastic scintillator[8]. Aluminum was found to be the major source of primordial contamination, with capacitors, resistors, field-effect transistors (FET), and rubber O-rings being small contributors[9]. The stainless steel screws contained ^{60}Co.

Samples from three types of copper were analyzed, and the one with the least amount of primordial radioactivity (<0.00003 d/m/g) was selected to replace the aluminum as a construction material. Brass screws replaced the stainless steel screws, and indium was used for the O-ring vacuum seal. The FET was modified to exclude the contaminated component, and the cryostat was modified to place the preamplifier outside the shield. The results of the extensive program of counting radioactive contamination in a variety of materials are summarized in Table II.

Table II: Primordial radionuclide concentrations in various materials used in fabrication of radiation detector systems

MATERIALS	RADIONUCLIDE CONCENTRATION IN dpm/kg		
	^{208}Tl	^{214}Bi	^{40}K
ALUMINUM	7 TO 200	<4 TO 2,000	<20 TO 1,000
BERYLLIUM	10	700	<1,000
COPPER	<0.3	<0.8 TO 3	<10
COPPER (GRADE 101)	<0.03	<0.05	<0.5
COPPER (OFHC)	<0.3	<0.8 TO 10	<10
EPOXY	50 TO 4,000	80 TO 53,000	<1,000 TO 72,000
GREASE, HIGH VACUUM	<1	<7	<8
INDIUM	<1	<3	<20
LEAD	<0.02	<0.04	<0.1
MOLECULAR SIEVE	400 TO 500	1,000 TO 3,000	8,000 TO 9,000
MYLAR, ALUMINIZED	100	200	<2,000
OIL, CUTTING	<0.4	<3	<20
PLASTIC, TUBING	4	<4	<800
PRINTED CIRCUIT BOARD	2,000	4,000	4,000
QUARTZ	6 TO 60	<20 TO 1,000	<200
REFLECTOR MATERIALS	<0.1 TO 100	<0.7 TO 200	<5 TO 300
RUBBER, SPONGE	50 TO 200	80 TO 1,200	<400 TO 2,000
SILICA, FUSED	<20	<10	<100
SILICONE, FOAM	20	50	<200
SODIUM IODIDE (Tl)	<3	<4	<30
SOLDER	<0.3	<0.8	<10
STEEL, STAINLESS	<2	<6	<200
STEEL, PRE WW II	0.5	0.8	<10
TEFLON	<0.3	<1 TO 7	<20
WIRE, TEFLON COATED	<4	<1	<20

The detector was rebuilt using radiopure materials identified in our measurements discussed above and then placed in the combination bulk and anticosmic ray shield described earlier[10]. Measurements verified that much of the radioactivity had been removed; however, the continuum due to cosmic rays was still too intense to allow a sensitive determination of the level of radioactivity remaining. The detector was moved to the Homestake Gold Mine, and the bulk shield reassembled about the detector, excluding the plastic, anticoincidence live shield and associated photomultiplier tubes. The experimental site is located near the Brookhaven solar neutrino and the University of Pennsylvania cosmic neutrino experiments.

The detector is inside of a lead shield having a minimum thickness 35.6 cm. The lead is surrounded by a layer of 0.050 cm thick cadmium sheet which is enclosed by a 10 cm thick liquid scintillator which can be used as a veto counter and which is in turn surrounded by a 20 cm thick layer of boron-loaded paraffin blocks to absorb neutrons. The lead shield sits on a 102 cm x 102 cm by 20.3 cm thick plastic scintillator which can also be used as part of the veto counter. The early results showed a high background directly attributable to ^{222}Rn and its daughters. This background was eliminated to a level below detectability by piping the boil-off nitrogen gas from the cryostat tank to the inside of the shield. In this way a slight positive pressure is maintained in the cavity containing the detector which provides a constant flow of N_2 gas outward through the narrow spaces between the bricks.

Data were collected for several months, and the results are shown in Table III for the dominant, well known background peaks. In addition these rates are compared to those of our benchmark experiments and show reduction factors of between 670 and 6600. The three gross spectra corresponding to a typical shielded detector without Compton suppression, the experiments with the radiopure cryostat above ground, and those conducted in the goldmine, are shown in Fig. 1. This impressive background reduction represents a new plateau in the technology of low background, γ ray counting.

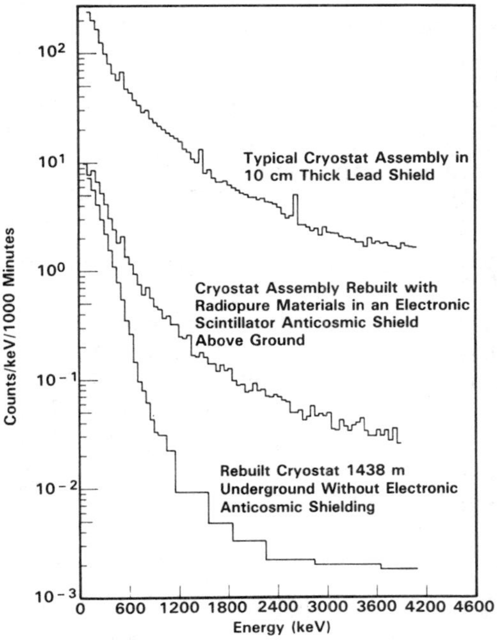

The most significant source of remaining background is the ^{210}Pb in the lead bricks themselves. This results in Pb x-rays, which are not troublesome, and a photon continuum from the bremstrahlung spectrum of ^{210}Bi which has a beta end point energy of 1161 keV. A significant reduction in this continuum has subsequently been observed by using a clean cylincrical copper liner 7.3 cm thick. The copper, however, has been found to contain approximately 0.00002 d/m/g of ^{54}Mn, 0.00005 d/m/g of ^{58}Co, and 0.00001 d/m/g of ^{60}Co, presumably induced by cosmic ray interactions while the copper is stored above ground. A lead shield being fabricated from 400

Fig. 1 Background spectra of the 135 cm^3 intrinsic Ge detector in three different shielded environments.

Table III: Comparison of primordial radioactivity levels in the background of the Ge spectrometer before and after rebuilding with radiopurity selected materials.

Primordial Radionuclide	Gamma ray energy (keV)	Count rate before rebuilding (c/hr)	Count rate after rebuilding (c/hr)	Improvement factor
^{235}U	185.72	73.0	<0.011	>6600
$^{228}Ac(^{232}Th)$	911.07	9.0	<0.0019	>4600
$^{234m}Pa(^{238}U)$	1001.03	3.4	<0.0014	>2500
^{40}K	1460.75	22.0	0.014	1600
$^{208}Tl(^{228}Th)$	2614.47	1.0	<0.0015	>670

year old Pb should substantially reduce this source of background if only the inner 10 cm of shielding is replaced by the old lead. After rebuilding the shield, with the old lead lining the detector cavity, direct observation of two-neutrino $\beta^-\beta^-$-decay should be possible. Our present data imply a conservative half life $T_{1/2}^{(2\nu)} \gtrsim 2.5 \times 10^{19}$ y (1σ) when all counts in the spectra, except those in the ^{60}Co peaks, between 0.95 and 1.45 MeV are attributed to this decay mode. This is, then, within about one order of magnitude of having reached the theoretically predicted, required sensitivity.

ANALYSIS OF THE DATA AND LIMITS ON $\beta^-\beta^-$-DECAY

Events from neutrinoless $\beta^-\beta^-$-decay would result in a full energy peak at 2040.71 ± 0.52 keV. In the present experiment, 98 days of data yielded less than one full count under an hypothesized peak at 2041 keV. The detector is an ensemble of N, ^{76}Ge atoms. If λ is the decay constant of each atom, then $(1-\lambda t)^N$ is the probability that the entire ensemble will survive the time interval t without a single decay. If $\lambda t < 1$, this expression can be easily expanded, and in the limit N >> 1, the binomial expansion of $(1-\lambda t)^N$ becomes $e^{-\lambda N t}$. The probability that there will be at least one decay in the ensemble during time t is then $(1-e^{-\lambda N t}) \equiv P$. If we wish to deduce the half life, below which we expect at least one count during the time interval t with a given level of confidence CL, we set P = CL and solve for $T_{1/2}$.

The result is,

$$T_{\frac{1}{2}}(CL) > \frac{(\ln 2)Nt}{\ln[1/(1-CL)]} \quad . \tag{6}$$

In the present detector $N = 4.314 \times 10^{23}$ ^{76}Ge atoms while the counting time was 0.2689 y. The limit on the half life for neutrinoless $\beta^-\beta^-$-decay from the 0^+ ground state of ^{76}Ge to the 0^+ ground state of ^{76}Se can be quoted as $>7.0 \times 10^{22}$ y (1σ). When this result is used with equation (4), the corresponding limit on the Majorana mass of the electron neutrino is 5.0 eV. To improve this limit significantly in a reasonable time, the number of ^{76}Ge atoms must be increased by about a factor of 10. Below, the design and development of our large detector called "Superdetector" and a projection of the ultimate sensitivity of such experiments are discussed.

DESIGN OF SUPERDETECTOR

The first step taken in the design of the large, 14 detector mosaic, composed of two 7 detector configurations[10], as shown in Fig. 2, was to calculate the efficiencies for a variety of multiple scattering and coincidence events. A large Monte Carlo code was built on the principle of an earlier effort[11]. The main phenomenon for which the efficiency was optimized was the neutrinoless double beta decay of ^{76}Ge to the first excited 2^+ state of ^{76}Se at 559 keV. The Monte Carlo calculation was used to optimize the geometry for observing coincidences between 1482 keV pulses in the detector in which the double beta decay occurs and a 559 keV pulse occuring in one other detector or shared among the others. Four probabilities defined below were calculated.

P1 is the probability that a 559 keV γ ray generated in any one of the crystals will deposit a sum of 559 keV in all 14 detectors; P2 is the probability that the

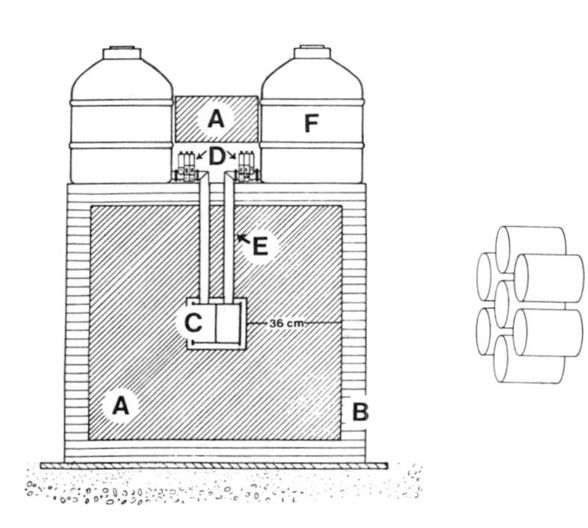

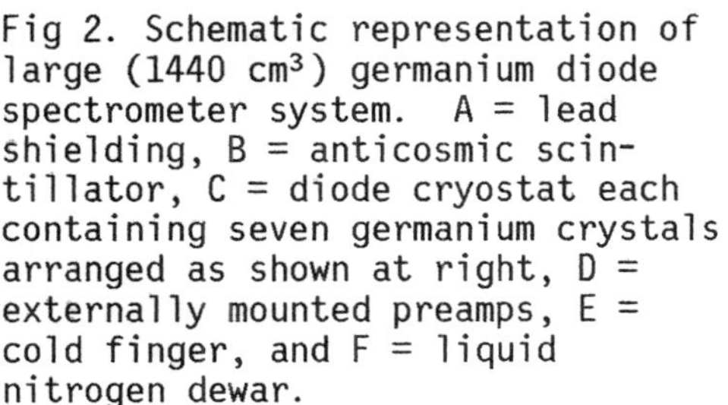

Fig 2. Schematic representation of large (1440 cm^3) germanium diode spectrometer system. A = lead shielding, B = anticosmic scintillator, C = diode cryostat each containing seven germanium crystals arranged as shown at right, D = externally mounted preamps, E = cold finger, and F = liquid nitrogen dewar.

559 keV γ ray will escape from the detector of origin and will have its energy deposited among all of the others; P3 is the probability that 559 keV is deposited in the detector of origin; and P4 is simply the probability that the 559 keV γ ray completely escapes the detector of origin. All of these probabilities are needed to interpret the data because a 1482 keV, $\beta^-\beta^-$-decay to the 2^+ first excited state, followed by the γ ray being absorbed in the same detector, can mimic a 2041 keV zero-neutrino $\beta^-\beta^-$-decay to the 0+ ground state.

A variety of distances between detectors, detector lengths and radii were used in a total of 243 Monte Carlo runs. A typical set of values for three different detector lengths is given in Table IV.

Table IV: Probabilities P1, P2, P3 and P4 calculated for three detector lengths.

Length(cm)	Radius(cm)	P1	P2	P3	P4
5.0	2.63	0.2762	0.0512	0.1517	0.4812
6.0	2.63	0.2952	0.0511	0.1684	0.4564
7.0	2.63	0.3099	0.0508	0.1824	0.4409

The data acquisition system has been designed and has separate preamplifiers for each germanium diode which feed their signals through separate linear amplifiers, ADC's and CAMAC parallel input registers. The CAMAC modules generate computer interrupts within a microsecond of the arrival of data. Upon receipt of an interrupt, an LSI-11/23 minicomputer traps to an interrupt service routine which gathers the information necessary to define an event packet. Each fixed length event packet contains: 1) an absolute event identification number, 2) the ADC the event came from, 3) the energy of the pulse, 4) the contents of both fast ($\sim$ 20 μsec) and slow ($\sim$ 100 μsec) coincidence masks, and 5) the arrival time of each event, namely, the year, month, day, hour, minute, second, and clock tick (1/60 second). When the packet is filled, the computer sets a data-ready flag and dismisses the interrupt. Final processing of the packet takes place in a completion routine running at background priority. The data are written to magnetic tape in a serial manner (list mode).

PROJECTED SENSITIVITY OF THE 1440 cm³ DETECTOR

The phase of our experimental work just completed, and that underway now, represent something new in the technology of low background Ge detectors. It is interesting to project the ultimate sensitivity of the 1440 cm^3 detector for the detection of the full energy double beta decay peak of ^{76}Ge at 2040.71 keV.

Let us estimate the expected half-life limit we could set with Superdetector when our present background rate 0.0025 counts/keV/1000 min is scaled by the increase in volume. There are a total of 4.97×10^{24} atoms of ^{76}Ge in the effective volume. By setting CL = 0.9 (90% confidence level), and assuming no observed peak at 2041 keV, a sensitivity of $T_{1/2} \gtrsim 1.9 \times 10^{24}$ yr will be reached in two years of counting. This would establish an upper limit less than one eV on the Majorana mass of the electron neutrino.

This work was supported by the U.S. Department of Energy under Contract No. DE-AC06-76RLO 1830 and the National Science Foundation under Grant No. PHY-8209562. The authors are grateful to R. Davis, A. Gilles, C.K. Lee, K. Lande, M. Cherry and R. Thompson for their valuable assistance in the experiments performed at Homestake.

References

1. E. Fermi, Nuov. Cim. 11, 1 (1934).
2. V.A. Lubimov, E.G. Novikov, V.Z. Nozik, E.F. Tretyakov, and V.S. Kosic, Phys. Lett. 94B, 266 (1980).
3. S. Borris, A. Golutvin, I. Laptin, V. Lubimov, V. Nagovizin, E. Novikov, V. Soloshenko, I. Tichomirov, E. Tretyakov, Proc. Int. Euro Physics Conf. on High Energy Physics, Brighton, 386 (1984).
4. Boris Kayser and Alfred Goldhaber, Phys. Rev. D28, 2341 (1983).
5. W.C. Haxton, G.J. Stephenson, Jr., and D. Strottman, Phys. Rev. Lett. 47, 153 (1981). For an updated review, see W.C. Haxton and G.J. Stephenson, Los Alamos reprint LA-UR-84396; Progress in Nuclear and Particle Physics (in press).
6. F.T. Avignone, III, R.L. Brodzinski, D.P. Brown, J. C. Evans, Jr., W.K. Hensley, J.H. Reeves and N.A. Wogman, Phys. Rev. Lett. 50, 721 (1983).
7. R.J. Ellis, B.J. Hall, G.R. Dyck, C.A. Lander, K. S. Sharma, R.C. Barber and H.E. Duckworth, Phys. Lett. 136B, 146 (1984).
8. N.A. Wogman, D.E. Robertson and R.W. Perkins, Nucl. Instr. and Methods 50, 1 (1967).
9. J.H. Reeves, W.K. Hensley, R.L. Brodzinski and Peter Ryge, PNL-SA-11358, IEEE Trans. Nucl. Sci. NS-31, No.1, 697 (1984).
10. F.T. Avignone, III, R.S. Moore, R.L. Brodzinski, D.P. Brown, J.C. Evans, Jr., W.K. Hensley, J.H. Reeves and N.A. Wogman, Proc. of the Miniconference on Low Energy Tests of Conservation Laws in Particle Physics, Blacksburg, VA, 1983, AIP Conf. Proc. 114, 206 (1984).
11. F.T. Avignone, III, Nucl. Instr. and Methods 174, 555 (1980).

SOLAR NEUTRINO EXPERIMENTS: THEORY

J. N. Bahcall
Institute for Advanced Study, Princeton, New Jersey 08540

ABSTRACT

The expected capture rates and their uncertainties are reevaluated for the chlorine and gallium experiments using improved laboratory data and new theoretical calculations. I also state a minimum value for the flux of solar neutrinos that is expected provided only that the sun is currently burning light nuclei at the rate it is emitting photons from its surface and that nothing happens to solar neutrinos on their way to the earth. These results are used - together with Monte Carlo simulations performed by Bruce Cleveland - to determine how much gallium is required for a solar neutrino experiment.

In this introductory talk, I will bring up to date the theoretical calculations by making use of the results of the many laboratory investigations of nuclear parameters that have occurred over the past few years. I will also describe a lower limit for the flux of p-p (and p-e-p) neutrinos that is valid provided only that 1) the sun is currently supplying energy by light element fusion at the rate it is losing energy from its surface by photon emission and 2) nothing happens to solar neutrinos on the way to the earth. I will then use the theoretical results - together with extensive Monte Carlo simulations - to determine how much gallium is required for a solar neutrino experiment. A number of the results described here have been obtained in collaboration with Bruce Cleveland, Ray Davis, and Keith Rowley and are adapted from the article by these authors.[1]

The bottom line for the ^{37}Cl experiment is that the standard solar model predicts - if nothing happens to the neutrinos on the way to the earth - about 6 SNU, with an effective 3-σ uncertainty of about 2 SNU. This is in conflict with the observations reported by Keith Rowley at this conference, which yield about 2 SNU (with a small 1-σ uncertainty of 0.3 SNU). There is no accepted solution for this problem, although many have been proposed and several of the more interesting suggestions will be discussed in the following talks. The discrepancy between theory and observation has remained approximately constant over the past 16 years, although there have been hundreds of careful and important papers refining the input data, the calculations, and the observations.

Table 1 lists the neutrino fluxes produced by the standard solar model and the corresponding capture rates for a ^{37}Cl and a ^{71}Ga detector. The neutrino fluxes were evaluated using Tables 8 and 11 of Bahcall et al. (1982),[2] in conjunction with the improved series

Table 1

Neutrino Fluxes and Expected Capture Rates

Neutrino Source	Flux at Earth ($10^{10}cm^{-2}s^{-1}$)	^{37}Cl Capture Rate (SNU)	^{71}Ga Capture Rate (SNU)
p - p	6.10	0	70.2
p - e - p	0.015	0.24	2.5
^{7}Be	0.40	0.95	27.0
^{8}B	0.00040	4.3	1.2
^{13}N	0.05	0.08	2.6
^{15}O	0.04	0.24	3.5
Total		5.8	107

The ^{71}Ga neutrino absorption cross sections that were used in compiling this table (neutrino source in parentheses) are, in units of $10^{-46}cm^2$: 11.5 (p-p), 166 (pep), 67.5 (^{7}Be), 3×10^3 (^{8}B), 56.7 (^{13}N), and 97.7 (^{15}O). The average neutrino cross sections for the possible calibrators are, in the same units, 55.1 (^{51}Cr) and 70.6 (^{65}Zn). The calculational details are described in Bahcall (1978); I have used here the improved Q-value for ^{71}Ge decay measured by Hampel and Schlotz (1984).

of measurements of g_A/g_V and the neutron lifetime[3,4] [$\tau_{\frac{1}{2}}$ = 629 s (1 $\pm$ 0.012)], of $S_{34}(0)$ (0.52 $\pm$ 0.05 keV-b) summarized by Alexander et al.[5] [see also Osborne et al.[6]], and of $S_{17}(0)$ (0.0238 $\pm$ 0.0023 keV-b) by Filippone et al.[7] A more accurate measurement by Hampel and Schlotz[8] of the Q-value for the ^{71}Ga decay (233.2 keV) changes by a few percent the neutrino absorption cross sections for ^{71}Ga calculated by Bahcall[9] with a less precise Q-value (235.7 keV). We have combined the new experimental results with the complete error analysis of Bahcall et al.[2] to find that the standard solar model predicts (effective 3-σ limits):

$$\Sigma(\psi\sigma)_{^{37}\mathrm{Cl}} = 5.8 \pm 2.2 \text{ SNU}, \tag{1a}$$

and

$$\Sigma(\psi\sigma)_{^{71}\mathrm{Ga}} = 107^{+11}_{-7} \text{ SNU}. \tag{1b}$$

The error estimates were also revised using the improved laboratory data.

I recalculated the uncertainties from each parameter discussed in Bahcall et al.[2] using the neutrino fluxes given in Table 1 of the present paper. I included an estimated uncertainty of 1.9 SNU from the lifetime[11] of ^{71}Ge.

The best-estimate for the capture rate for the gallium experiment given in equation (1b) does not include any contributions from excited states. I have continued to follow the phenomenological estimate of the likely magnitude of excited state transitions to ^{71}Ge that was obtained by Bahcall in 1978,[8,10] which corresponds to $\leq$ 8 SNU. This estimate was based upon the systematics of beta-decay rates for nuclei in the mass range near 81 that have measured transitions similar to the ones of interest for solar neutrino detection. The suggestion by Orihara et al.[12] that much stronger transitions occur is based upon data at too low a proton energy (35 MeV) to indicate reliably the strength of the Gamow-Teller transitions [see A. Baltz et al.[13]].

Keith Rowley has described at this conference the heroic experiment that Ray Davis and his collaborators have carried out over the past twenty years. The cumulative analysis of 59 experimental runs using ^{37}Cl over the period 1970-1983 (with a 1-σ error) yields:[14-17]

$$\Sigma(\psi\sigma)_{\mathrm{observed}} = 2.1 \pm 0.3 \text{ SNU} \tag{2}$$

The difference between equations (1a) and (2) is a quantitative statement of the solar neutrino problem. Note that the theoretical error estimates are designed to be effective 3-σ limits. The theoretical best-estimates have bounced around in a relatively narrow range (7 $\pm$ 2 SNU, see Fig. 4 of Bahcall et al.[2]) over the past 16 years, despite the many improved determinations of the parameters.

It is useful to evaluate the minimum rate of neutrino capture that is consistent with the assumption that the solar neutrino problem is caused by errors in the standard solar model (neutrino production). This minimum rate is achieved if the only neutrinos produced by the sun are from the p-p reaction (and the associated p-e-p reaction). The capture rate expected on ^{71}Ga in this case is - for the minumum rate -

$$\Sigma(\psi\sigma)_{\text{min,astronomical}} = 78 \text{ SNU}. \tag{3}$$

The corresponding neutrino fluxes are: p-p: 6.5×10^{10} cm^{-2} s^{-1} and p-e-p: 1.6×10^{8} cm^{-2} s^{-1}. Equation (3) is a conservative minimum since if only p-p (and p-e-p) neutrinos are produced in the sun, the expected capture rate in the ^{37}Cl experiment is 0.25 SNU. The difference between the observed rate (equation 2) and 0.25 SNU must be ascribed in this (minimum) scenario to unknown background sources. I have evaluated also the predicted capture rate for the other non-standard steady-state solar models for which detailed and complete calculations of neutrino fluxes are available (see Table 10 of ref. 9). For the model in which the heavy element is very low in the solar interior, I find 85 SNU and for the completely mixed sun I obtain 87 SNU. The recent study by Gilliland[18] of sudden perturbations of the nuclear abundances in the solar interior yields a model with an implied capture rate of 84 SNU.

Many authors[19-24] have suggested that oscillations of neutrinos between different eigenstates may reduce the flux of solar electron neutrinos that can be detected at earth. If one takes account of the continuous nature of the neutrino spectrum,[21] the maximum non-resonant reduction that can be expected - if all the mixing angles are optimal - is the production rate divided by the number of independent neutrino types.

Wolfgang Hampel has recently drawn special attention to the fact that, if all the mixing angles are optimal, there is a narrow range of neutrino mass differences in which the decrease due to oscillation can be larger than one over the number of independent neutrino types. Destructive interference can occur if the earth-sun distance corresponds to (n + 1/2) neutrino oscillation wavelengths, where n is a small integer. This resonant phenomenon, originally discussed by Gribov and Pontecorvo[20] for monoenergetic neutrinos, is illustrated for a continuous neutrino spectrum in Figure 1 of Bahcall and Frautschi's paper.[21]

A possible propagation solution that is consistent with the observed rate given in equation (2) is that the standard fluxes are all divided by 3 (for electron, muon, and tau neutrinos):

$$\Sigma(\psi\sigma)_{\text{oscillates}} = 36 \text{ SNU}. \tag{4}$$

Equation (4) represents the maximum rate that one would expect on the basis of a purely propagation solution for the solar neutrino problem since the explanation must be consistent with the difference between theory and observation in the ^{37}Cl experiment (cf. equations 1a and 2). Of course, the detected capture rate may be zero if all of the observed captures in the ^{37}Cl experiment are caused by background events and neutrinos decay in flight from the sun.

I will now summarize some calculations that have been carried out in collaboration with Cleveland, Davis, and Rowley. We have calculated the amount of gallium required to distinguish between the rates given in equations (3) and (4). We have carried out detailed Monte Carlo simulations, modeled upon the results that have been achieved with existing equipment and techniques.[25-27] The results of these calculations can be summarized in the following relation between the 1-σ percentage uncertainty, PU, and the number N of experimental runs, background rate b, and effective production rate p. We find that

$$PU \cong \frac{38\%}{(Np)^{1/2}} \left(1 + \frac{1.9b}{p}\right)^{1/2}. \qquad (5)$$

The effective production rate p is the product of the overall counting and extraction efficiencies times the saturation factor [1 - exp - (λ × exposure time)] times the true production rate. The following parameters were used in obtaining equation (5): duration of each exposure (21 days), counting efficiency (74%), extraction efficiency (95%); counting background rate (0.16 per day), and duration of counting (59 days). The counting efficiencies and background rates are close to the best values obtained by the Heidelberg group[25,26] and we believe that parameters similar to these can be achieved in a routine manner. The general conclusions reached below are not sensitive to the precise numerical values of the background and counting efficiencies (cf. eq. [5]). Equation (5) represents well the results of the numerical simulations for ^{71}Ge production rates between 0.2 and 3 atoms per day. The true production rate of ^{71}Ge atoms is

$$A = 3(\text{atoms/day})[(\text{tons of gallium}/100)(\text{SNU's}/100)], \qquad (6)$$

where SNU's is the solar neutrino capture rate on ^{71}Ga. For 15 tons of gallium, this true production rate is 0.35 per day for an assumed minimum capture rate of 78 SNU. This minimum rate is approximately equal to the observed production rate in the ^{37}Cl experiment.

Table 2 summarizes the principal results of our Monte Carlo simulations. The first 3 columns of the table give the number of tons (15, 30, or 45) for each simulated experiment, the duration of the experiment in years (2 or 4), and the assumed capture rate (either 78 or 36 SNU). The total number of atoms of ^{71}Ge that would be counted with the given conditions is listed in the fourth column. The PU value (percentage uncertainty in the inferred solar neutrino capture rate, see equation 5) is listed in the fifth column.

Table 2

Simulated Gallium Solar Neutrino Experiments
For Different Amounts of Gallium and Durations of the Experiment

The values given are for an assumed exposure time of 21 days, a counting efficiency of 74%, an extraction efficiency of 95%, a counting background rate of 0.16 day^{-1}, and a counting duration of 59 days. The results are taken from Bahcall, Cleveland, Davis, and Rowley (1985).

Tons of Gallium	Duration of Experiment (years)	SNU's	Total ^{71}Ge Atoms Counted	PU	Standard Deviations between 78 and 36 SNU	1 - σ error in lifetime
15	2	78	100	25	2.1	0.38
		36	46	49	2.4	0.44
	4	78	199	18	3.0	0.32
		36	92	34	3.4	0.40
30	2	78	199	15	3.7	0.30
		36	92	27	4.4	0.38
	4	78	398	10	5.2	0.25
		36	184	19	6.2	0.34
45	2	78	299	11	4.9	0.24
		36	138	19	6.0	0.33
	4	78	597	8	6.9	0.18
		36	276	14	8.5	0.28

The next-to-last column of Table 2 gives the number of standard deviations that separate the minimum astrophysically possible capture rate of 79.5 SNU from the maximum capture rate of 36 SNU attainable with a propagation solution. The last column of Table 2 gives the accuracy with which the radioactive lifetime of the counted atoms could be determined experimentally in the presence of a background rate of 0.16 atoms per day. In all circumstances, relatively large amounts of gallium are required in order to determine the lifetime of the counted atoms with good accuracy.

Table 2 shows that even moderate amounts of gallium are sufficient to distinguish in most cases between the counting rate expected on the basis of the most pessimistic astronomical explanation and the rate expected on the basis of the most optimistic propagation explanation. One can quantify this result by calculating the a priori probability that a true counting rate in the astronomically expected range of 110 $\pm$ 30 SNU might appear to be a 2σ up fluctuation of 36 SNU (i.e., be measured to be between 36 SNU + 2σ and 0 SNU). The largest predicted counting rate with which I am familiar is given by Sur and Boyd,[28] and is about 140 SNU. Assuming that the true production rate lies with equal probability between 80 and 140 SNU (standard solar models are wrong), one finds that - for 15 tons of gallium - there is an 8% chance that the measured rate is smaller than 36 SNU + 2σ after 2 years of the experiment. The chance of confusing a true capture rate in the astronomical range with a rate consistent with a propagation solution is only 1% after 4 years of running a 15 ton gallium experiment. If the true rate lies with equal probability between 0 and 36 SNU (a production solution), the chance of measuring an apparent capture rate larger than 79.5 SNU - 2σ is 14% for 2 years of running a 15 ton experiment and 2% for 4 years.

For a 30 ton experiment, the probability of a chance fluctuation causing a true capture rate in the astronomically allowed region to appear to be in the range of propagation solutions - or vice versa - is less than 1% after 2 years of running.

The differences between various non-standard steady-state solar models (see above) are relatively small (typically a few SNU's, comparable to or less than the errors due to input parameters) and are not distinguishable after 4 years of running with even a very large amount of gallium. If the correct explanation of the solar neutrino problem lies in the realm of neutrino production (solar models), then additional experiments with different detectors will be necessary.

After studying the results of the Monte Carlo simulations, we came to the conclusion that 15 tons is the minimum amount of gallium required to distinguish between production and propagation solutions of the solar neutrino problem. Thirty tons will yield a much sharper result than 15 tons, but for larger amounts of gallium the expected accuracy does not improve proportional to the total amount of detector material (cf. equation 5 and Table 2).

This work was supported in part by the National Science Foundation under grant no. PHY-8217352 to the Institute for Advanced Study. Much of this research was carried out in collaboration with Bruce

Cleveland, Ray Davis, and Keith Rowley. I am pleased to acknowledge their special contribution to this work and their sustained stimulation and support.

REFERENCES

1. J. N. Bahcall, B. T. Cleveland, R. Davis, and J. K. Rowley, Ap. J. Lett. (submitted) (1985).
2. J. N. Bahcall, W. F. Heubner, S. H. Lubow, P. D. Parker, and R. K. Ulrich, Rev. Mod. Phys. 54, 767 (1982).
3. B. Bopp, D. Dubbers, E. Klemt, J. Last, H. Schutze, W. Weibler, S. J. Freeman, and O. Scharpf, Journal de Physique 45, C3 (1984).
4. P. Parker, to be published (1984).
5. T. K. Alexander, G. C. Ball, W. N. Lennard, H. Geissel, and H. B. Mak, Nuclear Physics (to be published) (1984).
6. J. L. Osborne, C. A. Barnes, R. W. Kavanagh, R. M. Kremer, G. J. Mathews, J. L. Zyskind, P. D. Parker, and P. D. Howard, Phys. Rev. Lett. 48, 1664 (1982).
7. B. W. Filippone, A. J. Elwyn, C. N. Davids, and D. D. Koetke, Phys. Rev. 28, 2222 (1983).
8. W. Hampel and R. Schlotz, to be presented at the 7th International Conference on Atomic Masses and Fundamental Constants, Darnstadt-Seehim (September 1984).
9. J. N. Bahcall, Rev. Mod. Phys. 50, 881 (1978).
10. J. N. Bahcall, B. T. Cleveland, R. Davis, I. Dostrovsky, J. C. Evans, W. Frati, G. Friedlander, K. Lande, J. K. Rowley, R. W. Stoenner, and J. Weneser, Phys Rev. Lett. 40, 1351 (1978).
11. W. Hampel, private communication (1983).
12. H. Orihara, C. D. Zafiratos, S. Nishihara, K. Furukawa, M. Kabasawa, K. Maeda, K. Miura, and H. Ohnuma, Phys. Rev. Lett 51, 1328 (1983).
13. A. Baltz, J. Weneser, B. A. Brown, and J. Rapaport, Phys. Rev. (submitted) (1984).
14. B. T. Cleveland, R. Davis, and J. K. Rowley, in Proceedings of the Second International Symposium on Resonance Ionization Spectroscopy and Its Applications, Knoxville, Tennessee, April 16-20, 1984 (Adam Hilger Ltd., Bristol, England).
15. R. Davis, D. S. Harmer, and K. C. Hoffman, Phys. Rev. Lett. 20, 1205 (1968).
16. R. Davis, in Proceedings of Informal Conference on the Status and Future of Solar Neutrino Research, ed. by G. Friedlander (Brookhaven National Laboratory Report No. BNL 50879), Vol. 1, p. 1.
17. B. T. Cleveland, Nucl. Instr. Methods 214, 451 (1983).
18. R. L. Gilliland, Ap. J. (to be published) (1984).
19. B. Pontecorvo, Sov. Phys. -JETP 26, 984 (1968).
20. V. Gribov and B. Pontecorvo, Phys. Lett. B28, 495 (1969).
21. J. N. Bahcall and S. C. Frautschi, Phys. Lett. B29, 263 (1969).
22. L. Wolfenstein, Phys. Rev. D17, 2369 (1978).
23. S. M. Bilenky and B. Pontecorvo, Phys. Lett. 41c, 225 (1978).
24. A. K. Mann, Comment Nucl. Part. Phys., Vol. 10, 155 (1981).

25. T. Kirsten, in Proceedings of the Second International Symposium on Resonance Ionization, Spectroscopy and Its Applications Knoxville, Tennessee, April 16-20 (Adam Hilger Ltd., Bristol, England) (1984).
26. W. Hampel, private communication (1984).
27. G. Zatsepin, in Neutrino '82 (Proceedings of International Conference on Neutrino Physics, Balatonfured, Hungary), p. 53 (1982).
28. B. Sur and R. N. Boyd, Phys. Rev. Lett. (submitted) (1984).

MIXING VERSUS TURBULENT DIFFUSION MIXING AND THE SOLAR NEUTRINO PROBLEM.

Evry SCHATZMAN
Observatoire de Nice, B.P. 139, 06003 NICE Cedex, France.

ABSTRACT

A short account is given of the physical processus which can explain the presence of turbulent diffusion mixing (TDM) in the radiatively stable layers of rotating stars. The difference between full mixing and TDM is explained. In the case of the Sun, the excess of 3 He carried to the center of the Sun explains the efficiency of TDM in decreasing the central temperature and thereafter the small flux of Boron neutrinos.

Ohter consequences of TDM are described, concerning (1) the relation between the present abundance of 3 He in the galaxy and the primordial abundance, (2) the abundance of Lithium in main sequence stars, and (3) the possibility of using the solar sismology as a test of the effect of TDM on the μ-gradient.

INTRODUCTION

The idea of turbulent diffusion inside the stars is not a new one. It has been considered by Biermann already in 1937! The phenomenological theory of turbulent diffusion has been known for a long time. $\underline{l}$ being the correlation length and $\langle v^2\rangle^{1/2}$ the r.m.s. velocity, the quantity $D \simeq l \cdot \langle v^2\rangle^{1/2}$, which has the proper dimension, is the turbulent diffusion coefficient. Schatzman (1969a) introduced turbulent diffusion mixing (TDM) in order to explain the difference between Am and normal A stars and made (1969b) an unsuccessful attempt to explain the solar neutrinos deficiency by TDM. Baglin (1972) suggested that the turbulence which produces the TDM is generated by some shear flow instability related to rotation. The instability mechanism suggested by Baglin (1972) has been criticized (see especially Kippenhahn, 1981): the problem of the generation of turbulence in radiatively stable zones will be considered later. TDM turned out to give the possibility of explaining the Li deficiency in red giants (Schatzman, 1977, revised 1983), the ($^{12}C/^{13}C$) abundance ratio in giants of the first ascending branch (Genova, Schatzman, 1979; Bienaymé et al., 1984), the mass dependance of Li deficiency in the Hyades (Baglin, Morel, 1984), the solar spin-down (Endal, Sofia, 1981). Zahn (1984a, b) has shown that, for example, meridional circulation can drive a radial turbulent mixing; the mechanism which has been suggested by Zahn avoids the difficulty mentioned by Kippenhahn (1981), which comes from the Richardson stability criterion. The basic physics of the mechanism suggested by Zahn is that the shear flow generates a 2-D turbulence on equipotentials. The 2-D turbulence then, at small scales, as soon as inertia overtakes the Coriolis force, decays into a 3-D turbulence (the inertial force goes like $k^{1/3}$, whereas the Coriolis force goes like $k^{-1/3}$).

0094-243X/85/1260069-06 $3.00

Turbulent diffusion mixing (TDM) in radiatively stable regions is not an ad hoc assumption. It leads presently to a consistent modelization of a variety of abundance anomalies - or the absence of such.

MIXING AND TDM

It is well known that diffusive mixing is easily obtained by adding a second order term in the conservation equation,

$$\rho \frac{\partial x_i}{\partial t} = \quad K_{ij} \, X_i \, X_j \quad 2 + \frac{1}{r^2} \frac{\partial}{\partial r} \, r^2 \quad D \frac{\partial x_i}{\partial r} \tag{1}$$

where D is the turbulent diffusion coefficient, which is related to a characteristic length l and a characteristic velocity v of the turbulent flow by the relation:

$$D = (1/3) \; l \; . \; v \tag{2}$$

For the time being, we shall not consider any further the question of the value of l and v. It is quite convenient to write:

$$D = Re^{*} \; \nu \tag{3}$$

where ν is the microscopic viscosity (molecular plus radiative) and Re^{*} is a dimensionless measure of the efficiency of TDM.

Equation (1) is a reasonable approximation as long as the temperature fluctuations associated with the turbulence are small, such that the nuclear reaction rate can be linearized over the distance scale l. The linear fluctuations cancel each other and the nuclear reaction rate can be replaced by its local average,

$$\langle K_{ij} \, X_i \, X_j \rangle = \langle K_{ij} \rangle \, \langle X_i \rangle \, \langle X_j \rangle \tag{4}$$

The vertical scale of the motions has also to be small compared to the vertical scale height for the diffusion approximation to be valid. In the case of a deep convective zone, as it was considered for example by Sackman (1974) in order to explain the Li anomalies in red supergiants, none of these conditions are fulfilled and eq. 1 is probably a poor approximation.

If the time scale of the mixing is shorter than the thermonuclear reaction rate, then it is possible to consider space averages over the mass fraction q M = M(r) and to ignore the gradients:

$$\frac{1}{qM} \int 4\pi \, r^2 \, \rho \, \frac{\partial x_i}{\partial t} \, dr = \Sigma \, \frac{1}{qM} \int 4\pi \, r^2 \, \rho^2 \, K_{ij} \, X_i \, X_j \, dr \tag{5}$$

or

$$\frac{\partial x_i}{\partial t} = \Sigma \, \frac{1}{qM} \, X_i \, X_j \int 4\pi \, r^2 \, \rho^2 \, K_{ij} \, dr \tag{6}$$

However, if the thermonuclear reaction rate is very large (e.g. the D(p, γ) ^{3}He reaction), then, as usual, the local concentration has to be introduced. For further discussion of the papers of Ezer and Cameron (1968), Shaviv and Beaudet (1968), Bahcall et al. (1968), Shaviv and Salpeter (1968), see Schatzman and Maeder (1981).

What shows up immediately from the study of the standard model is the very pronounced ^{3}He peak (fig. 2 of Schatzman - Maeder, 1981). Diffusion generates a flow of ^{3}He downhill both sides of the top. For the time being, we are interested by the ^{3}He flow towards the center.

If we assume that TDM is a perturbation of the standard model, it is possible to obtain an estimate of the effect of TDM. When the system of equations (1) is written to the first order in D, it becomes evident that the usual balance between the different reactions cannot be used anymore. Close to the center, the perturbation in the ^{3}He concentration can be written

$$\frac{\delta x_3}{x_3} = (7/2)\ D\ b\ \tau_{nucl}\ (t/\tau_{nucl})^2\ (1 + (t/\tau_{nucl})) \qquad (7)$$

or

$$\frac{\delta x_3}{x_3} = (7/2)\ D\ b\ \tau_{nucl} \left| \frac{x_{10}}{x} - 1 \right|^2 \frac{x_1}{x_{10}} \qquad (7')$$

where $\tau_{nucl} = 2\ K_{11}\ \rho\ x_1$ is the time scale of Hydrogen burning. K_{11} has been assumed to vary like T^4 close to the center, ρ like T^3. Close to the center, the temperature has been assumed to be given by $T = T_c\ (1 - b\ r^2)$. With the model $Re^* = 100$ of Schatzman and Maeder (1981), x_{10} = 0,73, x_1 = 0,585, this gives

$$\frac{\delta x_3}{x_3} = 0.09$$

which is actually what is obtained from the numerical solution of the diffusion equation. Formulae (7, 7') express the fact that the disturbance to the ^{3}He balance is due to the diffusive flow which has a non-zero divergence.

The slight excess in ^{3}He increases with the efficiency of the TDM.

RESULTS

The numerical results (Schatzman, Maeder, 1981) which have been obtained with the phenomenological relation (3) with Re^* = constant shows a strong contribution of the ^{3}He (^{3}He, 2p)α reaction to the energy generation in the central regions of the Sun (fig. 4 of Schatzman, Maeder, 1981). The hydrogen concentration is increased by TDM just as it would be by homogeneization over a mass fraction q, but the increase of the amount of nuclear fuel in the central regions, due to the ^{3}He flow favors an extra-decrease of the central temperature if compared with ordinary mixed models. We can consider that it is the flow of ^{3}He, downhill, which makes the whole difference between mixing and TDM. With a value of Re^* around 100 it is easy to match the solar neutrino flux.

ABUNDANCE OF LIGHT ELEMENTS

The obvious counterpart of the solar neutrino flux is the surface abundance of ^3He. The observed abundance ratio ($^3He/^4He$) is around $5 \cdot 10^{-4}$ with extreme values $4 \cdot 10^{-4}$ to $8 \cdot 10^{-4}$ (Hall, 1975). Geiss (1973) has derived from abundance measurements in lunar samples a slow increase of the abundance ratio of at most 20 % during the last 1 or 2 billion years. The rate of increase of the abundance ratio in the lunar samples is certainly difficult to interpret: but the sign of the rate of change of the abundance is positive and this corresponds to the slow arrival of ^3He at the surface of the Sun.

The abundance ratio ($^3He/^4He$) at the surface of the Sun is very sensitive to the efficiency of the TDM. The following table (Table I), obtained by interpolation from the models $Re^* = 100$ and $Re^* = 200$ show that the value of Re^* differs from the value which fits the solar neutrino flux.

Table I

Re^*	$(^3He/^4He)_{surface}$
30	$2.8 \cdot 10^{-4}$
31	$3.12 \cdot 10^{-4}$
32	$3.44 \cdot 10^{-4}$

With these values of Re^* the abundance ratio of ^3He increased by 39 % during the last billion year. This does not seem really contradictory with the observations.

As noticed by Fowler (1981), the meaning of the abundance ratio depends strongly on the assumed abundance of deuterium at the epoch of formation of the Sun. According to the recent discussion of Laurent (1983) on the cosmic abundance of ^2D, it comes out that the abundance ratio ($^2D/^1H$) can be as low as 10^{-6} (then generating in the Sun a ratio ($^3He/^4He$) $\simeq 10^{-5}$, or as high as a few 10^{-5} (thus generating in the Sun the whole observed ^3He).

Turbulent diffusion mixing in other stars, similarly, brings ^3He in the outer layers. If we accept the results of Köster and Weidemann (1983) and Weidemann (1984) that the mass of the progenitors of white dwarfs may be as high as 8 to 10 solar masses, we are facing the situation where up to 85 % of the mass of a star has been expelled into the interstellar matter, this fraction being enriched into ^3He. There is a general argument that the mass of interstellar matter in the galaxy is slowly decreasing, but we must accept the idea that it may have been more nuclearly processed than it is usually assumed to. An estimate (Schatzman, 1984) of the rate of change of the abundance of ^3He in interstellar matter shows that it comes almost entirely from the combination of nuclear processing and TDM in the outer layers of the stars. Therefore, it is not possible to give from the present abundance of ^3He any reliable value of the primordial abundance of ^3He.

CONSISTENCY WITH OTHER TESTS

It is quite obvious that stellar models with $D = Re^* \nu$, and Re^* = const throughout the whole star, and time and angular velocity independant are very rough approximations. The solar tests (Lithium deficiency, ^3He surface abundance, neutrino flux) show clearly that Re^* is not a constant. Furthermore, the evolution towards the giant branch shows that TDM must stop some time. The μ-gradient which is generated by the nuclear reactions may have the stabilizing effect which presents the 3-D turbulence from taking place.

It is probable that solar sismology will turn out to be a good test of the presence of TDM, as mixing generates a μ-gradient which differs from the μ-gradient of the standard models. However, the different authors disagree on the value of the frequency separation of the low degree pressure modes, $\delta\nu_{n,\ 1} = \nu_{n,\ 1} - \nu_{n-1,\ 1+2}$. For example, with the same chemical composition, Shibahashi et al. (1983) and Ulrich and Rhodes (1983) give the following values (Table II) (taken from a paper of Gough, 1983).

Table II

Frequency separation

Theory.	$\delta\nu_0$	$\delta\nu_1$
Shibahashi et al. (1983)	6.7 μhz	11.7
Ulrich et al. (1983)	8.3	16.3
Observations.		
Claverie et al. (1981)	8.3	-
Grec et al. (1983)	9.4	15.3

The change of the μ-gradient due to TDM can draw the Shibahashi values towards the values of Grec et al., whereas it draws the Ulrich et al. values away from the observations.

If the interpretation of the g-modes is correct, the period spacing according to Berthomieu et al. (1983) can be explained also by the change of the μ-gradient compared to standard models.

CONCLUSION

The models of stars with TDM are certainly very promising, but the theory of TDM is still in the infancy. The behaviour of the turbulent diffusion coefficient D, as a function of the space variable r, as a function of the μ-gradient, as a function of the angular velocity, is not yet well understood.

The model of Zahn (1984a, b) which we have mentioned above is very encouraging and is likely to bring us from phenomenology to a real theory.

I think that there is a good hope of building during the coming years a consistent model of the generation of TDM, of its effect on stellar evolution and stellar structure, leading then to reliable observational tests of its presence inside the stars.

The measure of the solar neutrino flux will be a crucial test of our stellar models, and will probably tell us about the presence of TDM down to the deep interior of the Sun. In that respect, the competing experiments presently in project (e.g. Gallium) are crucial.

REFERENCES

A. Baglin, Astron. Astrophys. 19, 45, 1972.
A. Baglin, P. Morel, E. Schatzman, to be published, 1984.
J.N. Bahcall, N.A. Bahcall, R.K. Ulrich, Astrophys. L. 2, 91, 1968.
G. Berthomieu, J. Provost, E. Schatzman, Nature 308, 254, 1984.
O. Bienaymé, A. Maeder, E. Schatzman, Astron. Astrophys. 131, 316, 1984.
L. Biermann, Astron. Narchr. 263, 185, 1937.
A. Claverie, G.K. Isaak, C.P. McLeod, H.B. van der Ray and T. Roca-Cortes, Solar Phys. 74, 51, 1981.
A. Endal, S. Sofia, Astrophys. J. 243, 625, 1981.
D. Ezer, A.G.W. Cameron, Astrophys. L. 1, 177, 1968.
W. Fowler, private communication, 1981.
J. Geiss, 13th International Conference Denver, 1973.
F. Genova, E. Schatzman, Astron. Astrophys. 78, 323, 1979.
D. Gough, ESO Workshop on Primordial Helium, eds. P.A. Shaver, p. 117, 1983.
D. Kunth, K. Kjär, ESO Garching RFA, p. 335.
G. Grec, E. Fossat, M.A. Pomerantz, Solar Phys. 82, 55, 1983.
D.N.B. Hall, Astrophys. J. 197, 509, 1975.
R. Kippenhahn, H.C. Thomas, Fundamental Problems in the Theory of Stellar Evolution, IAU n° 93, eds. D. Sugimoto, D.Q. Lamb, D.N. Schramm, p. 237, 1981.
C. Laurent, ESO Workshop on Primordial Helium, ed. P.A. Shaver, 1983.
I.J. Sackman, R.L. Smith, K.H. Despain, Astrophys. J. 187, 555, 1974.
E. Schatzman, Astron. Astrophys. 3, 331, 1969a.
E. Schatzman, Astrophys. L. 3, 189, 1969b.
E. Schatzman, Astron. Astrophys. 56, 211, 1977.
E. Schatzman, Observational Tests of the Stellar Evolution, eds. A. Maeder, A. Renzini, published 1984 by the I.A.U., p. 491.
E. Schatzman, A. Maeder, Astron. Astrophys. 96, 1, 1981.
G. Shaviv, G. Beaudet, Astrophys. L. 2, 17, 1968.
H. Shibahashi, A. Noels, M. Gabriel, Astron. Astrophys. 123, 293, 1983.
R.K. Ulrich, E.J. Rhodes Jr., Astrophys. J. 265, 551, 1983.
V. Weidemann, D. Koester, Astron. Astrophys. 121, 77, 1983.
V. Weidemann, Astron. Astrophys. 134, L 1, 1984.
J.P. Zahn, Cours de Saas Fée, Société Suisse d'Astronomie, ed. A. Maeder, 1984a.
J.P. Zahn, Cours de Goutelas, Société Française des Spécialistes d'Astronomie, ed. A. Baglin, 1984b.

PARTICLE TRANSPORT IN SOLAR TYPE STARS

Georges Michaud
Département de Physique, Université de Montréal
C.P. 6128, Succ. A, Montréal, Québec, Canada H3C 3J7

ABSTRACT

The potential effects of three particle transport processes on solar evolution are discussed. Even though its effect is not very large, atomic diffusion should in principle be included in the standard solar model. It increases the neutrino flux. Turbulent diffusion could significantly reduce the neutrino flux. No reliable estimate of the strength of turbulence has been made from first principles. Its effect on ^{3}He, Li and Be abundance is discussed. A model with $\overset{*}{Re} \simeq 50$ may reduce sufficiently the neutrino flux while being compatible with the constraints imposed by the abundances and the solar oscillation spectrum. This remains to be shown by detailed calculations.

Such a low value of turbulence implies that the solar core is spinning at least three times faster than the surface. To reduce significantly the rotation rate of the solar core would require more turbulence than needed to reduce the neutrino flux to the observed value.

The parametrization of turbulence by a constant $\overset{*}{Re}$ is arbitrary. Its real value presumably varies in space and time. Whatever its value, it will make worse the disagreement between the cosmological and the evolutionary age of globular clusters.

INTRODUCTION

Many aspects of the stellar hydrodynamics input into stellar models are still uncertain. The mixing length that is used to parametrize convection zone is an arbitrary parameter. The surface rotation of the Sun is known but not the central rotation. The degree of turbulence of the stellar interior is unknown.

In order to use the abundances of certain elements to put constraints on stellar hydrodynamics, we review three particle transport processes that are potentially important in stars. We show how the slowing down of the inner core is related to particle transport. Constraints on turbulence from observations of the Sun, of white dwarfs and of peculiar stars are reviewed. Finally we show how the Li abundance observed on Population II stars constitutes a test that is not passed by the current evolutionary models for Population II stars with effective temperatures similar to that of the Sun.

PARTICLE TRANSPORT IN SOLAR TYPE STARS

We briefly discuss three particle transport processes: atomic diffusion, meridional circulation and turbulent diffusion. The

0094-243X/85/1260075-13 $3.00

last two processes are related in that rotation feeds the turbulence but differential rotation is limited by the turbulent transport of angular momentum.

Atomic diffusion is the "bottom line" particle transport process. It is always present. Its effects can only be wiped out by a more efficient particle transport process such as turbulent diffusion or convection.[1] The velocity of diffusion may be written[2,3]:

$$v_{12} = - D_{12} \left[\frac{\partial \ell nc}{\partial r} - (2A-Z-1) \frac{\partial \ell np}{\partial r} + \alpha_T \frac{\partial \ell nT}{\partial r} - \frac{Am_p g_k}{kT}\right], \qquad (1)$$

where

$$c \equiv n(A,Z) / n(H)$$

assuming

$$n(A,Z) \ll n(H).$$

The first term ($\partial \ell nc/\partial r$) is the "usual" diffusion term. It comes from the tendency of a spike in concentration to spread out in the gas.

The second term ($\partial \ell np/\partial r$) leads essentially to gravitationnal settling. It comes from the tendency of each type of ion to come to its own hydrostatic equilibrium in the prevailing gravity and electric field (see e.g. Michaud[4]).

The third term ($\partial \ell nT/\partial r$) is the thermal diffusion due to the interaction of particles (A,Z) with the energy flux. The results of Chapman[5] (see also Chapman and Cowling[2]) lead to thermal diffusion dominating the $\partial \ell np/\partial r$ term for most elements in the Sun. It has however recently been shown by Paquette et al.[6] that this was a gross overestimate caused mainly by neglecting the energy dependence of the Coulomb logarithm.[7] In the Sun, thermal diffusion is 5 to 10 times overestimated by the Chapman and Cowling[2] formulas. Thermal diffusion is then less important in the Sun than gravitational settling and we neglect it in what follows. Similarly Michaud et al.[8] have shown that the radiative acceleration term was small in the Sun and we neglect it here.

To evaluate meridional circulation, we use the recent results of Tassoul and Tassoul.[9,10] Using a boundary layer technique with all appropriate boundary conditions, they solved a system of equations for meridional circulation. They obtained an angular velocity distribution which may be written

$$\Omega = \Omega_o\left[\Theta + \varepsilon\omega(r,\theta)\right] \qquad (2)$$

with

$$\varepsilon = \Omega_o^2 R^3 / GM. \qquad (3)$$

They found Θ to be a function of both r and t. For instance if they started with uniform rotation and a ratio of turbulent to molecular velocity given by

$$\nu_T \;/\; \nu_{mol} = 150, \qquad (4)$$

they found that the center is rotating four times faster than the surface after 4.5×10^9 yr. They assumed that the surface slowed down by a factor of five. If ν_T is smaller, the center is completely decoupled from any slowing down of the surface. It keeps its original velocity. For other initial distributions of the angular momentum, the center of the Sun can rotate 10 times faster than the surface. Only for values of ν_T/ν_{mol} larger than 1000 does the interior rotation of the Sun approach uniform rotation. If, as we shall see below, such large values of turbulence are ruled out, one must conclude that the solar core is rotating faster than the surface, unless one were to assume that initially the solar core rotated more slowly than the surface.

Their results on meridional circulation confirm those of Mestel[11] that the μ-gradient considerably reduces the circulation. Values for the meridional circulation velocity are shown in Table I, for $\varepsilon = 0.013/\tau_o^2 \simeq 10^{-4}$, corresponding to an initial rotation period, τ_o, of 10 days for the Sun. They will be compared below to other transport velocities.

Table I Meridional Circulation Velocities in the Sun

$r/R_\odot$	Velocity (cm s^{-1})
0.04	10^{-14}
0.08	3×10^{-14}
0.2	10^{-13}
0.4	10^{-11}
0.8	8×10^{-10}
0.96	10^{-9}

On the Earth, it is found necessary to use turbulent viscosity and particle transport to model both the oceans (e.g. Marshall[12]) and the atmosphere (e.g. White and Green[13]). This is not true only close to the boundaries but also in the center of the oceans. There are various causes for the turbulence, one being simply that the difference in surfaces of constant pressure and those of constant density feed growing disturbances in azimuthal flows. This is the barocline instability. At the present time this turbulence cannot be calculated from first principles. It can only be measured. In rotating stars, the surfaces of constant pressure and density do not coincide and so the dynamical barocline instability (and related instabilities) should also exist (see also Tassoul and Tassoul,[10] Sect. III). It presumably depends on the rotation rate and so should vary in space and time. We cannot, at the present time, estimate the value of the turbulent viscosity and of the turbulent diffusion coefficient it leads to.

Turbulence adds to the diffusion velocity a term[1]

$$v_T = - D_T \frac{\partial \ell nc}{\partial r} \tag{4}$$

proportional to the abundance gradient. In the absence of reliable estimates for D_T, Schatzman et al.[14] have parametrized D_T using

$$D_T = \overset{*}{Re}\ \nu_{mol} \tag{5}$$

where $\overset{*}{Re}$ is an arbitrary constant to be fitted to the observations of abundances, the neutrino flux and the oscillation spectrum of the Sun.

To relate the internal rotation rate of the Sun to the constraints on abundances, it is useful to relate D_T and ν_T, as used by Tassoul and Tassoul. Consider a mixture composed mainly of protons plus a trace element of charge Z. The coefficients D_{11} and ν_{mol} are determined by very similar expressions. Except for a factor of ρ, the density, which appears because different units are used by Chapman and Cowling for ν_{mol} and D_{11}, they mainly differ by the averaging over the deflection angles (see eqs. [10.34.11] and [10.34.12] of Chapman and Cowling[2]) occuring during the atomic collisions. This is not too surprising since the molecular viscosity is concerned with the exchange of momentum between different layers caused by the thermal motion of atoms or ions while D_{11} is concerned with the exchange of momentum between the same particles. The coefficient of atomic diffusion, D_{12}, between two types of particles then differs mainly from D_{11} by a factor $1/Z_2^2$, appearing in the cross section between particles 1 and 2.

A naive interpretation of turbulence could be that the presence of macroscopic motions increases both the exchange of momentum between stellar layers (the viscosity from ν_{mol} to the coefficient of turbulent viscosity ν_T) and the particle transport (from D_{11} to D_T) by roughly similar factors. Both are increased because the macroscopic random motions are added to the molecular thermal agitation. One then expects:

$$\nu_T / \nu_{mol} \simeq D_T / D_{11} \tag{6}$$

to hold in order of magnitude. If one defines D_T using equation (5) one then has also

$$\nu_T \simeq \overset{*}{Re}\ \nu_{mol} \qquad (7)$$

where we used $\nu_{mol} \simeq D_{11}$.

Table II Transport Velocities (cm s^{-1})

$r/R_\odot$	$v_{12}(^4He)$	$v_{12}(^3He)$	v_T^a	v_{Mcr}
0.05	2×10^{-10}	——	$- 2 \times 10^{-8}$	10^{-14}
0.125	-	$- 10^{-9}$	$- 2 \times 10^{-7}$	10^{-13}

[a]For $\overset{*}{Re} = 100$.

The particle transport velocities are compared in Table II for two cases of interest. They were chosen where ^3He and ^4He have substantial gradients in the standard solar model so where particle transport can have most effect on stellar evolution. To have substantial effect on stellar evolution, particles must be transported over a significant fraction of the solar radius, say over 10^{10} cm. It is then easy to verify, using the velocities in Table II, that only turbulent diffusion can modify the element distribution in the 4.6×10^9 years lifetime of the Sun. Meridional circulation leads to a time scale of order 10^{15} years. Atomic diffusion leads to time scales of 10^{11} to 10^{12} years meaning that it could modify abundances by a few percent. Turbulent diffusion with $\overset{*}{Re} = 100$, however leads to time scales of 10^9 to 10^{10} years and so could completely modify the abundance distribution. For this evaluation, the atomic diffusion coefficient of Paquette et al.[6] was used ($D_{12} \simeq 1.5$ for the hydrogen helium mixture at $r \sim 0.1\ R_\odot$). Using the Chapman and Cowling[2] values would change the diffusion velocities by a factor of less than two in these cases (though the change in the thermal diffusion contribution is by a greater factor than that, as discussed above). In the evaluation of the atomic diffusion velocity, equation (1) was used. That ^4He is not a trace element was neglected. Again this cannot modify the order of magnitude of the result (see Montmerle and Michaud[15]). It should also be noted that in our evaluation of the diffusion velocity for ^4He, the $\partial \ell nc/\partial r$ and $\partial \ell np/\partial r$ terms cancel partially in equation (1), the first one being about a third of the second.

From the above results, it should be clear why the only value of potential interest for $\overset{*}{Re}$ are between 10 and 100. Above that range the star is completely mixed by turbulences which is clearly incompatible with the existence of giant stars. Below that range turbulent diffusion becomes in competition with atomic diffusion and the effects are very small.

As mentioned above, atomic diffusion is a bottom line transport process which can only be wiped out by more efficient processes. As such, it should be included in the standard solar model. In practice, it has generally been neglected probably because of the complications it leads to. Noerdlinger[16] has obtained that it increased the helium content at the center of the Sun by about 6%. This agrees with the rough orders of magnitude obtained here from the diffusion velocity. These calculations should probably be repeated with the atomic diffusion calculations of Paquette et al.[6] It will lead to some increase of the neutrino flux.

CONSTRAINTS ON TURBULENCE AND SOLAR NEUTRINOS

Through its effect on the abundances of ^3He, Li, Be, ^{13}C and ^4He the mechanism of turbulent diffusion lends itself to a number of observational tests. The results are summarized on Figure 1.

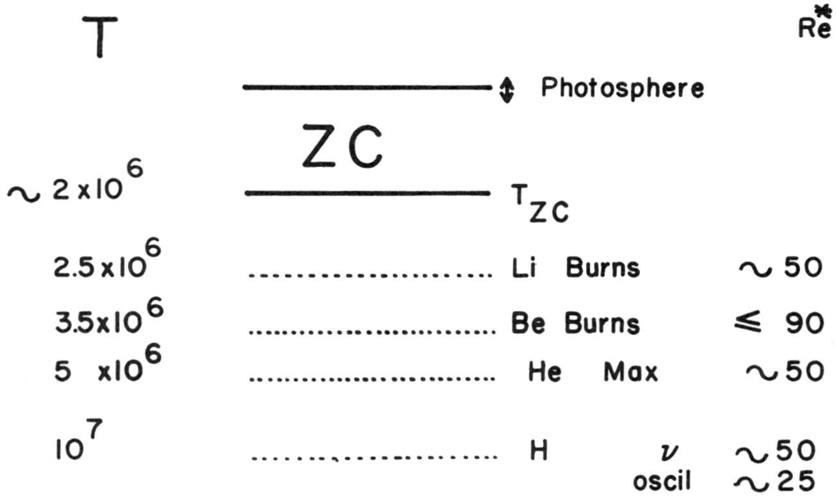

T			Re^*
	————	Photosphere	
	ZC		
$\sim 2 \times 10^6$	————	T_{ZC}	
2.5×10^6	....	Li Burns	~ 50
3.5×10^6	....	Be Burns	$\leqslant 90$
5×10^6	....	He Max	~ 50
10^7	....	H ν	~ 50
		oscil	~ 25

Fig. 1. Constraints on the turbulence parameter Re^*. Turbulence transports particles both toward the surface and the interior. Constraints on the turbulence between the surface and certain zones in the Sun (here referred to by their temperature) are imposed by the surface abundance of Li, Be and ^3He. Reducing the neutrino flux requires $Re^* \simeq 50$, while the solar oscillation spectrum suggests a lower value.

The first and most rigid evaluation of $\overset{*}{Re}$ comes from the observation that Li is 100 to 200 underabundant in the Sun while Be is about normal,[17,18] say not more than two times underabundant. Because Li burns at $T = 2.5 \times 10^6$ K while Be burning requires 3.5×10^6 K, the likely explanation is that the surface convection zone of the Sun is connected to the Li burning region while it is disconnected from the Be burning region. The situation, however, becomes more complicated once it is noted that the standard solar models do not have deep enough convection zones. It is very difficult and perhaps impossible to construct solar models with deep enough convection zones to burn Li (see e.g. Spiegel[19]). Considering further the gradual decrease of the Li abundance with T_{eff}, it has rather been suggested that the connection between the Li burning region and the surface occurred via turbulent mixing.[20] If one uses the standard solar models, one obtains $\overset{*}{Re} \sim 50$ below the convection zone if Li is to be burned. This value is however quite uncertain since it depends sensitively on the choice of α.[21] What is better established however is that Be should not burn. This means that turbulence should not be strong enough to transport Be between the region where it burns ($T = 3.5 \times 10^6$K) and the bottom of the convection zone. Since there must be a link between the Li burning zone and the surface, the Be constraint implies no link between $T = 3.5 \times 10^6$ K and $T = 2.5 \times 10^6$ K which determines a precise upper limit of $\overset{*}{Re} \lesssim 90$ in that region. While the value obtained from the Li abundance is very model dependent, this upper limit is well established.

The ^3He abundance in the solar wind has been used by Schatzman et al.[14] to determine a value of $\overset{*}{Re} \simeq 50$, at $T = 5 \times 10^6$K where the maximum of the ^3He abundance is. The observation that ^4He is two to three times underabundant in the solar wind[22] weakens this argument by suggesting separation of He in the solar wind. Schatzman et al.[14] tried to avoid this difficulty by considering the ^3He / ^4He ratio. It could also have been affected however. The two isotopes need not have been affected by the same factor.

Another constraint could come from the ^{13}C abundance. Genova and Schatzman[23] have shown how certain problems with ^{13}C abundances in giants could be solved by turbulent transport in the main sequence phase. They have not however calculated how this would change the ^{13}C abundance on the solar surface and so it is not possible to determine an upper limit on $\overset{*}{Re}$ from the observed ^{12}C and ^{13}C abundances on the Sun and in the solar wind. Such calculations should however be done.

The solar oscillation spectrum offers another test of element distribution in the Sun. As applied by Ulrich and Rhodes[24] to the Schatzman et al.[14] solar models, it seems to exclude the models with $\overset{*}{Re} = 100$ and 200. However, it is not clear to us, that Ulrich and Rhodes used exactly the same model as obtained by Schatzman et al. Further, as suggested by the above discussion, a value of $\overset{*}{Re} \simeq 50$ seems more likely than 100 from a number of observational tests. No model with Re^* ≃ 50 has been tested for solar oscillation for the simple reason that no such model has yet been calculated in detail. Berthomieu et al.[25] have obtained that Re^* = 25 gave the best fit to the solar g modes. This is uncertain since it is based on interpolated models. From the published analyses of solar oscillations, we conclude tentatively that a value of Re^* = 25 is suggested.

The solar model obtained by Schatzman et al.[14] had a solar neutrino flux of 11.6 SNU and they required Re^* = 100 to reduce it to 2.39 SNU. However the standard solar model leads to a neutrino flux of about 6 SNU which, if scaled linearly, would require an Re^* ≃ 50 to reduce the neutrino flux to 2.4 SNU. This is very uncertain since stellar evolution models are highly nonlinear.

To sum up, it appears that a solar model with Re^* ≃ 50 is a possibility though this remains to be confirmed by detailed evolutionary calculations.

Considering the uncertainties of the value of Re^* in the Sun, it is useful to consider other astrophysically determined values of Re^*.

The most strict upper limit comes from the very existence of hydrogen rich white dwarfs. Michaud and Fontaine[26] have shown that the superficial hydrogen is transported by atomic and turbulent diffusion to a zone where hydrogen burns ($T \simeq 1.6 \times 10^7$K), in white dwarfs with $T_{eff} \gtrsim 10000$ K. When there is any turbulence the process becomes so efficient that all hydrogen burns rapidly and no hydrogen rich white dwarfs should be observed. This is contrary to observation and sets an upper limit of $Re^* \lesssim 1$. It applies to the region between the convection zone and $T = 1.6 \times 10^7$K.

In Am-Fm stars, it is generally accepted that most anomalies are caused by atomic diffusion. Michaud et al.[27] have suggested that this occurred in the presence of mass loss and that only an upper limit could be set to Re^*. The most stringent upper limit comes from the underabundances of Ca and Sc which give $Re^* \lesssim 5$. The overabundances of heavy elements would only lead to upper limits of around 10^3. Only upper limits are obtained from Am-Fm stars because mass loss and meridional circulation seem much more important than turbulence in limiting abundance anomalies.

THE Li ABUNDANCE: A CONSTRAINT ON STELLAR HYDRODYNAMICS

The recent observation by Spite and Spite[28] of Li in old Population II stars poses a constraint that the stellar evolution models for such objects do not pass. They observed that the Li abundance is the same, independent of T_{eff} in the range $3.80 \geq \log T_{eff} \geq 3.74$ (see their Fig. 5). Lithium is destroyed only in stars with $\log T_{eff} < 3.74$. The problem comes from the Li abundance decreasing because of atomic diffusion if $\log T < 6.2$ while it decreases because of Li burning for $\log T > 6.4$. This only leaves a very small interval $6.2 \leq \log T \leq 6.4$ where the convection zone or more precisely the mixed zone must end throughout the stellar evolution for all those stars, if the Li abundance is to be constant over the observed T_{eff} range.

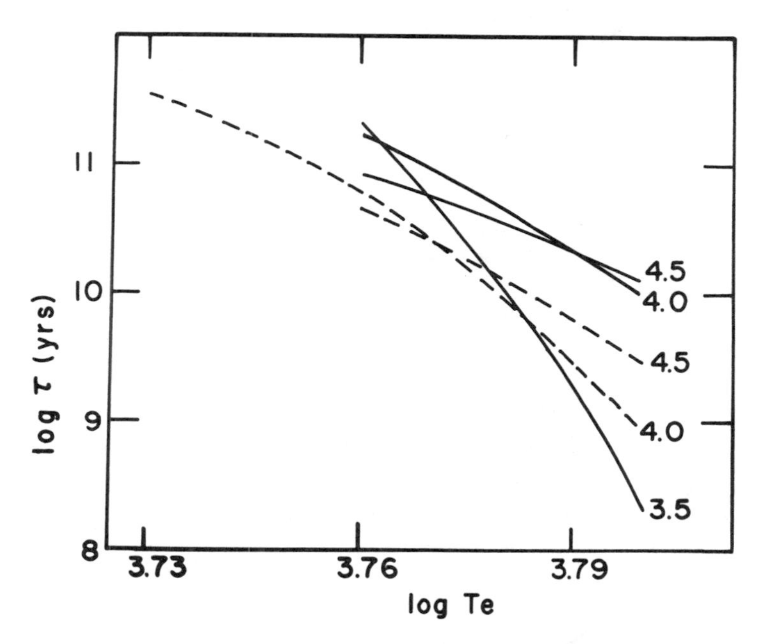

Fig. 2. Time scale of lithium diffusion as a function of effective temperature, for various gravities. The curves are labelled by log g. The full lines are for $\alpha = 1.5$ and the dashed lines for $\alpha = 1.1$.

On Figure 2 is shown the diffusion time scale of Li (from Michaud et al.[21]) at the bottom of the convection zone over the effective temperature range of interest for a few values of α and log g. Clearly that time scale varies considerably in that range. One could hope to force no diffusion by increasing α sufficiently so that the diffusion time scale is large for all cases. Figure 3 shows what happens then in the 0.7 $M_{\odot}$ model. The convection zone extends down to temperatures higher than log T = 6.4 early in the evolution. As a consequence, lithium burning becomes efficient and the lithium abundance is reduced by a factor of 3.3 by nuclear reactions and by a factor of 1.22 by diffusion. The total reduction factor is thus 4.1. The smallest reduction of the Li abundance in any of the Population II evolutionary models of Sweigart and Gross[29] and of VandenBerg[30] has been a factor of 4. We have tried without success to reduce this by arbitrarily extending the convection zone by a fixed factor. Furthermore if one assumes that the same value of α should be used for all masses covering the observed range of Li abundances, the observed reduction is then at least a factor of ten. Only by arbitrarily as

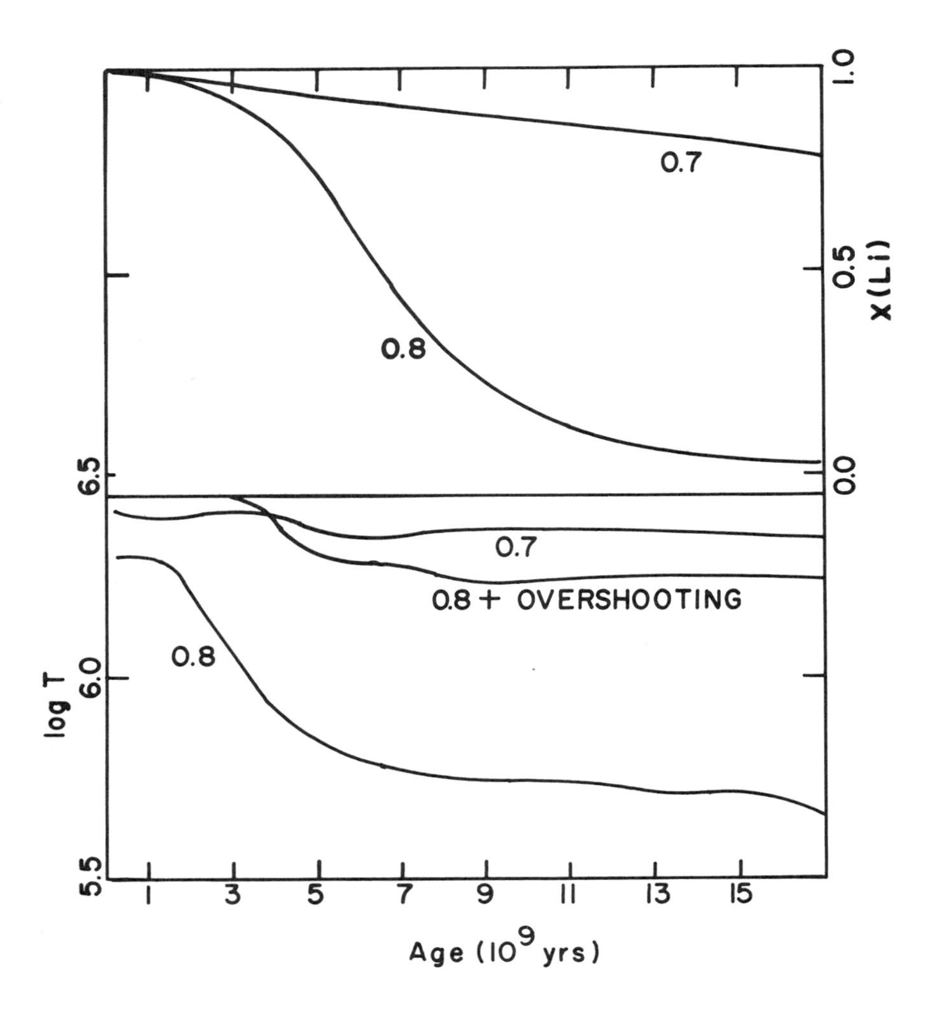

Fig. 3. The upper part of the figure gives the Li abundance, in the superficial convection zone as a function of time. They were obtained using standard evolutionary Population II models. The curves are labelled by the mass of the models in solar units. They were chosen to span the range of observations of Spite and Spite[28]. The lower part of the figure gives the temperature at the bottom of the convection zone in the same models during their evolution. The one labelled 0.8 + OVERSHOOTING had been arbitrarily extended to minimize diffusion. The variation of T at the bottom of the convection zone, for a given α, makes it impossible to eliminate Li burning and Li diffusion without <u>arbitrarily</u> imposing to the convection zone to end within $6.2 \lesssim \log T \lesssim 6.4$.

suming that the mixed zone always ends in the range $6.2 \lesssim \log T \lesssim 6.4$ can the reduction of the Li abundance be kept within a factor of two. This requires not only that α be different in models with different masses but also that α vary with time during the evolution.

CONCLUSION

Because of the presence of at least the barocline instability, one expects some turbulence in stellar interiors. It is not currently possible to evaluate how large that turbulence should be nor how it should vary with time and radius. We only know it should be there. We cannot prove it must be large enough to have observable effects on stellar models.

From the effect it has on the chemical abundances of Li and ^3He, it appears that a value of $Re^* \simeq 50$ is favored. It is assumed here constant only for convenience since it should vary in time and radius. From all other astrophysical determinations, only upper limits on Re^* are obtained. They are as low as 1 in white dwarfs.

A value of $Re^* \simeq 50$ appears sufficient to explain the observed neutrino flux. This however remains to be confirmed by detailed solar evolution models since the only models currently available used $Re^* \simeq 100$. They ran into problems with the solar oscillation spectrum but this need not apply to models calculated with $Re^* \simeq 50$.

Furthermore, a constant value of Re^* is used only for convenience, to allow explicit calculations depending on only one parameter. The understanding we have of turbulence does not require Re^* to be a constant. However, if turbulence is to have any effect on the solar neutrino problems, it must reduce the hydrogen abundance in the center, whatever be the space and time dependence of Re^*. This implies that the main sequence stage of stellar evolution is prolonged, since the hydrogen reservoir becomes larger. It does not imply that no giants are produced specially if one takes into account that Re^* is likely to decrease as the rotation decreases during the solar evolution. However it seems the globular cluster age would necessarily be increased by at least 30%. This number needs to be confirmed by detailed calculations. This increase of the age of globular clusters is a problem in that the age of globular clusters, even without this effect, is larger than the cosmological age. The problem being there in any case should we consider this to exclude turbulent diffusion?

Finally we are perhaps close to a final confirmation or infirmation of the solar evolution models. The problems with the neutrino flux, observed with Cl, may or may not be solved by modifications to the standard solar model as descibed in this paper. Such an explanation may be excluded or confirmed by the solar oscillation spectrum. But the final confirmation will come only when an experiment, such as the gallium experiment, has confirmed the flux of low energy neutrinos. Otherwise it will require a change in fundamental physics.

I thank G. Beaudet, G. Fontaine, W.A. Fowler and I. Iben for useful discussions.

REFERENCES

1. E. Schatzman, Astron. Astrophys. 3, 331 (1969).
2. S. Chapman and T.G. Cowling, The Mathematical Theory of Non-Uniform Gases, 3rd ed. (Cambridge University Press, Cambridge, 1970).
3. G. Michaud, Astrophys. J. 160, 641 (1970).
4. G. Michaud, in Highlights of Astronomy, edited by E.A. Müller (D. Reidel, Boston, 1977), Part 2, 4, p. 177.
5. S. Chapman, Proc. Phys. Soc., London 72, 353 (1958).
6. C. Paquette, C. Pelletier, G. Fontaine, and G. Michaud (in preparation).
7. G. Fontaine and G. Michaud, in IAU Colloquium 53, White Dwarfs and Variable Degenerate Stars, edited by H.M. Van Horn and V. Weidemann (University of Rochester, Rochester, 1979), p. 192.
8. G. Michaud, Y. Charland, S. Vauclair, and G. Vauclair, Astrophys. J. 210, 447 (1976).
9. M. Tassoul and J.L. Tassoul, Astrophys. J. 279, 384 (1984).
10. M. Tassoul and J.L. Tassoul, Astrophys. J. 286, 000 (1984).
11. L. Mestel, in Stars and Stellar Systems, Vol. 8, Stellar Structure, edited by L.H. Aller and D.B. McLaughlin (University of Chicago Press, Chicago, 1965), p. 465.
12. J.C. Marshall, Journal of Phys. Ocean. 11, 257 (1981).
13. A.A. White and J.S.A. Green, Quart. J. R. Met. Soc. 108, 55 (1982).
14. E. Schatzman, A. Maeder, F. Angrand, and R. Glowinski, Astron. Astrophys. 96,1 (1981).
15. T. Montmerle and G. Michaud, Astrophys. J., Suppl. Ser. 31, 489 (1976).
16. P.D. Noerdlinger, Astron. Astrophys. 57, 407 (1977).
17. A.M. Boesgaard, Astrophys. J. 210, 466 (1976).
18. A.M. Boesgaard, Publ. Astron. Soc. Pac. 88, 353 (1976).
19. E.A. Spiegel, in Highlights of Astronomy, edited by L. Perek (D. Reidel, Dordrecht, 1967), p. 261.
20. S. Vauclair, G. Vauclair, E. Schatzman, and G. Michaud, Astrophys. J. 223, 567 (1978).
21. G. Michaud, G. Fontaine, and G. Beaudet, Astrophys. J. 282, 206 (1984).
22. W.C. Feldman, J.R. Ashbridge, S.J. Bame, and J.T. Gosling, in The Solar Output and its Variation, edited by O.R. White (Associated University Press, Boulder, 1977), p. 351.
23. F. Genova and E. Schatzman, Astron. Astrophys. 78, 323 (1979).
24. R.K. Ulrich and E.J. Rhodes, Astrophys. J. 265, 551 (1983).

25. G. Berthomieu, J. Provost, and E. Schatzman, in Proceedings of the Study Conference, Oscillations as a Probe of the Sun's Interior, edited by G. Belvedere and L. Paterno, in press (1984).
26. G. Michaud and G. Fontaine, Astrophys. J. 283, 787 (1984).
27. G. Michaud, D. Tarasick, Y. Charland, and C. Pelletier, Astrophys. J. 269, 239 (1983).
28. F. Spite and M. Spite, Astron. Astrophys. 115, 357 (1982).
29. A.V. Sweigart and P.G. Gross, Astrophys. J., Suppl. Ser. 32, 367 (1976).
30. D.A. VandenBerg, Astrophys. J., Suppl. Ser. 51, 29 (1983).

INSTABILITIES, MIXING AND SOLAR NEUTRINOS

IAN W. ROXBURGH

Theoretical Astronomy Unit
School of Mathematical Sciences
Queen Mary College
University of London

ABSTRACT

Instabilities driven by differential rotation during spin down of a rotating solar model are analysed and it is shown that with a very small composition gradient, the first unstable mode is the Axisymmetric Baroclynic Diffusive (ABCD) instability. It is argued that if this instability occurs, it leads to an almost horizontal re-adjustment of chemical composition and only very slight mixing.

Mixing due to the ^3He instability is energetically possible but it is argued that finite amplitude oscillations lead to a quasi-steady state without mixing with ^3He being burnt to ^4He during such oscillations.

INTRODUCTION

One potential resolution of the Solar Neutrino Problem is the mixing of chemical elements in the inner part of the sun. Such mixing maintains a higher central abundance of hydrogen than in the case for unmixed 'standard model', this in turn, requires a lower temperature for a given rate of energy generation and consequently, fewer reactions follow the Be-B branch of the p-p chain, resulting in a lower predicted flux of neutrinos. A fully mixed model results in a predicted neutrino flux below that observed.

What can cause such mixing? One possibility is that as the sun's rotation decreased, due to angular momentum loss in the solar wind, the differential rotation became unstable. The resulting instabilities transport angular momentum to the surface and cause chemical mixing.

Recent analysis of the stability of differential rotation by myself, (Roxburgh 1984a) and by Knobloch and Spruit (1983) [see also

Knobloch, Spruit and Roxburgh 1983], supported these ideas by showing that the Axisymmetric Baroclynic Diffusive Instability was not stabilised by chemical composition gradients and so, could continue during solar evolution. Models with solar neutrino fluxes of 3 SNU were developed, but they have difficulty in explaining the recent data on solar oscillations. However, it is one thing to show that instabilities can exist and, quite another to understand their non-linear development. I here present an argument that the instability redistributes chemical composition almost horizontally, resulting in marginal stability and negligible mixing.

What about the ^{3}He instability (Christensen Dalsgaard, Dilke and Gough 1974)? The non-linear development of this instability is unknown - but whilst it is energetically possible for this to drive mixing, simple 'back of the envelope' calculations suggest that relatively low amplitude oscillations are sufficient to destroy the ^{3}He at the rate that it is produced. After something like 3.10^9 years the build-up in compositon gradients suppresses the instability and the solar evolution is not much different from the standard model.

INSTABILITIES DUE TO DIFFERENTIAL ROTATION

I do not wish to present here the details of the stability analysis, but just to pick out the relevant points for the present discussion on the ABCD instability. The condition for instability can be expressed in the approximate form:

$$[R_1 \sin \Gamma - R_2 \sin(\lambda-\Phi)]^2 > 8 R_1 \sigma [\sin \theta - R_2 \sin \Phi] \qquad (1)$$

where $\sigma = \nu/k$ is the Prandtl number ($\approx 10^{-6}$), and λ, θ, Φ, Γ, are respectively the angles between the equatorial plane and the effective gravity, and between the equatorial plane and the normals to surfaces of constant entropy, molecular weight and angular momentum. R_1 and R_2 are defined in terms of the boyancy N_T, composition N_μ and rotation N_Ω frequencies by

$$R_1 = N^2{}_\Omega/N^2{}_T, \; R_2 = N_\mu{}^2/N_T{}^2.$$

If $\Theta = \Phi$, (surfaces of constant entropy and composition coincide), then for a radially varying angular velocity $\Omega = \Omega(r)$ the instability condition reduces to (Knobloch, Spruit and Roxburgh 1983)

$$\left[r \frac{d\Omega}{dr} \right]^2 > 8 \sigma N_T^2 \qquad (2)$$

At marginal stability, the angular velocity varies very little for the outer 50% of the sun and then increases monotonically to a central value of about 13 times the surface value.

The argument was then advanced that due to angular momentum in the solar wind, the surface layers are spun down (c.f. Gill and Roxburgh 1984), leading to instability. Resulting 'turbulent diffusion' then transports angular momentum from the interior to the surface down the marginally stable angular velocity gradient and also produces chemical mixing, leading to a lowering of the predicted neutrino flux, (Roxburgh 1984).

However, this model has difficulties; as pointed out by the author (Roxburgh 1984b), in order to keep the star mixed, the energy supplied by the instability has to be sufficient to raise the Helium rich material through the potential well of the star. In the absence of any mixing, the rate of change of gravitational energy is of the order of:-

$$\frac{dE_g}{dt} \sim E_g \left[\frac{1}{\mu} \frac{d\mu}{dt} \right] \sim \frac{E_g}{t_{nuc}} \qquad (3)$$

where $t_{nuc} \sim 10^{10}$ years is the nuclear evolution time scale and E_g the gravitational energy. If the instability is driven by differential rotation, the energy ultimately comes from the kinetic energy of rotation, so that for mixing, we need:-

$$\frac{dE_{rot}}{dt} > \frac{dE_g}{dt}, \quad \text{ie: } \left[\frac{E_{rot}}{E_g} \right] \left[\frac{t_{nuc}}{t_{rot}} \right] > 1 \qquad (4)$$

where t_{rot} is the spindown time of the sun. Since E_{rot}/E_g is small, ($\sim 10^{-5}$ in the present Sun), this requires a very rapid spindown time of the order of 10^5 years. Such a short spindown time is incompatible with observations of the rotation of stars and of the present Sun.

If mixing cannot take place, what does the ABCD instability do? I think it redistributes chemical compositon by almost horizontal mixing to achieve a marginal state. To see this we note from condition (1) that the differential rotation is stable if:-

$$\sin(\lambda - \Phi) = \frac{R_1}{R_2} \sin\Gamma = \frac{|\nabla\,\Omega^2\,\omega^4|}{\omega^3|\nabla\,ln\quad\mu|}\,g\,\sin\Gamma \qquad (5)$$

where ω is the distance from the rotation axis. With $R_1 \gg R_2$ this is not possible, but as chemical evolution proceeds and Ω decreases $R_1/R_2 \ll 1$ and an angle Φ exists such that, were the surfaces of constant μ so distributed, the instability would be surpressed. For a standard unmixed model of the present sum
$R_1/R_2 \sim 10^{-5}$.

This particular stable distribution of μ is not that produced by evolution in an unmixed star; for example, if the energy generation rate is $\propto \rho\ T^{4.5}$ then $(\lambda-\Phi) \simeq -\ 0.35\ (\lambda-\theta)$. Thus, to achieve marginal stability there has to be some redistribution of chemical compostion, requiring almost horizontal mixing along surfaces of constant entropy. Since the ABCD mode is itself almost horizontal, such a redistribution may be easily achieved. If this is correct then substantial differential rotation can be sustained without effective vertical mixing and the standard unmixed models of solar evolution are basically correct.

INSTABILITIES DRIVEN BY ^{3}HE ABUNDANCE

Christensen Dalsgaard, Dilke and Gough (1974) first demonstrated that a standard solar evolution model becomes unstable to an $\ell = 1$ or 2, g mode after some 2 - 3.10^8 years. The instability is driven when the energy input from the ^{3}He burning exceeds radiative damping; in unmixed models ^{3}He abundance increases away from the centre and eventually reaches the critical value for instability. Such an instability can occur in solar type stars, due in part to the temperature sensitivity of ^{3}He burning and in part, to the high amplitude of the perturbation in the region of the peak of ^{3}He production. Can this overstable oscillation lead to mixing?

Such mixing is energetically possible. The steady state burning of this ^{3}He would produce about 3% of the sun's luminosity, this is considerably in excess of the rate of change of gravitational energy due to the growing chemical inhomogenity.

$$\Delta L_{34} \sim 0.03\ L_\odot \sim 0.03\ \frac{V_g}{t_{th}} > \frac{V_g}{t_{nuc}} \qquad (6)$$

since $t_{th} \sim 0.003\ t_{nuc}$.

This does not, however, mean that mixing takes place, since with finite (but small) amplitude oscillations, the ^{3}He is converted to ^{4}He, thus reducing the abundance of ^{3}He that drives the instability. Consider the simplified model of the pp chain, where the number density of ^{3}He satisfies the equation:-

$$\frac{dn_3}{dt} = \frac{1}{2} n_1^2\, r_{11} - n_3^2\, r_{33} \qquad (7)$$

In equilibrium $n_3^2/n_1^2 = r_{11}/2r_{33}$, the abundance of ^{3}He is such that the creation and destruction just balance. If we are near the peak of the ^{3}He, there is no longer an equilibrium situation which we may approximate by:-

$$\frac{dn_3}{dt} = \frac{1}{2} n_1^2 r_{11} - n_3^2 r_{33} = \frac{1}{4} n_1^2 r_{11} \qquad (8)$$

Now consider the average over a cycle and take $r_{11} \propto T^5$, $r_{33} \propto T^{20}$, $T = T_o + \Delta T \sin \omega t$, then averaging over a cycle gives

$$\frac{dn_{3o}}{dt} = \frac{1}{4} n_1^2 r_{11o} \left[1 - 85 \left[\frac{\Delta T}{T}\right]^2\right] \qquad (9)$$

which is zero if $\Delta T/T \simeq 0.11$. Thus with a finite amplitude oscillation it is possible to destroy ^{3}He at the rate at which it is produced.

The resulting solar model will differ slightly from the standard model, since more energy is released with the ^{3}He burning to ^{4}He. To a first approximation this is equivalent to taking the age of the sun some 2 - 3% less than in the standard model. After some 3.10^9 years the model becomes stable and the resulting neutrino flux is only reduced by a few percent.

If the above argument is basically correct, then we should expect many solar type stars to be undergoing non-radial oscillations. Whilst the amplitude at the peak of the ^{3}He abundance is significant $\Delta r/r \sim 5\%$. This does not necessarily imply a large surface amplitude. Until we have a better understanding ofthe effect of outer convective zones on such oscillation amplitudes it is not possible to make sound predictions. Nevertheless, it would be of considerable interest to look for such oscillations in solar type stars.

REFERENCES

1. Christensen Dalsgaard J, Dilke FWW and Gough DO (1974) Mon Not R Astro Soc 169 p429
2. Gill RS and Roxburgh IW (1984), Space Research Prospects in Stellar Activity and Variability,p 335, Obs. de Paris
3. Knoblock E, Spruit HC and Roxburgh IW (1983) Nature 304 p320
4. Knoblock E and Spruit HC (1983) Astron and Astrophys 125, p59
5. Roxburgh IW (1984a) Mem Soc Ast Ital 55, p273
6. Roxburgh IW (1984b) Observation Tests of The Stellar Evolution Theory, ed A Maeder & A Renzini p519, D. Reidel.

CONSTRAINTS OF OBSERVATIONS OF SOLAR OSCILLATIONS ON SOLAR MODELS WITH MIXING BY TURBULENT DIFFUSION

Arthur N. Cox
Russell B. Kidman
Michael J. Newman
Los Alamos National Laboratory, Los Alamos, New Mexico 87545

ABSTRACT

The pulsational behavior of solar models with mixing produced by turbulent diffusion has been investigated, for both p (pressure) modes and g (gravity) modes. It is found that while a small amount of mixing may be allowed by the observed data on solar oscillations, values of the pseudo-Reynolds number large enough to reduce the predicted neutrino counting rate to the level seen by the ^{37}Cl experiment are in conflict with published solar oscillation data.

INTRODUCTION

Schatzman and Maeder[1] have investigated solar models with mixing induced by turbulent diffusion with diffusion coefficient

$$D_T = R_e^* \nu \quad , \tag{1}$$

where ν is the microscopic molecular viscosity and R_e^* is a pseudo-Reynolds number. They found, as shown in Table I, that increasing the value of the pseudo-Reynolds number increased the amount of mixing induced and consequently increased the amount of hydrogen remaining at the center of the sun in the present epoch, decreased the central temperature, and reduced the counting rate predicted for the ^{37}Cl experiment, as is typical for solar models with significant mixing of the central regions. Models with pseudo-Reynolds number as large as about 100 or greater produce counting rates for the ^{37}Cl experiment in agreement with the observations.

Table I. Central conditions for solar models with mixing from Reference 1.

R_e^*	X_c	$T_c, 10^6$ K	N_ν, SNU
0	0.359	15.65	11.66
100	0.569	14.73	2.38
200	0.616	14.46	1.43

P MODE OBSERVATIONS

We have constructed solar models of the Schatzman and Maeder type by matching the composition profiles given in Fig. 2 of Reference 1, for $R_e^* = 0$, 100, and 200, and have performed pulsational stability analyses for such models. In Fig. 1 we display the difference in μHz between the theoretical frequency calculated from our models and the measured[2] frequency for p modes of degree $\ell = 1$ through 5 as a function of order n. A satisfactory solar model would produce a horizontal line at zero deviation of theory from observation on this plot. We see from Fig. 1 that although there is some drift from good agreement with experiment for the standard unmixed model $R_e^* = 0$, the discrepancy between theory and observation becomes much worse as the pseudo-Reynolds number R_e^* is increased.

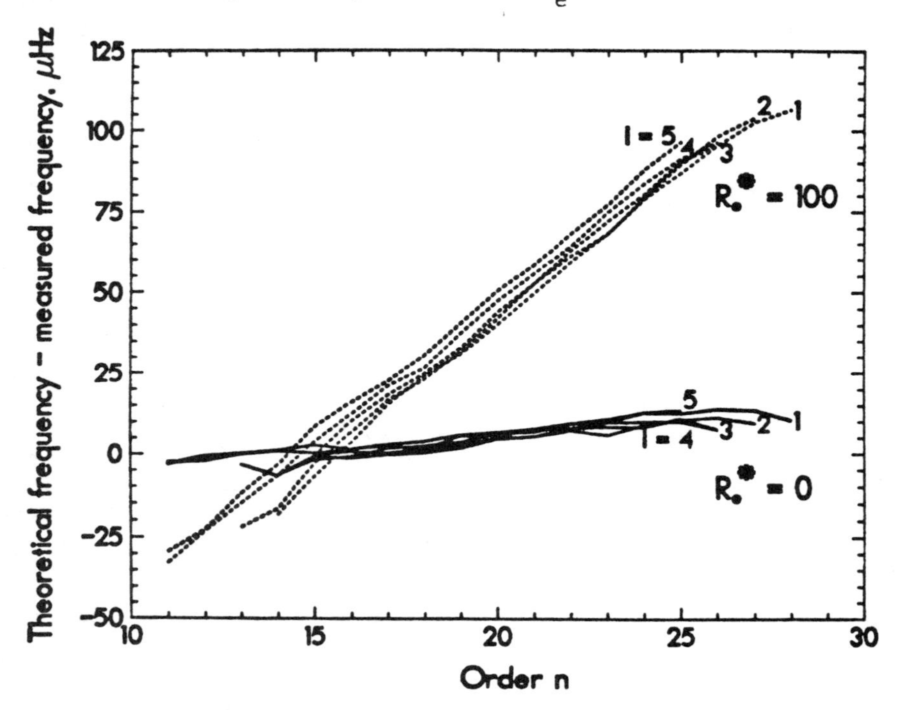

Figure 1. Deviation of theoretical frequencies from the measured frequencies for low degree p modes.

Simple pulsation theory predicts that the frequency splitting

$$\Delta\nu(\ell,n) = \nu_{\ell,n} - \nu_{\ell+2,n-1} \tag{2}$$

for low degree p modes differing by 1 in n and by 2 in ℓ should be approximately zero. Thus the deviation of this quantity from zero is a sensitive probe of the detailed structure of a solar model. Fig. 2

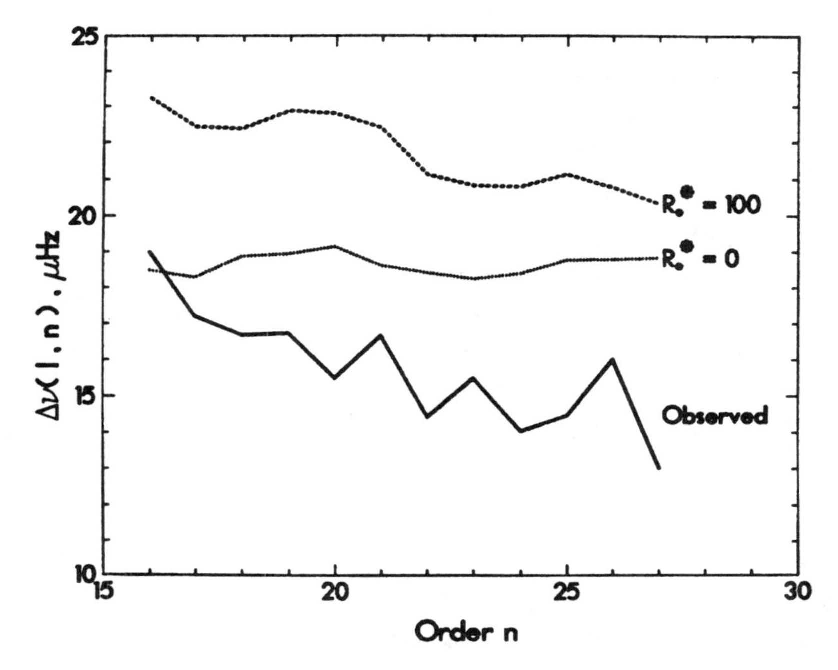

Figure 2a. p mode frequency splitting from $\ell = 1$ to $\ell = 3$.

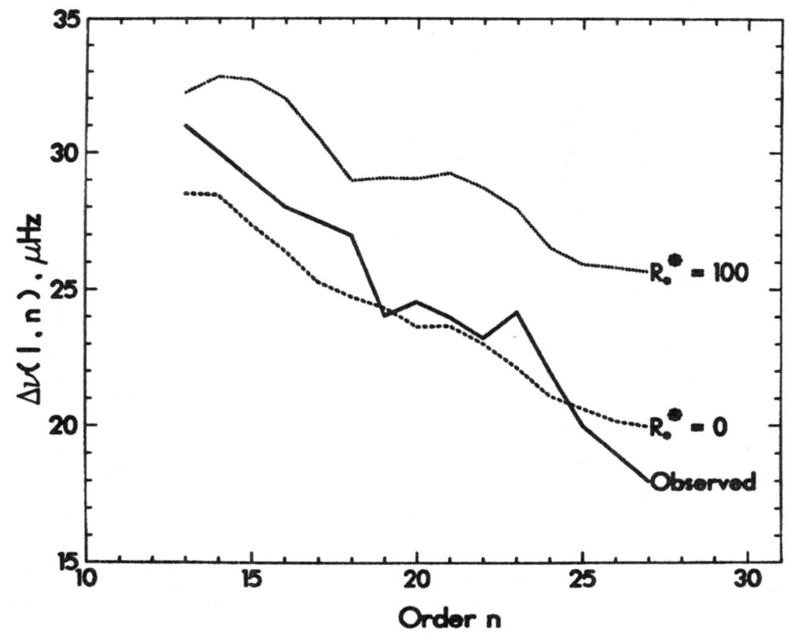

Figure 2b. p mode frequency splitting from $\ell = 2$ to $\ell = 4$.

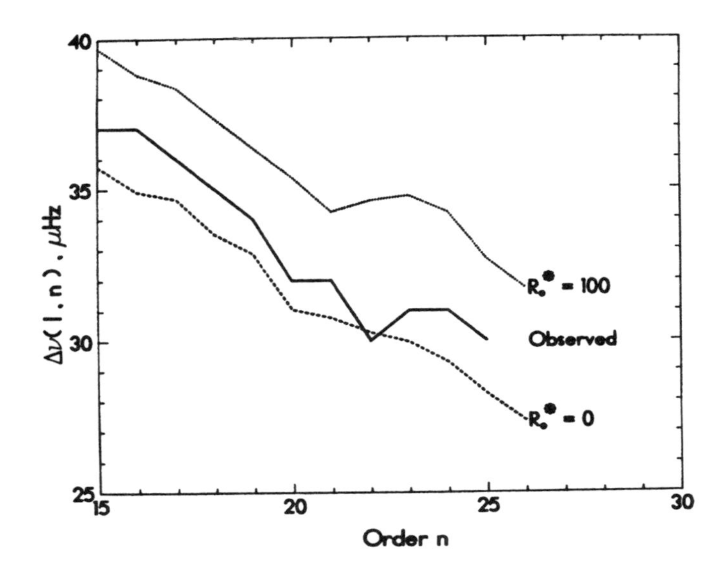

Figure 2c. p mode frequency splitting from ℓ = 3 to ℓ = 5.

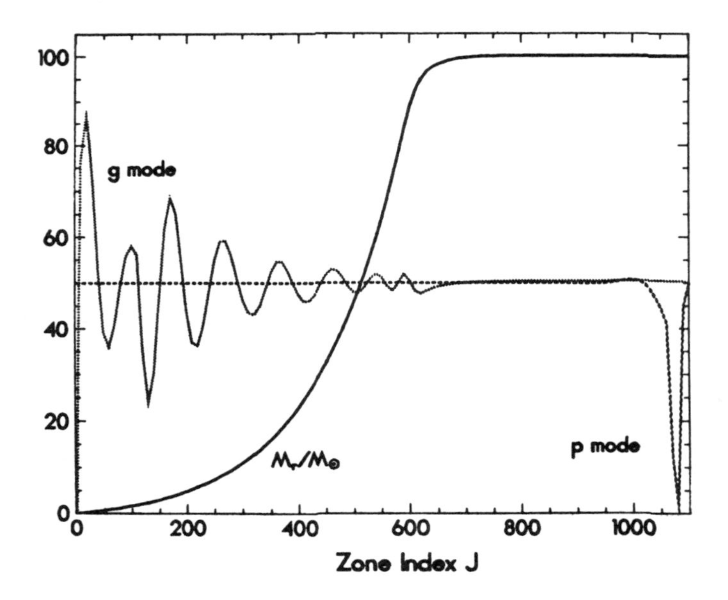

Figure 3. Schematic amplitude of oscillation for ℓ = 3, n = 17 p and g modes. The center horizontal line indicates zero amplitude of these modes. Deep amplitudes for the p modes are typically 0.001 of the surface amplitude.

shows the frequency splitting for the p modes of Fig. 1. For $\ell = 1$ and $\ell = 3$ modes in Fig. 2a, even the standard model shows too much splitting at higher orders, and mixed models lie much higher than the observations in this plot. For $\ell = 2$ and $\ell = 4$ frequency splitting in Fig. 2b the standard model $R_e^* = 0$ is much closer to the observed splitting, and for $\ell = 3$ and $\ell = 5$ splitting in Fig. 2c it begins to appear that small intermediate values of pseudo-Reynolds number may offer a better fit to the observations than the standard model, although values of R_e^* as large as 100 still predict much too much frequency splitting for these modes.

Thus the p mode observations of solar oscillations do not seem to allow sufficient mixing by the Schatzman and Maeder mechanism to produce neutrino counting rates compatible with the ^{37}Cl experiment. However, it is well known that p mode oscillations are more sensitive to conditions nearer the solar surface, and that g mode oscillations are more sensitive to conditions deep in the solar interior, as shown schematically in Fig. 3.

Here the amplitude of a typical p mode and a typical g mode are shown plotted on an arbitrary scale as a function of zone number J in our standard model, with the mass fraction (in percent) interior to zone J for comparison. On this scale the p modes show most of their structure exterior to most of the mass of the sun, and the g modes show most of their structure interior to most of the mass of the sun.

G MODE OBSERVATIONS

The difficulty in using gravity modes to diagnose the solar interior is that pressure mode observations have been on a much sounder observational footing for many years. Consequently most analyses of constraints of solar oscillation observations on models of solar structure have been in terms of p modes. Recently, however, published accounts of g mode observations have begun to appear in the astrophysical literature. For example, Delache and Scherrer[3] have made identifications of oscillations with periods from 171.9 minutes to 328.6 minutes with g modes of $\ell = 1$ and $\ell = 2$ and n ranging from 6 to 20. Such modes are conveniently analyzed in terms of the asymptotic formula

$$P = P_0 \frac{(n + \ell/2 - 1/4)}{\sqrt{\ell(\ell + 1)}} \quad . \tag{3}$$

Thus if $P\sqrt{\ell(\ell + 1)}$ is plotted as a function of $n + \ell/2 - 1/4$, a straight line of slope P_0 should result. A plot of this type for the Delache and Scherrer identifications is shown in Fig. 4.

The best fit to the data shown in Fig. 4 gives a period spacing $P_0 = 38.6$ minutes. However, Gabriel[4] has reinterpreted the Delache and Scherrer data to find $P_0 = 35.98$ minutes or 35.64 minutes, depending on the particular mode assignments chosen, in good agreement with the results for his standard solar model $P_0 = 35.52$ minutes. The ACRIM experiment on the Solar Maximum satellite produced data which has been interpreted to give $P_0 = 38.9$ minutes.

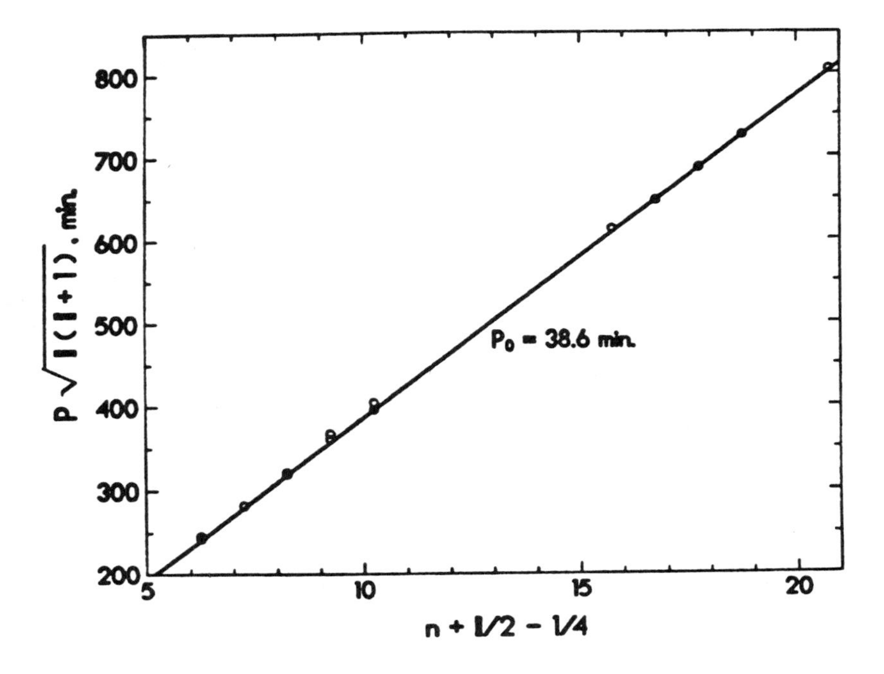

Figure 4. g mode periods from Reference 3.

Table II. Period spacing (P_0) for g modes; Theory vs Observation

Observed (min.)				
Gabriel	Severny[6] et al.	Delache and Scherrer	ACRIM	van der Raay[7] et al.
36.0	37.4	38.6 ± 0.5	38.9	41.2

Theory (min.)				
	Berthomieu et al.	Gough[8]	Gabriel	This work
$R_e^* = 0$	33.9	34.5	35.5	37
$R_e^* = 100$	54.6	-	-	58
$R_e^* = 200$	62.9	-	-	-

Our standard solar model with $R_e^* = 0$ gives $P_0 = 37$ minutes.

Berthomieu, Provost, and Schatzman[5] have calculated g mode period spacings for solar models with mixing induced by turbulent diffusion, and display P_0 as a function of pseudo-Reynolds number R_e^*. They find that P_0 increases from 33.9 minutes for their $R_e^* = 0$ model to 54.6 minutes at $R_e^* = 100$ and 62.9 minutes at $R_e^* = 200$.

DISCUSSION

The observational and theoretical situation as regards g modes is sumarized in Table II. Berthomieu et al.[5] have argued that since their $R_e^* = 0$ model produces a period spacing smaller than the observations, small intermediate values of R_e^* are indicated. However, the observed period spacings seem to lie in the range of about 36 to 41 minutes, and the theoretical period spacing for $R_e^* = 0$ lies in the range of about 34 to 37 minutes. The two ranges overlap, and there seems to be no serious disagreement between the standard solar models without mixing and the g mode observations reported in the literature. On the other hand, values of the pseudo-Reynolds number large enough to significantly reduce the counting rate predicted for the ^{37}Cl solar neutrino experiment are not compatible with the period spacings reported for g modes.

The interpretation of g mode observations is controversial at present, and indeed the very existence of gravity modes in the solar oscillation spectrum is still under question in some circles. Thus the g mode observations discussed above may not provide a strong constraint for solar models with mixing produced by turbulent diffusion. Nonetheless, it is perhaps significant that both the well established p mode oscillations of the sun and the g mode oscillations which have been reported support the same conclusion. Although small values of R_e^* cannot be ruled out by the data, mixing produced by turbulent diffusion in amounts large enough to produce neutrino counting rates consistent with the ^{37}Cl experiment does not seem to be allowed by observations of solar oscillations.

REFERENCES

1. Schatzman, E. and Maeder, A., Astron. and Astrophys. 96, 1 (1981).
2. Woodard, M. and Hudson, H.S., Nature 305, 589 (1984).
   Harvey, J. and Duvall, T., in Proc. 'Solar seismology from space' (ed. R.K. Ulrich, NASA, Washington, DC, 1984), in press.
3. Delache, P. and Scherrer, P.H., Nature 306, 651 (1983).
4. Gabriel, M., Astron. and Astrophys. 134, 387 (1984).
5. Berthomieu, G., Provost, J., and Schatzman, E., Nature 308, 254 (1984).
6. Severny, A.B., Kotov, V.A., and Tsap, T.T., Nature, 307, 247 (1984).
7. van der Raay, H.B., Claverie, A., Isaak, G., McLeod, J.M., Roca Cortes, T., Palle, G., and Delache, P., Oscillations as a probe of the Sun's interior, (ed. G. Belvedere and L. Paterno), Mem. Soc. astr. Italiana (1984), in press.
8. Gough, D.O., in Proc. 'Solar seismology from space' (ed. R.K. Ulrich, NASA, Washington DC, 1984), in press.

REVIEW OF NUCLEAR INPUT TO SOLAR NEUTRINO CALCULATIONS

B. W. Filippone
California Institute of Technology, Pasadena, CA 91125

ABSTRACT

A review of the experimental nuclear physics data that is necessary in the calculation of the solar neutrino flux is presented. The emphasis is on recent experiments which have attempted to reduce the uncertainties in the relevant astrophysical S-factors. The effects of the remaining uncertainties in the experimental input is discussed for three solar neutrino detectors: ^{37}Cl, ^{71}Ga, and ^{81}Br.

INTRODUCTION

The calculated flux of solar neutrinos is directly related to the rates of the nuclear reactions that produce the neutrinos. The rates of the nuclear reactions in turn depend on the energy (i.e. temperature) dependence of the relevant cross sections. Because the temperature at the solar core is $\sim 16 \cdot 10^6\ ^0K$ the reactions take place,for the most part, way below the Coulomb barrier and are for all practical purposes unmeasureable. The solution to this dilemma is to parameterize the cross section in terms of the nearly energy independent S-factor

$$S(E_{c.m.}) = \sigma(E_{c.m.})E_{c.m.} \exp[\ (E_g / E_{c.m.})^{1/2}\ ] \tag{1.0}$$

where

$$E_G = (2\pi\alpha Z_1 Z_2)^2 \mu c^2 / 2$$

(here μ is the reduced mass of the incident channel and α is the fine structure constant.) This factor has most of the non-nuclear energy dependence removed. It is then up to the experimenters to determine the normalization of the S-factor through experiments at higher energies and to check the energy dependence of S to as low an energy as is feasible.

The obvious question to address here is then " Have there been any experimental developments in the determination of the important S-factors since the last solar neutrino conference in 1978?" The response is a resounding " YES"; no less than 30 papers have been published on this subject since that conference. What has resulted from all this activity, aside from a few false starts and an occasional scare, is a better handle on many of key cross sections. This has been achieved through new experiments using novel techniques combined with independent cross checks to reduce systematic uncertainties and in some cases through measurements at lower energies than had been previously possible.

Our roadmap for this discussion will be the proton-proton chain of nuclear reactions shown in Fig. 1. We will focus on the three S-factors where most of the recent activity has been concentrated: S_{11}--the pp and pep S-factor, S_{34}--the $^3He(\alpha,\gamma)^7Be$ S-factor, and S_{17}-- the $^7Be(p,\gamma)^8B$ S-factor. The impact of these new results on three detectors will be considered: ^{37}Cl (for which ^{8}B neutrinos provide > 75% of the calculated capture rate), ^{71}Ga (pp neutrinos give 70% of the signal), and ^{81}Br (^{7}Be neutrinos give ~70% of the signal).

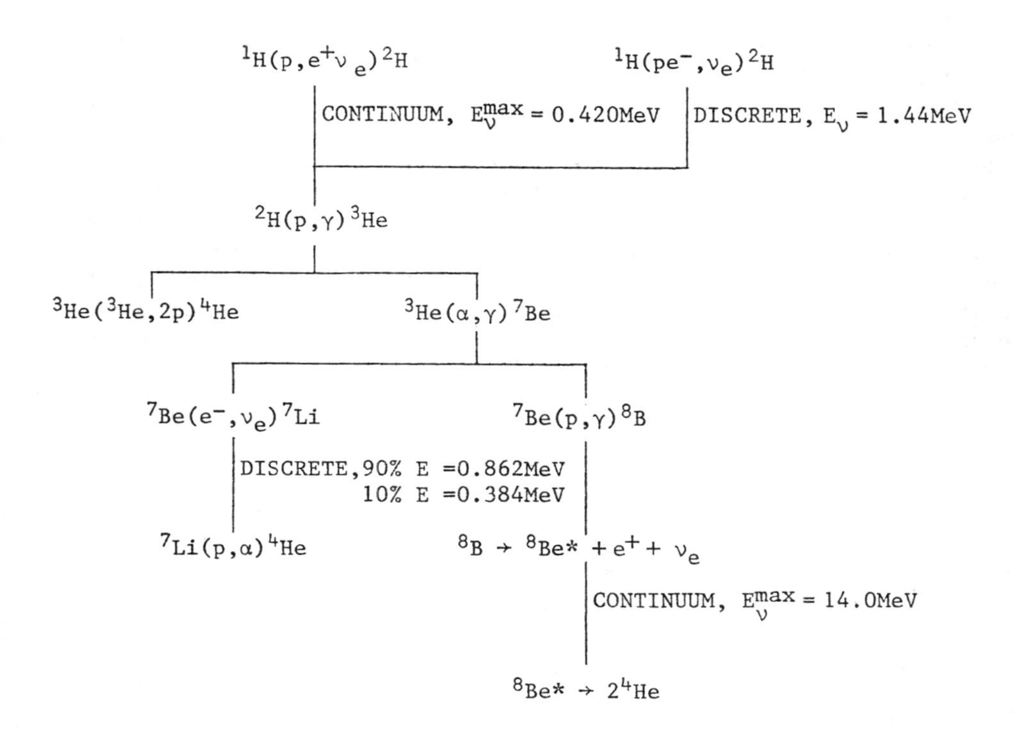

Fig. 1. The proton-proton chain of nuclear reactions.

S_{11}

The pp and pep reactions essentially determine the rate of energy generation in the sun. This is because they are the slowest reactions in the pp chain due to the weak interaction. Because of this they are also practically inaccessible to experiment. Our knowledge of their cross sections is determined through calculation, but experimental input is needed here to determine the weak matrix element. The matrix element can be written in terms of the square of the axial vector coupling constant (G_A). This constant can be determined from the neutron half-life (which depends on both G_A and the vector coupling constant G_V) and super-allowed Fermi transitions in nuclei (which give G_V). While G_V can be determined with an accuracy of < 0.1% from the Fermi transitions, recent direct measurements of the neutron half-life with quoted errors of 1-2% differ from each other by ~ 5%, presumably due to hidden systematic errors likely to occur in such a difficult measurement.

There are however other ways of determining G_A through determinations of $\lambda = |G_A/G_V|$ which use correlation observables in neutron decay. By measuring the beta asymmetry in polarized neutron decay or by measuring the angular correlation between the electron and the $\bar{\nu}$ (inferred from the momentum of the recoiling proton) λ can be determined. New measurements using both of these techniques have been reported over the last several years, with good agreement between the experiments. We can compare the three different types of experiments by converting all the results to a value for the neutron half-life, as shown in table I. The poor agreement between the direct measurements is evidenced by a chi-squared per degree of freedom (χ^2_ν) of 7.1, while the inferred values appear to agree too well. It is interesting however that the weighted means from the direct and inferred measurements are in good agreement. The safest approach is probably to adopt the value from the inferred measurements but scale the error by the square root of χ^2_ν = 2.7 from the direct measurements. The value obtained is then $t_{1/2} = 622 \pm 11$. For a similar analysis see ref. 8.

Table I

Recent direct and inferred values of the neutron half-life (errors given for the mean values are internal errors; see text for discussion of adopted values)

Experiment	Method	$t_{1/2}$(sec)
Christensen *et al.*[1] (1972)	Direct	637 ± 10
Krohn and Ringo[2] (1975)	β-asymmetry	629 ± 12
Bondarenko *et al.*[3] (1978)	Direct	608 ± 5
Stratowa *et al.*[4] (1978)	$\bar{\nu}_e - e^-$ correlation	625 ± 14
Erozolimskii *et al.*[5] (1979)	β-asymmetry	627 ± 10
Byrne *et al.*[6] (1980)	Direct	649 ± 12
Bopp *et al.*[7] (1984)	β-asymmetry	616 ± 7
$\bar{t}_{1/2}$ (direct), $\chi^2_\nu = 7.1$		618 ± 4
$\bar{t}_{1/2}$ (inferred), $\chi^2_\nu = 0.46$		622 ± 5
$t_{1/2}$ (adopted)		622 ± 11

The impact of this uncertainty on the predicted capture rates in the three solar neutrino detectors discussed in the introduction is most severe in the case of ^{37}Cl. This is because of the extreme sensitivity of the ^8B ν's to temperature. The relationship between temperature and S_{11} is the result of the constraint to fit the calculated energy generation rate to the observed solar constant. An increase in S_{11} results in a decrease in the temperature because the reaction can now produce energy at the same rate (determined by the solar constant) but at a lower temperature. Using the solar model code of ref. 9 we find that a

2% uncertainty in the neutron half-life results in a 5%, 1%, and 3% uncertainty in the neutrino capture rate for ^{37}Cl, ^{71}Ga, and ^{81}Br respectively. This does not include errors in the estimation of the meson-exchange corrections and in the overlap integral for pp and the deuteron which also contribute to the error in S_{11}.

S_{34}

A great deal of interest was sparked in the ^{3}He(α,γ)^{7}Be S-factor by an indication (ref. 10) that previous measurements may have overestimated S_{34} by nearly a factor of two. Such a modification of S_{34} would reduce the predicted neutrino capture rate in the ^{37}Cl experiment by nearly one-half, going a long way towards solving the solar neutrino problem. This suggestion initiated a whole series of new independent experiments, in some cases using novel techniques, which hoped to confirm or disprove this result. To make a long story short, none of the new measurements were able to substantiate the very low value for S_{34}. A summary of these results is displayed in table II. The value quoted for ref. 10 is the published value. There are indications[11] that this value includes some measurements with a gas-jet target that used an incorrect value for the target thickness. This apparently increases the extracted value of S_{34} to 0.40 keV-barns.

Table II

Measurements of the ^{3}He(α,γ)^{7}Be S-factor
(errors given for the mean are internal errors;
see text for discussion of adopted value)

Experiment	Method	S_{34} (keV-b)
Parker *et al.*[13] (1963)	Capture γ-rays	0.47 ± 0.05
Nagatani *et al.*[14] (1969)	Capture γ-rays	0.57 ± 0.06
Kräwinkel *et al.*[10] (1982)	Capture γ-rays	0.30 ± 0.03
Osborne *et al.*[15] (1982,1984)	Capture γ-rays	0.52 ± 0.03
"	^{7}Be activity	0.56 ± 0.04
Robertson *et al.*[16] (1983)	^{7}Be activity	0.63 ± 0.04
Volk *et al.*[17] (1983)	^{7}Be activity	0.56 ± 0.03
Alexander *et al.*[18] (1984)	Capture γ-rays	0.47 ± 0.05
$\bar{S}_{34}$ (capture γ-rays), $\chi^2_\nu = 0.8$		0.51 ± 0.02
$\bar{S}_{34}$ (activity), $\chi^2_\nu = 1.1$		0.58 ± 0.02
S_{34} (adopted)		0.54 ± 0.06

Of the new techniques tried, three of the experiments detected the reaction by counting the 478 keV γ-ray from the decay of ^{7}Be($t_{1/2}$ = 53 days). An interesting sidelight to this revolves around the γ-ray branching ratio of ^{7}Be to the 478 keV state in ^{7}Li. The branching ratio is needed for this technique in order to determine the number of ^{7}Be nuclei produced in the ^{3}He(α,γ)^{7}Be reaction. An unpublished suggestion that the accepted value for this branching ratio could be in error by 50% (thereby altering the inferred S_{34} by a similar amount) was met, within six months, by no less than 15 remeasurements by vastly different techniques which substantiated the earlier number. The interested reader is referred to ref. 12 for a summary of these new measurements.

In adopting a value for S_{34} it would seem best not include the value from ref. 10 until clarification of the corrected value is published. Table II shows the mean value for S_{34} for the capture γ-ray data (without ref. 10) as well as for the activity measurements. Although both techniques seem to be internally consistent, there appears to be a systematic difference between the two methods, with the activity measurements somewhat higher. As there are no compelling reasons to choose one technique over the other, we adopt a value of $S_{34} = 0.54 +-0.06$. The uncertainty here results from an additional systematic error of 5%, added in quadrature, because of the difficulty in separating statistical from systematic errors in each individual experiment. Also added in quadrature is an error of 10% (see ref. 16) resulting from the uncertainty in the theoretical extrapolation to solar energies. The uncertainty in S_{34} results in an uncertainty in neutrino capture rate of 8%, 2%, and 7% respectively for ^{37}Cl, ^{71}Ga, and ^{81}Br.

S_{17}

The ^{7}Be$(p,\gamma)^{8}$B reaction is responsible for the high energy neutrinos which provide most of the signal in the ^{37}Cl experiment. The measurement of S_{17} is complicated by the fact that the target nucleus, ^{7}Be, is radioactive. This activity, however can be put to good use. The 478 keV γ-rays from the decay can be used to determine the number of ^{7}Be nuclei present in the target. Scanning the activity with a collimated γ-ray detector determines the area of the target and thus the areal density can be determined. The same quantity can be measured independently by monitoring the build-up of ^{7}Li (the ^{7}Be decay product) as a function of time, using a reaction sensitive to ^{7}Li. The $^{7}Li(d,p)^{8}Li$ reaction is ideally suited for this because the reaction product - ^{8}Li - decays by high energy β emission to ^{8}Be* followed by emission of two α particles. This decay closely mimics the decay of ^{8}B so that the same apparatus used to detect the occurrence of the ^{7}Be$(p,\gamma)^{8}$B reaction can be used to monitor the build-up of ^{7}Li. In order for this technique to be useful though the cross section (σ_{dp}) for ^{7}Li$(d,p)^{8}$Li must be known; actually the cross section at the peak of the broad resonance at 0.77 MeV is used as the normalization point. There has been considerable work on this cross section since 1978, as illustrated in table III, where a summary of the modern measurements is displayed. As discussed in ref. 26 the best strategy here is probably to remove ref. 20 from the data set and double the unjustifiably small error bars of ref. 22 to obtain the weighted mean. When this is done a reasonable $\chi^2_\nu = 1.1$ is obtained.

Table III

Measurements of the $^7Li(d,p)^8Li$ cross section at the peak of the 0.77 MeV resonance (see text for discussion of the adopted value)

Experiment	Method	σ_{dp} (mb)
Kavanagh[19] (1960)	^8Li β^-	163 ± 15
Parker[20] (1966)	β^- delayed α's	211 ± 15
McClenahan and Segel[21] (1975)	^8Li β^-	138 ± 20
Schilling *et al.*[22] (1976)	β^- delayed α's	181 ± 8
Mingay[23] (1979)	^8Li β^-	174 ± 16
Elwyn *et al.*[24] (1982)	reaction protons	146 ± 13
Filippone *et al.*[25] (1982)	β^-delayed α's	148 ± 12
σ_{dp} (adopted)		157 ± 10

There is also new data for the $^7Be(p,\gamma)^8B$ reaction itself. In this experiment (ref. 26) the two techniques for determining the ^7Be areal density were combined for the first time. The agreement between the two techniques was excellent when the mean value for σ_{dp}, discussed above, was used. In addition, the measurements were extended to lower energies than had been previously possible. The cross section and S-factor from this new experiment are shown in figs. 2 and 3. The non-resonant calculation of ref. 27 has been normalized to the off-resonance data. The resonance at 630 keV gives a negligible contribution at solar energies.

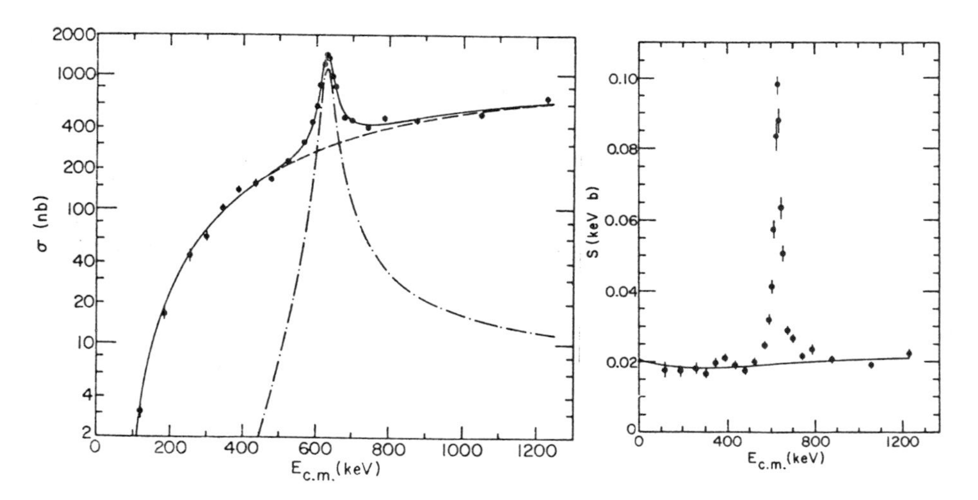

Fig. 2. Total cross section for $^7Be(p,\gamma)^8B$. Fig. 3. S-factor for $^7Be(p,\gamma)^8B$.

All measurements of S_{17} are summarized in table IV. Two values are quoted for ref. 26 corresponding to the use of two independent normalization methods but the uncertainties from the two methods are not independent. Thus in determining an adopted value we use a single number from that experiment of $S_{17} = 0.0216 \pm 0.0025$keV-b. and take a weighted mean. The uncertainty in the adopted value is obtained from the internal error scaled by the square root of χ_ν^2 (note: $\chi_\nu^2 = 2.3$) and including a $\pm$ 5% uncertainty in the theoretical extrapolation (see ref. 26). The resulting uncertainty in S_{17} gives an uncertainty of 7%, 0%, and < 1% in the predicted capture rates for the ^{37}Cl, ^{71}Ga, and ^{81}Br detectors.

Table IV

Measurements of the ^{7}Be$(p,\gamma)^{8}$B S-factor normalized to the same σ_{dp}(0.77 MeV) = 157 ± 10 mb (see text for discussion of adopted value)

Experiment	Normalization	S_{17} (keV-b.)
Kavanagh [19] (1960)	σ_{dp}	0.016 ± 0.006
Parker [20] (1966)	σ_{dp}	0.028 ± 0.003
Kavanagh *et al.* [28] (1969)	σ_{dp}	0.0273 ± 0.0024
Vaughn *et al.* [29] (1970)	σ_{dp}	0.0214 ± 0.0022
Wiezorek *et al.* [30] (1977)	^{7}Be activity	0.045 ± 0.011
Filippone *et al.* [26] (1983)	σ_{dp}	0.0221 ± 0.0028
"	^{7}Be activity	0.0206 ± 0.0030
S_{17} (adopted)		0.024 ± 0.002

CONCLUSIONS

A large body of work, performed over the last several years, has provided us with a substantial improvement in the nuclear physics input to solar neutrino calculations. As a result of this work it would appear that we can find no error in the nuclear input to the calculations large enough to resolve the solar neutrino problem, although uncertainties in the nuclear input still contribute a significant amount to the uncertainty in the predicted neutrino capture rates. Presently we appear limited by the large theoretical extrapolations in attempting to determine more precisely the relevent nuclear parameters. Experimentally the only hope might be striving for lower energies where the theoretical extrapolations are less severe.

REFERENCES

1. C. J. Christensen, A. Nielsen, A. Bahnsen, W. K. Brown, B. M. Rustad, Phys. Rev. **D5**, 1628 (1972).
2. V. E. Krohn, G. R. Ringo, Phys. Lett. **55B**, 175 (1975).
3. L. N. Bondarenko, V. V. Kurguzov, Yu. A. Prokofiev, E. V. Rogov, P. E. Spivak, JETP Lett. **28**, 303 (1978).
4. C. Stratowa, R. Dobrozemsky, P. Weinzierl, Phys. Rev. **D18**, 3970 (1978).
5. B. G. Erozolimskii, A. I. Frank, Yu. A. Mostovoi, S. S. Arzumanov, L. R. Voitzik, Sov. J. Nucl. Phys. **30**, 356 (1979).
6. J. Byrne, J. Morse, K. F. Smith, F. Shaikh, K. Green, G. L. Greene, Phys. Lett. **92B**, 274 (1980).
7. P. Bopp, D. Dubbers, E. Klemt, J. Last. H. Schütze, W. Weibler, S. J. Freedman, and O. Schärpf, J. Phys. **45**, 21 (1984).
8. Particle Data Group,Rev. Mod. Phys. **56**, S1 (1984).
9. B. W. Filippone and D. N. Schramm, Astrophys. J. **253**, 393 (1982).
10. H. Kräwinkel, H. W. Becker, L. Buchmann, J. Görres, K. U. Kettner, W. E. Kieser, R. Santo, P. Schmalbrock, H.-P. Trautvetter, A. Vlieks, C. Rolfs, J. W. Hammer, R. E. Azuma, and W. S. Rodney, Z. Phys. **A304**, 307 (1982).
11. C. Rolfs, private communication.
12. R. T. Skelton and R. W. Kavanagh, Nucl. Phys. **A414**, 141 (1984).
13. P. D. Parker and R. W. Kavanagh, Phys. Rev. **131**, 2578 (1963).
14. K. Nagatani, M. R. Dwarakanath, and D. Ashery, Nucl. Phys. **A128**, 325 (1969).
15. J. L. Osborne, C. A. Barnes, R. W. Kavanagh, R. M. Kremer, G. J. Mathews, J. L. Zyskind, P. D. Parker, and A. J. Howard, Phys. Rev. Lett. **48**, 1664 (1982), and Nucl. Phys. **A419**, 1664 (1984).
16. R. G. H. Robertson, P. Dyer, T. J. Bowles, R. E. Brown, N. Jarmie, C. J. Maggiore, and S. M. Austin, Phys. Rev. **C27**, 11, (1983).
17. H. Volk, H. Kräwinkel, R. Santo, and L. Walleck, Z. Phys. **A310**, 91, (1983).
18. T. K. Alexander, G. C. Ball, W. N. Lennard, H. Geissel, and H.-B. Mak, Nucl. Phys. **A427**, 526 (1984).
19. R. W. Kavanagh, Nucl. Phys. **15**, 411 (1960).
20. P. D. Parker, Phys. Rev. **150**, 851 (1966), and Astrophys. J. **153**, L85 (1968).
21. C. R. McClenahan and R. E. Segel, Phys. Rev. **C11**, 370 (1975).
22. A. E. Schilling, N. F. Mangelson, K. K. Nielson, D. R. Dixon, M. W. Hill, G. L. Jensen, and V. C. Rogers, Nucl. Phys. **A263**, 389 (1976).
23. D. W. Mingay, S. Afr. J. Phys. **2**, 107 (1979).
24. A. J. Elwyn, R. E. Holland, C. N. Davids, and W. Ray, Jr., Phys. Rev. **C25**, 2168 (1982).

25. B. W. Filippone, A. J. Elwyn, W. Ray, Jr., and D. D. Koetke, Phys. Rev. **C25**, 2174 (1982).

26. B. W. Filippone, A. J. Elwyn, C. N. Davids, and D. D. Koetke, Phys. Rev. Lett. **50**, 412 (1983) and Phys. Rev. **C28**, 2222 (1983).

27. T. A. Tombrello, Nucl. Phys. **71**, 459 (1965).

28. R. W. Kavanagh, T. A. Tombrello, J. M. Mosher, and D. R. Goosman, Bull. Am. Phys. Soc, **14**, 1209 (1969); R. W. Kavanagh, in *Cosmology, Fusion, and Other Matters*, edited by F. Reines (Colorado University Press, Boulder, 1972),p. 169.

29. F. J. Vaughn, R. A. Chalmers, D. Kohler, and L. F. Chase, Jr., Phys. Rev. **C2**, 1657 (1970).

30. C. Wiezorek, H. Kräwinkel, R. Santo, and L Wallek, Z. Phys. A. **282**, 121 (1977).

CAN (p,n) REACTIONS HELP US DETERMINE NEUTRINO CROSS SECTIONS?

C. D. Goodman
Indiana University, Bloomington, Indiana 47405

ABSTRACT

The absorption of a neutrino on a nucleus induces a transition in which a nucleon within the nuclear structure changes its isospin and an electron is emitted. It would be desirable to determine the nuclear matrix elements of this process from beta decay measurements, but some of the transitions involved for existing and proposed neutrino detectors are not accessible to beta decay. The (p,n) reaction can induce the appropriate transitions with almost no momentum transfer to the nucleus. Thus, (p,n) measurements hold a promise of providing needed information to predict neutrino absorption cross sections. The use of this technique is reviewed.

INTRODUCTION

The elementary process by which a neutrino is detected is

$$\nu + n \longrightarrow p + e^- \tag{1}$$

Unfortunately, we cannot use free neutrons as the detector if only because the neutron mass is greater than the proton mass and the n --> p transition takes place spontaneously. In reality we must use neutrons bound in nuclei as the detectors. This fact makes our prediction of the neutrino absorption cross section dependent on our knowledge of the relevant nuclear structure and, therefore, introduces a complication that makes it impossible to calculate the neutrino absorption cross sections starting from only a knowledge of the elementary process.

A practical alleviation of the difficulty is to start from measured beta decay ft values as fiducial points and use model calculations only to extrapolate the predictions to closely similar transitions. Even this limited use of a nuclear model may be inaccurate. Ideally, one would like to be able to make measurements of the actual transitions that will occur in the neutrino detection process. If one could measure the Gamow-Teller (GT) matrix elements to the spectrum of states accessible to the neutrino absorption process, then one could calculate the expected neutrino counting rate by folding the neutrino energy spectrum into the GT strength function. The spectrum cannot be measured directly with beta decay in the forward direction because it is energetically "uphill." Some GT matrix elements can be obtained from beta decay in the inverse direction leading to the neutrino detection target as the final state.

In this talk I wish to examine the possibility of using a

combination of (p,n) data and beta decay data to determine the GT strength function. Initial success in finding a proportionality between GT matrix elements and (p,n) cross sections has been quite encouraging,[1] and the successful application of the technique to the ^{37}Cl case[2] has given even further encouragement to such a program. The general scheme is to normalize (p,n) cross section measurements to GT matrix elements known from beta decay and then to use the reaction measurements to extend our knowledge somewhat, but not too far, beyond the transitions that can be measured with beta decay. This procedure bypasses the rather difficult problem of understanding the reaction process in great enough detail to extract accurate structure information directly from reaction cross sections.

THE NUCLEAR STRUCTURE ASPECTS OF THE GT MATRIX ELEMENTS

We may picture a neutrino being absorbed by a neutron in a nucleus (see fig 1). An electron is emitted and the neutron is transformed to a proton. Since this process occurs with almost no momentum transfer to the nucleus, the proton must be created with the same angular momentum and radial wave functions that the neutron had prior to the transformation. A nuclear state meeting those conditions in general does not exist as an eigenfunction of the final nucleus. We can, however, think of this fictitious state expressed as a linear combination of the real states of the nucleus. In this language, the amplitudes for exciting real final states by neutrino absorption are the projections of the fictitious state on the energetically accessible eigenstates of the final nucleus.

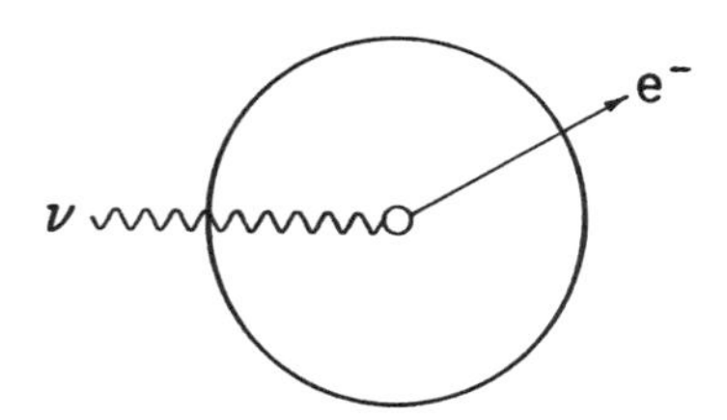

Fig. 1. Absorption of a neutrino on a neutron in a nucleus.

It is convenient to think of the description of the nucleus in terms of state vectors. We shall denote the ground state of the target nucleus by the state vector, $|P\rangle$, where P is for "parent" state. We symbolize the changing of protons to neutrons and neutrons to protons with isospin raising, t^+, and isospin lowering, t^-, operators. For nucleons:

$$t^+|p\rangle = |n\rangle \qquad (2)$$

$$t^-|n\rangle = |p\rangle \qquad (3)$$

These transitions correspond to what are generally called Fermi transitions in beta decay. In a nucleus in its ground state neutrons and protons will pair off to fill the available single particle states and most neutrons will be prevented from changing to protons because the corresponding proton states are already occupied. However, in a nucleus with a neutron excess, all neutron to proton transitions cannot be Pauli blocked, simply because there are fewer protons than neutrons. Then, operating on the nuclear state vector with the total isospin lowering operator cannot yield a zero vector. We may write:

$$\sum_i t_i |P\rangle = |IAS\rangle \qquad (4)$$

where IAS stands for "isobaric analog state," which is an eigenstate of the daughter nucleus. This is a way of symbolizing the discovery of Anderson, Wong and McClure[3] that all of the Fermi transition strength lies in a sharp state of the daughter nucleus. The left member of the equation represents the total transition strength and the right member represents the sharp state that has come to be called the isobaric analog state. The Fermi strength distribution is represented pictorially in fig. 2.

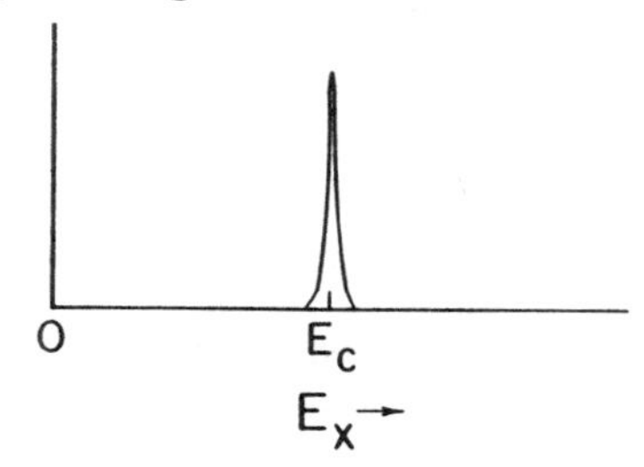

Fig. 2. Pictorial representation of the Fermi transition strength function.

Changing neutrons to protons in a nucleus costs coulomb energy because the new proton interacts with the other protons. Thus, if neutrino detection were to be accomplished only through analog state transitions, the threshold energy would have a very steep Z dependence, and only very light nuclei could be used for detecting solar neutrinos. The functional dependence of the coulomb energy difference is approximately $Z/A^{1/3}$. Fortunately, neutrons can be changed to protons with simultaneous spin flips -- the so-called Gamow-Teller transitions. Here, too, the coulomb energy must be supplied, but some energy can be gained back through the spin interaction.

$$st^- |p\rangle = |n\rangle \qquad (5)$$

Here s represents the three component spin operator. The difference between this and eq. (3) is that the spin orientation of the neutron may be different from that of the proton.

An easy way to picture how the spin interaction might help the process of neutrino detection is to imagine first that the spin interaction is turned off. Then the energy required to change a proton to a neutron is simply the coulomb energy difference between the parent and the daughter state, regardless of whether the spin flips. In this case Gamow-Teller transitions as well as Fermi transitions would yield sharp states at the same energy. These sharp states might be called giant Gamow-Teller and giant Fermi resonances since they contain all of the GT or F strength built on the parent state. If we now turn on the spin dependence of the nuclear force, it will smear and fragment the giant GT resonance. A little piece of the GT strength may be pushed down so far that the energy difference between the parent state and that state containing the fragment of strength is nearly zero. Then we have an energetically useful neutrino detector, but we have only a small bit of the total GT strength contributing. This is, in fact, the situation that we actually encounter in such detectors as gallium, indium, and bromine, for example.

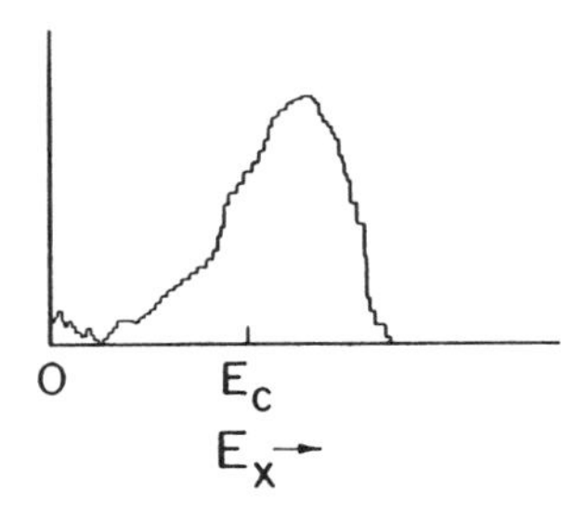

Fig. 3. Pictorial representation of the Gamow-Teller transition strength function.

Let us return to the symbolic representations. The GT operator acting on the parent state vector yields a new vector that we might call the collective Gamow-Teller vector, $|CGT\rangle$. In the light of the previous paragraph we can expect that the $|CGT\rangle$ is not an eigenstate of the daughter nucleus. The question we must ask is how big are the projections of that vector on the low lying levels of the daughter nucleus.

$$\sum_i s_i t_i |P\rangle = |CGT\rangle \neq \text{eigenstate} \qquad (6)$$

In principle we can set down a procedure for a shell model calculation to find the projections we want. We calculate $|P\rangle$. We operate on it with the GT operator. We calculate the first few eigenstates of the daughter nucleus and we form the projections. Thinking about it in this way, we can see the accuracy problems that arise. The total GT strength, $\langle CGT|CGT\rangle$, is constrained by a simple sum rule (see ref. 4 for a qualitative and quantitative description of the sum rule). In fact only a little over half of the calculated sum strength is actually seen in experiments, so in a way, the model is already wrong in predicting the strength in the major part of the giant resonance. We might worry, then, about whether the model might be wrong *a fortiori* in predicting the strengths of the little peaks. For the neutrino cross sections we are asking the model to tell us how much of the GT strength is in the little bumps on the wings of the GT giant resonance. This depends on fine details of the residual force used in the shell model calculation.

The discussion above illustrates the importance of measuring the relevant matrix elements for neutrino detection rather than relying simply on calculations.

THE (p,n) REACTION AS A GAMOW-TELLER PROBE

We wish to induce with a strongly interacting probe the same transformations that take place in neutrino absorption. That is, we want to cause a neutron in the nucleus to change into a proton, gently, without imparting much momentum to the nucleus. We shall do this with high energy protons, and the experimental challenge will be to select out of all the things that happen the events that belong to the process of interest. If we shoot a high energy proton into a nucleus and observe a high energy neutron coming out with the same momentum that the proton had, then we can infer that there has been no change in the orbital or radial wave function of the nucleus, but a neutron must have changed to a proton. We must, of course, make the observations at zero degrees. Otherwise we will have imparted transverse momentum.

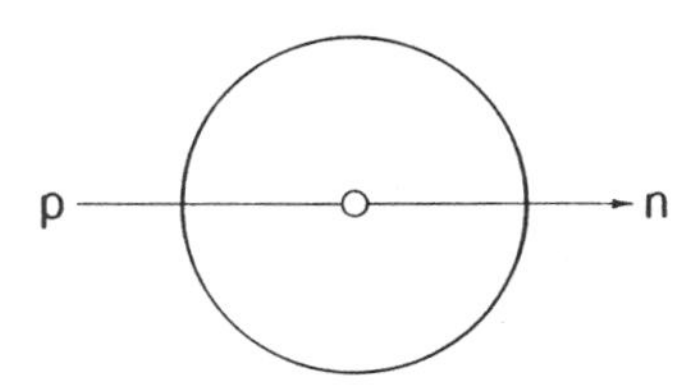

Fig. 4. Changing a neutron to a proton in the nucleus with a (p,n) reaction. Because of the large mass of the projectile, we must make the observation at zero degrees to avoid transferring momentum to the bound nucleon.

The relevance of the high energy is twofold. First, we must transfer a little bit of energy to the nucleus to reach the state of interest. The corresponding momentum transfer at zero degrees gets smaller as the bombarding energy is increased. This is illustrated in eq. (7), which is not the full kinematic equation but simply the relativistic momentum-energy relationship. The square-root factor has the values 5.27 at 35 MeV and 2.39 at 200 MeV.

$$E^2 = p^2c^2 + m^2c^4 \; ; \quad \Delta pc = [1+1/(E-1)]^{1/2}\Delta E \qquad (7)$$

Second, we want to emphasize single step reactions. This dictates the use of protons more or less in the 100-300 MeV range, where the nucleon-nucleon interaction is weak.

The theoretical rationale for believing that one might be able to extract Gamow-Teller matrix elements from (p,n) cross sections comes from the belief that the expression for the (p,n) cross section in the distorted wave impulse approximation (DWIA)) can be factorized into three parts, a nuclear structure factor, an interaction strength factor, and a distortion factor. The argument is discussed in ref. 1. The relevant equation is

$$d\sigma/d\omega(0^{\circ}) = (\mu/\pi\hbar^2)^2(k_f/k_i)[N_{\tau}J_{\tau}^2B(F)+N_{\sigma\tau}J_{\sigma\tau}^2B(GT)] \qquad (8)$$

The N's are the distortion factors, the J's are the factors that contain the interaction strengths, and B(F) and B(GT) are the Fermi and Gamow-Teller structure factors. These are the squares of the overlap matrix elements with the spin multiplicity factors already built in. We define the units such that B(F) = 1 and B(GT) = 3 for a free neutron to proton transition. The three comes from the fact that the total spin operator has three spatial components. Note that the only structure matrix elements that appear in this approximation for the zero degree cross section are the Fermi and Gamow-Teller matrix elements. These are the only transitions that do not involve spatial rearrangements. Transitions that involve changes in angular momentum or radial quantum numbers imply momentum transfers. In making the measurements at zero degrees we suppress the transitions requiring momentum transfer, but the accuracy of the approximation in which we ignore them is something that has to be examined for each case. Since the energy transfer for the transitions involved in neutrino detection is very small, the approximation is likely to be better here than it is for measuring GT strength at high excitation.

In eq. (8) we are assuming an impulse approximation; that is, we assume that the proton interacts with a single neutron through the free n-n interaction. Thus, the factors J and J may depend on the bombarding energy but should not depend on the target nucleus. The distortion factors, the N's, however, represent the distortion of the incoming and outgoing waves by the nucleon-nucleus potential and will depend on Z and A of the target nucleus.

Looking at the form of eq. (8), one may think of several ways to apply it to the problem of extracting the structure factor from the measured cross section. The suggestion in ref. 1 is to calculate the distortion factor using a distorted wave code, to measure the cross sections for fiducial cases where the structure factors are known from beta decay, and solve for the J values. Then, with the J values already known, one can calculate the N's for an unknown case, but in the measured cross section and solve for B(GT).

The method of normalizing to fiducial transitions is illustrated in the level scheme in fig. 5. Here the beta decay ft value for the ^{42}Ti decay has been measured. In the (p,n) reaction the isospin mirror of that transition is observed. The value of B(GT) is assumed to be the same as for the isospin mirror. With the measured (p,n) cross section and the B(GT) deduced from the beta decay measurement, the product $N_{\sigma\tau}J_{\sigma\tau}$ can be determined from eq. (8). The quantities B(F) and B(GT) are related to the beta decay ft values as follows:

$$1/ft = G_V^2 B(F) + G_A^2 B(GT) \tag{9}$$

$$6163.4/ft = B(F) + 1.56\ B(GT) \tag{10}$$

For a pure GT transition

$$B(GT) = 3951/B(GT) \tag{11}$$

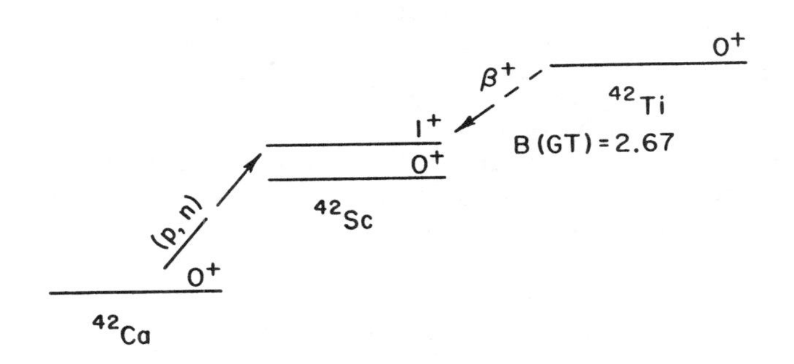

Fig. 5. Level scheme illustrating the method of normalizing (p,n) cross sections to beta decay measurements.

The data presented in ref. 2 and the application of this technique to the ^{37}Cl case led us to believe that we could use this procedure more generally to find the B(GT) for unknown transitions involved in neutrino detection. Then some dark clouds rose on the horizon that made us question the accuracy of this approach. One cloud was the report by Orihara et al.[5] of a measurement of $^{71}Ga(p,n)^{71}Ga$ suggesting that the ground state and first excited state transitions have about equal GT strengths. This result is contrary to the estimate by Bahcall et al.[6] and also seems contrary to some preliminary (p,n) evidence in an experiment underway at Indiana. If the Orihara result is true, the spectral sensitivity of ^{71}Ga is shifted more towards higher energy neutrinos than had been previously thought, and the discrimination between p+p neutrinos and ^{8}B neutrinos is degraded.

The other cloud was the observation by our group at Indiana, in work in progress and not yet published, that in some worst cases the calculated (p,n) cross sections, calculated with what we thought were reasonable optical model parameters, are smaller than the measured cross sections by as much as 30 percent. This caused us to question the reliability of distortion factors calculated with distorted wave codes.

The clouds seem to be clearing. The Orihara experiment has now been reanalyzed by Baltz et al.[7] who conclude that at 35 MeV, the energy of the Orihara experiment, contributions from L=2 transfer are not negligible and, in fact, dominate the excited state transition. This is to say that eq. (8) is not a good enough approximation at 35 MeV. The calculations of Baltz et al., however, indicate that the approximation is much better at higher energy. The question of the ^{71}Ga spectral sensitivity may be answerable by further experiments at Indiana which are underway.

The question of distortion factors is more involved, but we can also dodge the issue to a large extent. It is apparent from eq. (8) that we should be able to extract B(GT) for an unknown transition in a given spectrum quite accurately if we know B(GT) for another transition in the same spectrum. That idea is too limited in its application to be generally useful. However, Taddeucci et al.[8] made a very important empirical observation, expressed in eq. (12), that the double ratio of cross section to GT strength divided by cross section to F strength has a very simple functional dependence on bombarding energy in the energy range of about 50-200 MeV.

$$\left(\frac{\sigma(\mathrm{GT,0\ deg})/B(\mathrm{GT})}{\sigma(\mathrm{F,0\ deg})/B(\mathrm{F})}\right)^{1/2} = E_p/(55 \pm 1\ \mathrm{MeV}) \qquad (12)$$

This allows us to normalize to an isobaric analog state transition in a given spectrum, which is a good deal less restrictive in its application than normalizing to known GT transitions.

We attempted to apply this kind of normalization to the (p,n) spectra from ^{13}C and ^{15}N, and the difficulties that showed up illustrate potential problems that we thought might plague the extraction of the GT matrix elements for the neutrino detectors. In both cases beta decay ft values are available for the ground state mirror transitions, which are mixed F and GT transitions. We determined B(F) and B(GT) for the ground state transitions form the beta decay ft values and then used eqns. (8) and (12) to determine GT the fractions of the ground state cross sections. Once the GT fraction is known, B(GT) for any other transition in the spectrum can be determined from the ratio of that cross section to the ground state cross section, and one does not need to use the absolute cross section.

Using this procedure, we found that in both mass 13 and mass 15 the strongest GT transitions had values of B(GT) that were about a factor of three smaller than the predictions of a simple shell model.[9] The result seemed especially disturbing for mass 15 where the nucleus in the simple model is just a single p-shell hole state, and one could argue that if the model works anywhere, it should work there. The relevance of this result to the question of neutrino detectors is that we had here a case of an odd nucleus where we normalize to a transition that has both F and GT strength, as in the ^{71}Ga case, for example, and we found a large discrepancy with a structure calculation. Can we convince ourselves that our reaction analysis is correct?

Suppose we question the method by which we arrive at the GT fraction of the ground state cross section. Suppose for some unexplained reason the parameter in eq. (12) is somewhat different for mixed F and GT transitions than it is for pure transitions. Let us check the sensitivity of the observed cross-section ratio of the ground state to the excited state to changes in the parameter of eq. (12) and to changes in the GT matrix elements. In figs. 6 and 7 we have plotted the double ratio defined in eq. (12) vs. the cross section ratio that would be observed. We have used two different assumptions about the matrix elements. One curve applies to the simple shell model values, the other curve is generated by using the parameter value stated here in eq. (12), the ground state matrix element from beta decay, and an excited state matrix element fitted to the observed cross-section ratio. In other words, that curve is drawn to fit the data. At 160 MeV the double ratio has the value 8.46 and at 200 MeV it is 13.22 and these values seem well determined from measurements of separate F and GT transitions in ^{14}C. The point to notice is that the curves are very steep in the vicinity of those values. To try to account for the measured cross section ratios by assuming the shell model structure matrix elements and altering the double ratio seems out of the question. The curves in figs. 6 and 7 indicate that for proton energies in the 160-200 MeV range the cross-section ratios of the transitions in question in masses 13 and 15 are determined almost entirely by the structure.

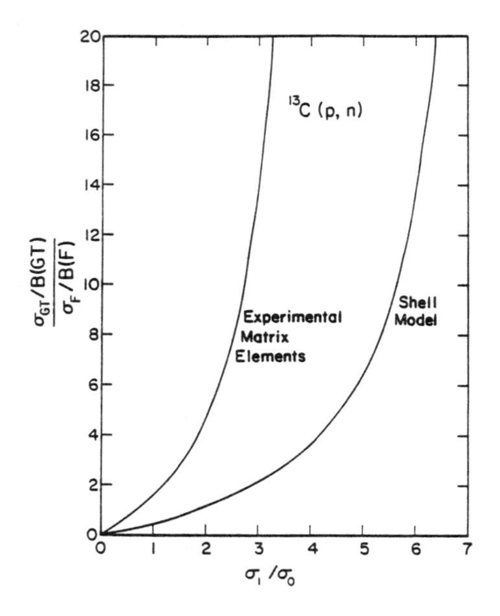

Fig. 6. Reaction strength ratio vs. observed peak ratio for $^{13}C(p,n)$. See text for explanation.

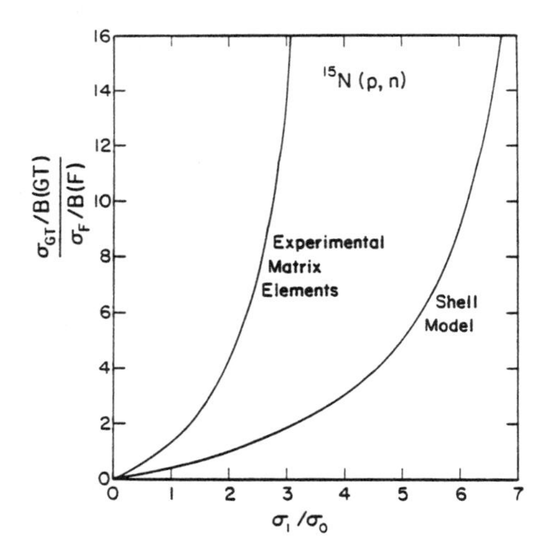

Fig. 7. Reaction strength ratio vs. observed peak ratio for $^{15}N(p,n)$. See text for explanation.

In the cases of those nuclei we have performed an additional check by measuring the spin-flip probabilities using a polarized proton beam and a neutron polarimeter.[9] The spin-flip probability is zero for the Fermi component and is 2/3 for the Gamow-Teller component. The spin-flip probability for a mixed transition is the weighted mean of these values weighted by the relative cross sections for the two components. The spin-flip measurement thus yields an independent determination of the GT fraction in the ground state transition. The result of the spin-flip measurement is in agreement with the result of the analysis stated above.

Fig. 8 shows a spectrum from $^{71}Ga(p,n)^{71}Ge$ at E_p = 120 MeV. Neither the resolution nor the statistics are good enough to resolve the question of the relative values of B(GT) for the two lowest states. The data do suggest, however, that an improved experiment could provide the desired information on the GT strength function.

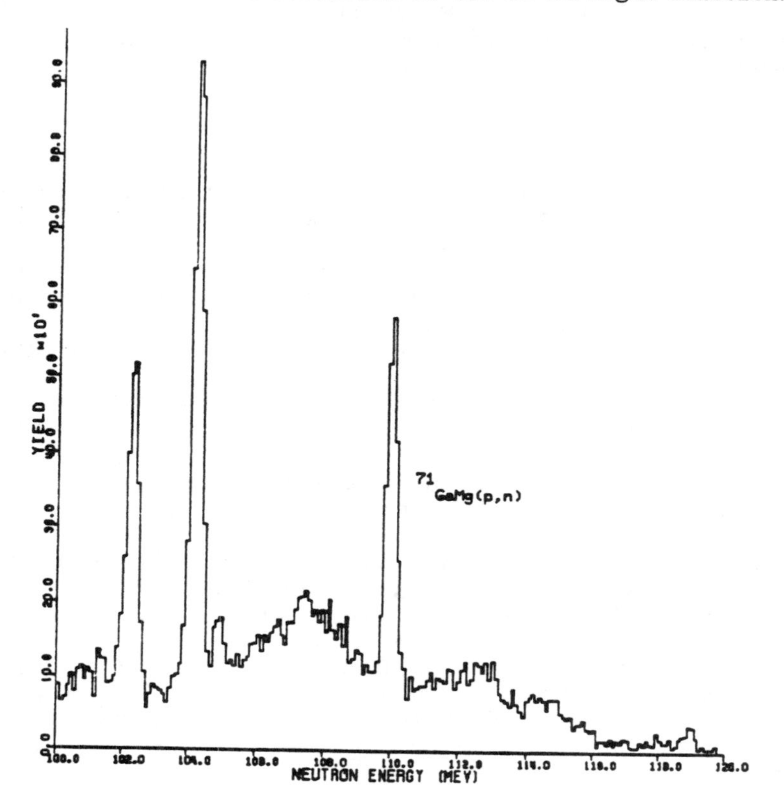

Fig. 8. Spectrum from $^{71}Ga(p,n)^{71}Ge$. The target was made by amalgamating ^{71}Ge with ^{24}Mg. The strong peak at neutron energy 110 MeV is the gallium IAS transition. The very strong peaks at lower neutron energies are from $^{24}Mg(p,n)$.

CONCLUSIONS

Although we have not yet provided definitive information on the Gamow-Teller strength functions of interest for neutrino detectors, we have examined many questions concerning the technique of using the (p,n) reaction as a GT probe. The prospects seem good for extracting GT matrix elements for specific transitions and for obtaining a broad view of the GT strength function with the (p,n) reaction. The main problems seem to be the preparation of adequate targets and allocatin enough accelerator time to obtain good statistics in the spectra.

ACKNOWLEDGEMENTS

I am especially indebted to J. Rapaport and T. N. Taddeucci for help in the preparation of this work. Our (p,n) group, including R. C. Byrd, C. C. Foster, C. Gaarde, D. J. Horen, J. S. Larsen, J. Rapaport, E. Sugarbaker, T. N. Taddeucci, I. J. van Heerden, and T. P. Welch, gets the credit for performing the experiments which generated the data on which this talk is based.

REFERENCES

1. C. D. Goodman, C. A. Goulding, M. B. Greenfield, J. Rapaport, D. E. Bainum, C. C. Foster, W. G. Love, and F. Petrovich, Phys. Rev. Lett. 44, 1755 (1980).
2. J. Rapaport, T. Taddeucci, P. Welch, C. Gaarde, J. Larsen, C. Goodman, C. C. Foster, C. A. Goulding, D. Horen, E. Sugarbaker, and T. Masterson, Phys. Rev. Lett. 47, 1518 (1981).
3. J. D. Anderson, C. Wong, and J. W. McClure, Phys. Rev. 126, 2170 (1962).
4. C. D. Goodman, in "Proceedings of the International Conference on Nuclear Physics," eds. P. Blasi and R. A. Ricci, Florence, Italy, (1983), p. 165.
5. H. Orihara, C. D. Zafiratos, S. Nishihara, K. Furukawa, M. Kabasawa, K. Maeda, K. Miura, and H. Ohnuma, Phys. Rev. Lett. 51, 1328 (1983)
6. J. N. Bahcall, W. F. Huebner, S. H. Lubow, P. D. Parker, and R. K. Ulrich, Revs. Mod. Phys. 54, 767, (1982).
7. A. J. Baltz, J. Weneser, B. A. Brown, and J. Rapaport, Phys. Rev. Lett. 53, 2078 (1984).
8. C. D. Goodman, R. C. Byrd, I. J. Van Heerden, T. A. Carey, D. J. Horen, J. S. Larsen, C. Gaarde, J. Rapaport, T. P. Welch, E. Sugarbaker, and T. N. Taddeucci, submitted to Phys. Rev. Lett. (1984).

NUCLEAR PROCESSES AND NEUTRINO PRODUCTION IN SOLAR FLARES

R.E. Lingenfelter
Center for Astrophysics and Space Sciences
University of California, San Diego, La Jolla, CA 92093

R. Ramaty, R.J. Murphy* and B. Kozlovsky**
Laboratory for High Energy Astrophysics
Goddard Space Flight Center, Greenbelt, MD 20771

ABSTRACT

We briefly review recent observational and theoretical studies of nuclear processes in solar flares and use these studies to determine the flare neutrino flux.

INTRODUCTION

Recent observational and theoretical studies of gamma rays and neutrons from nuclear interactions of solar-flare accelerated particles in the solar atmosphere have provided a wealth of new information on particle acceleration in flares and on the nature of the flare process itself. These results have been discussed in a number of recent papers[1-14]. Here we consider only those aspects which bear directly on the estimation of neutrino production in flares, which is of interest because of suggested[15-17] temporal correlations between neutrino fluxes and solar flare activity. In particular, we consider the energy spectra and total numbers of accelerated particles in flares and their resulting production of β^+ emitting radionuclei and pions which should be the primary sources of neutrinos

*Also Physics Department, University of Maryland, College Park, MD 20742

**Permanent Address: Physics Department, Tel Aviv University, Ramat Aviv, Tel Aviv, Israel.

The possibility of detecting neutrons and gamma ray line emission from solar flares was first suggested some thirty years ago by Biermann[18] and Morrison[19]. In detailed calculations of the expected fluxes it was subsequently shown [20] that the principal gamma ray lines should be those at 2.223 MeV from neutron capture on ^{1}H, at 0.511 MeV from positron annihilation, and at 4.438 and 6.129 MeV from deexcitation of nuclear levels in ^{12}C and ^{16}O, respectively. It was further shown that measurements of both the 2.223 and 4.438 or 6.129 MeV line intensities from flares could give the first measure of both the spectrum and total number of accelerated particles in flares, since the ratios of these line intensities were strongly dependent on the particle spectrum. From calculations of the expected neutron production it was also shown [20,21] that measurement of the time dependence of the solar flare neutron flux alone could give a second, independent measure of the particle spectrum and number. This was possible because the transit time of neutrons from the sun to the vicinity of the earth should generally be much longer than their production time, thus allowing a time-of-flight determination of their spectrum, which was in turn strongly dependent on the accelerated particle spectrum.

Gamma ray line emission was first observed with a detector on OSO-7 from the solar flare of 4 August 1972 by Chupp et al.[22] at the predicted line energies of 0.51, 4.4 and 6.1 MeV. These and other weaker lines have since been observed from more than 30 flares by detectors on HEAO-1[23], HEAO-3[24], HINOTORI[6,10] and most extensively SMM[1,5,8,12]. Neutrons were first observed with the SMM detector by Chupp et al.[3,12] from the flare of 21 June 1980. They have subsequently been detected[11] from the flare of 3 June 1982, where relativistic neutrons were also observed[25] in the ground based cosmic ray neutron monitor at Jungfraujoch and neutron-decay protons were observed[26] with detectors on ISEE-3 in interplanetary space.

These observations have stimulated much more extensive theoretical studies[4,7,9,14,27-35].

NUCLEAR PROCESSES

These studies of the gamma ray line and neutron measurements have provided much new information on the nature of nuclear processes in solar flares and the surrounding solar atmosphere. They have shown[35] that most of the nuclear interactions are caused by accelerated particles that remain trapped in the magnetic fields of the flare region and interact as they slow down in the solar atmosphere and not by those accelerated particles that eventually escape into interplanetary space. This is most clearly seen by the fact that, if the escaping particles were responsible for the observed gamma ray line emission, they should also show great enrichments in spallation products, such as Li, Be and B, which were not observed[36] in the flare particles in interplanetary space.

The measurements of both the neutron-capture line at 2.2 MeV and the C and O deexcitation lines between 4 and 7 MeV allow us to determine[4,7,35] both the spectrum and total number of accelerated particles trapped at the sun. The spectral determination is possible because the effective threshold for neutron production is significantly higher than that for C and O excitations. Thus the line emission from the two processes samples different portions of the accelerated particle spectrum. The spectral indices and total numbers of accelerated protons determined from these gamma ray line measurements of a number of flares are listed in Table 1. These are given for two possible spectral forms, a Bessel function in momentum and a power-law in kinetic energy, expected[35] from different acceleration processes.

Although these observations alone do not allow us to distinguish between the spectral forms, measurements of the spectra of flare accelerated particles that escaped into interplanetary space are generally best described[37] by a Bessel function in momentum. These measurements strongly

TABLE 1

SOLAR FLARE ACCELERATED PARTICLES
(From Ref. 35)

	IN SOLAR ATMOSPHERE				IN INTERPLANETARY SPACE	
	Bessel Function		Power Law			
FLARE	αT	N_p(>30MeV)	s	N_p(>30MeV)	Spectral Index	N_p(>30MeV)
	Determined from Gamma Ray Line Measurements					
4 Aug. 1972	0.029±0.004	1.0×10^{33}	3.3±0.2	7.2×10^{32}	-	4.3×10^{34}
11 Jul. 1978	∿0.032	1.6×10^{33}	∿3.1	1.3×10^{33}	-	-
9 Nov. 1979	0.018±0.003	3.6×10^{32}	3.7±0.2	2.6×10^{32}	-	-
7 Jun. 1980	0.021±0.003	9.3×10^{31}	3.5±0.2	6.6×10^{31}	$\alpha T \simeq 0.015$	8×10^{29}
1 Jul. 1980	0.025±0.006	2.8×10^{31}	3.4±0.2	1.9×10^{31}	-	$<4\times10^{28}$
6 Nov. 1980	0.025±0.003	1.3×10^{32}	3.3±0.2	1.0×10^{32}	-	3×10^{29}
10 Apr. 1981	0.019±0.003	1.4×10^{32}	3.6±0.2	1.0×10^{32}	-	-
	Determined from Neutron and Gamma Ray Line Measurements					
21 Jun. 1980	0.025±0.005	7.2×10^{32}	INCONSISTENT		$\alpha T \simeq 0.025$	1.5×10^{31}
3 Jun. 1982	0.034±0.005	2.9×10^{33}	INCONSISTENT		$s \simeq 1.7$	3.6×10^{32}

suggest that the trapped particles also have a Bessel function spectral form because the spectral indices, αT, of the trapped and escaping particles are quite similar, in cases where they have been determined for particles from the same flare, as can be seen for those 7 and 21 June 1980 (Table 1). Moreover, the ranges of indices, $0.018 < \alpha T < 0.034$ for the trapped particles[35] and, $0.014 < \alpha T < 0.036$, from the escaped particles[37] are essentially the same. This not only suggests that they are accelerated together but that their escape is generally not strongly dependent on rigidity, with the notable exception of the 3 June 1982 flare. But even when the spectrum of escaping particles is the same as that of the trapped particles, the fraction of escaping particles can still vary greatly from less than 0.1% in the flare of 1 July 1980 to as much as 98% in that of 4 August 1972.

Measurements[12] of the time dependent neutron flux also permit an independent determination of both the energy spectrum and total number of accelerated particles, as can be seen in Figure 1 for the limb flare of 21 June 1980.

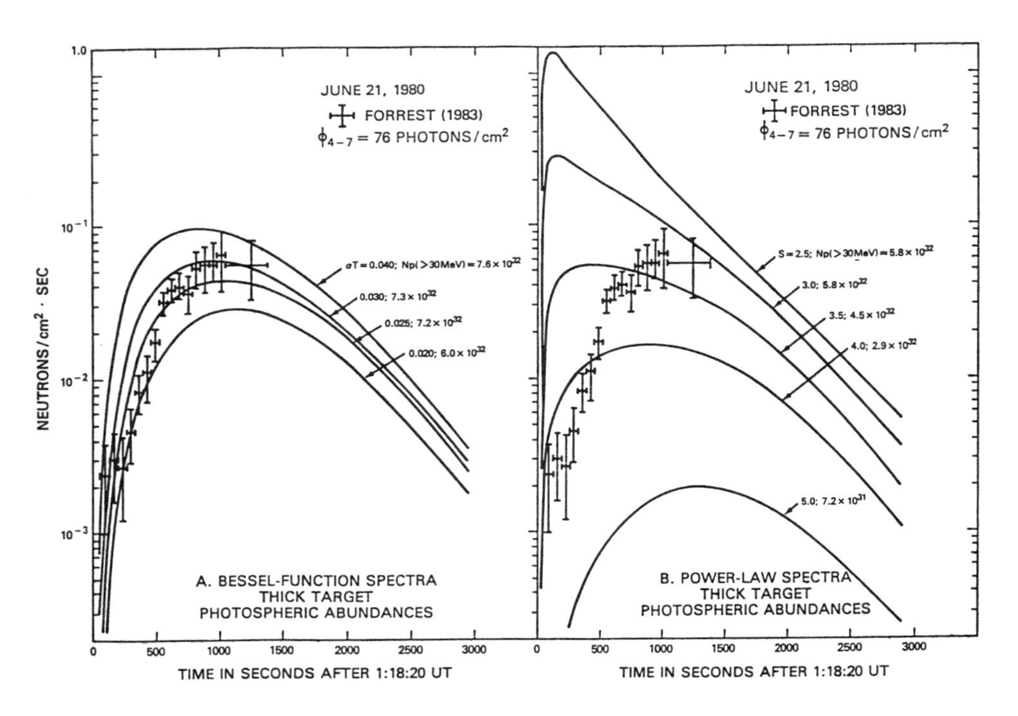

Fig. 1. Determination of the number and spectrum of flare accelerated protons at the sun from observations of the time dependent neutron flux[12] and the gamma ray line emission in the 4-7 MeV range[38]. (From Ref. 35)

Moreover, when the measured[38] fluence of 76 photons/cm^2 in deexcitation lines in the 4-7 MeV range is also considered, it is evident that the combined neutron and gamma ray emission cannot result from particles with a power law spectrum. For, as we see from Figure 1b none of the combinations of power law spectra and total particle numbers that could produce the observed 4-7 MeV intensity can also produce a neutron flux consistent with that which was measured. As can be seen in Figure 1a, however, both observations are quite consistent with accelerated particles having a Bessel function spectrum with $\alpha T \sim 0.025$ and a total number of 7×10^{32} protons > 30 MeV.

There is, however, one further test of consistency in the measured[13] time dependent flux of 0.511 MeV positron annihilation radiation. This can be compared[35] with that expected from positrons produced by trapped accelerated particles with the spectrum and total number determined from the neutron and 4-7 MeV gamma ray fluxes from the 21 June 1980 flare. And, as can be seen in Figure 2, the expected flux is in excellent agreement with that observed, strongly

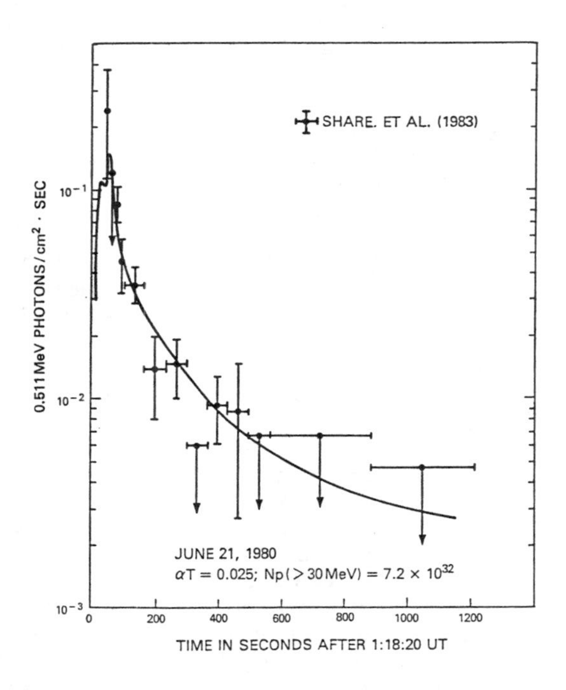

Fig. 2. Observed[13] 0.511 MeV line flux for the 21 June 1980 flare compared with that expected from the number and spectrum of accelerated particles determined in Fig. 1. (From Ref. 35)

suggesting that the spectral form of accelerated particles at the sun is indeed well described by a Bessel function in momentum.

NEUTRINO PRODUCTION

This agreement between the expected and observed 0.511 MeV is especially important for the estimation of the neutrino production in solar flares, since β^+ and π-μ-e decays of accelerated-particle-produced radionuclei and pions, which are the sources of most of the annihilating positrons, are also the main sources of neutrinos.

We can thus directly estimate the fluence of solar flare produced ν from the measured fluence of 0.511 MeV positron annihilation radiation:

$$F_\nu \approx F_{511} (\nu/e^+) (e^+/\gamma_{511})$$

where ν/e^+ is nearly 1, since most of the positrons are produced by β^+ and π-ν-e decay and e^+/γ_{511} lies between 0.5 and 2, depending on whether the positrons annihilate directly into two 0.511 MeV photons or via positronium, where two-photon annihilation results in one-fourth of the time and the rest of the time the annihilation results in a three-photon continuum.

A measurable solar flare contribution to the neutrino flux was first suggested[15] as the possible cause of the apparent excess of ^{37}Ar measured by Davis et al[39] in run 27 of the Brookhaven neutrino experiment, since this run from 7 July to 5 November 1972 included the period of the great flares of 4 and 7 August 1972. The measured ^{37}Ar production in this run was 44±10 atoms of ^{37}Ar, 2.6 times the average over the period from 1970 to 1983. This would imply an apparent excess of ~27 atoms of ^{37}Ar at the end of the run and if due to neutrinos produced in the 4 August flare, would require a flare-induced production of ~170 atoms of ^{37}Ar when corrected for decay.

This ^{37}Ar production can be directly compared with that expected from flare neutrinos, since the positron annihilation line flux was measured[22] during that flare. The measured 0.511 MeV flux at 1AU of (6.3± 2.0) x10^{-2} photons/cm^2-sec over the first 300 sec of the flare of 4 August 1972, implies a total fluence of ~30 photons/cm^2 at 1AU, including contributions from longer lived (>300 sec) β^+ emitters. From the simple relationship given above this implies a total flare-produced neutrino fluence of no more than 60 neutrinos/cm^2 at the earth. For the spectrum determined from the gamma ray line ratios (Table 1) roughly 97% of these neutrinos would have maximum energies ranging from ~ 0.9 to 16 MeV from ^{11}C, ^{12}N and other radio nuclide decay; and 3% would have average energies of the order of 100 MeV from π-μ-e decay. Since the average cross sections for ^{37}Ar production from ^{37}Cl by these neutrinos is ~10^{-44} cm^2 and 10^{-39} cm^2 respectively, high-energy neutrino interactions are clearly dominant. But in the Brookhaven experiment tank, containing 2×10^{30} atoms of ^{37}Cl, the total ^{37}Ar production thus expected from neutrinos made in the flare of 4 August 1972 is only 3×10^{-9}, over ten orders of magnitude less that needed to account for the apparent excess.

ACKNOWLEDGEMENTS - We wish to acknowledge support from NSF through the Solar Terrestrial Program and NASA through the Solar Terrestrial Theory Program.

REFERENCES

1. E.L. Chupp in R.E. Lingenfelter et al., eds, Gamma-Ray Transients and Related Astrophysical Phenomena (Am. Inst. Phys., N.Y., 1982) p. 363.
2. R. Ramaty and R.E. Lingenfelter, Ann. Rev. Nucl. Part. Sci., 32, 235 (1982).
3. E.L. Chupp, et al., Astrophys. J., 263, L95 (1982).
4. R. Ramaty, R.J. Murphy, B. Kozlovsky and R.E. Lingenfelter, Solar Phys., 86, 395 (1983).
5. E.L. Chupp, Solar Phys., 86, 383 (1983).
6. M. Yoshimori et al., Solar Physics, 86, 375 (1983).
7. R. Ramaty, R.J. Murphy, B. Kozlovsky and R.E. Lingenfelter, Astrophys. J., 273, L41 (1983).
8. T.A. Prince et al., 18th Internat. Cosmic Ray Conf. Papers, 4, 79 (1983).
9. R.E. Lingenfelter, R. Ramaty, R.J. Murphy and B. Kozlovksy, 18th Internat. Cosmic Ray Conf. Papers, 4, 101 (1983).
10. M. Yoshimori et al., 18th Internat. Cosmic Ray Conf. Papers, 4, 85 (1983).
11. E.L. Chupp et al., 18th Internat. Cosmic Ray Conf. Papers, 10, 334 (1983).
12. D.J. Forrest in M.L. Burns et al., eds. Positron-Electron Pairs in Astrophysics (Am. Inst. Phys., N.Y., 1983) p. 3.
13. G.H. Share, E.L. Chupp, D.J. Forrest and E. Rieger, in M.L. Burns, et al., eds., Positron-Electron Pairs in Astrophysics (Am. Inst. Phys., N.Y., 1983) p. 15.
14. E.L. Chupp, Ann. Rev. Astron. Astrophys., 22, 359 (1984).
15. A. Subramanian, Curr. Sci., 48, 705 (1979).
16. K. Sakurai, Nature, 278, 146 (1979).
17. G.A. Bazilevskaya et al., 18th Internat. Cosmic

Ray Conf. Papers, 4, 218 (1983).

18. L. Biermann, O. Haxel and A. Schluter, Z. Nature, 6A, 47 (1951).
19. P. Morrison, Nuovo Cim., 7, 858 (1958).
20. R.E. Lingenfelter and R. Ramaty in B.S.P. Shen, ed., High-Energy Nuclear Reactions in Astrophysics (W.A. Benjamin, N.Y., 1967) p. 99.
21. R.E. Lingenfelter, E.J. Flamm, E.M. Canfield and S. Kellman, J. Geophys. Res., 70, 4077 and 4087 (1965).
22. E.L. Chupp et al., Nature, 241, 333 (1973).
23. H.S. Hudson et al., Astrophys. J., 236, L91 (1980).
24. T.A. Prince et al., Astrophys. J., 255, L81 (1982).
25. H. Debrunner, E. Fluckiger, E.L. Chupp, and D.J. Forrest, 18th Internat. Cosmic Ray Conf. Papers, 4, 75 (1983).
26. P. Evenson, P. Meyer and K.R. Pyle, Astrophys. J., 274, 875 (1983).
27. R. Ramaty and R.E. Lingenfelter, 13th Internat. Cosmic Ray Conf. Papers, 2, 1590 (1973).
28. B. Kozlovsky and R. Ramaty, Astrophys. J., 191, L43 (1974).
29. H.T. Wang and R. Ramaty, Solar Phys., 36, 129 (1974).
30. R. Ramaty, B. Kozlovsky and R.E. Lingenfelter, Space Sci. Rev., 18, 341 (1975).
31. R. Ramaty, B. Kozlovsky and A.N. Suri, Astrophys. J., 214, 617 (1977).
32. I.A. Igragimov and G.E. Kocharov, Soc. Astron. Lett., 3, 221 (1977).
33. R. Ramaty, B. Kozlovsky and R.E. Lingenfelter, Astrophys. J. Supp. 40, 487 (1979).
34. R. Ramaty, R.E. Lingenfelter and B. Kozlovsky in R.E. Lingenfelter et al., eds, Gamma Ray Transients and Related Astrophysical Phenomena (Am. Inst. Phys., N.Y., 1982) p. 211.
35. R.J. Murphy and R. Ramaty, Adv. Space Res. (in press 1985).
36. R.E. McGuire, T.T. Von Rosenvinge and F.B. McDonald, 16th Internat. Cosmic Ray Conf. Papers, 5, 61 (1979).
37. R.E. McGuire, T.T. Von Rosenvinge and F.B. McDonald, 17th Internat. Cosmic Ray Conf. Papers, 3, 65 (1981).
38. E. Rieger, et al., 18th Internat. Cosmic Ray Conf. Papers, 4, 83 (1983).
39. R. Davis, Jr. and J.C. Evans, 13th Internat. Cosmic Ray Conf. Papers, 3, 2001 (1973).

SHORT-TIME VARIATIONS OF THE SOLAR NEUTRINO LUMINOSITY

(Fourier analysis of the argon-37 production rate data)

H.J.Haubold and E.Gerth

Zentralinstitut für Astrophysik
Akademie der Wissenschaften der DDR
DDR - 1502 Potsdam-Babelsberg
German Democratic Republic

Abstract

We continue the Fourier analysis of the argon-37 production rate for runs 18-80 observed in Davis' well known solar neutrino experiment. The method of Fourier analysis with the unequally-spaced data of Davis and associates is described and the discovered periods we compare with our recently published results for the analysis of runs 18-69 (Haubold and Gerth, 1983). The harmonic analysis of the data of runs 18-80 shows time variations of the solar neutrino flux with periods π = 8.33; 5.26; 2.13; 1.56; 0.83; 0.64; 0.54; and 0.50 years, respectively, which confirm our earlier computations.

1. Introduction

The thermonuclear reactions occuring in the interior of the Sun involve the production of electron neutrinos with a roughly estimated flux at Earth of $N_\nu = (2L_\odot)/(4\pi [AU]^2 E) \approx 6.5 \times 10^{10} \nu\, cm^{-2} s^{-1}$, where E is the liberated energy in the fusion of four protons into one nucleus of helium-4. It is well known that the neutrino flux observed by R.Davis, Jr. and his associates in the chlorine-argon experiment is about one-third of the value estimated using the standard solar model : $\sum(\sigma\Phi)_{obs} = 2.1 \pm 0.3$ SNU (Davis, Cleveland, and Rowley, 1984), $\sum(\sigma\Phi)_{theor} = 7.6 \pm 3.3$ SNU (Bahcall, Huebner, Lubow, Parker, and Ulrich, 1982). This discrepancy constitutes one of the most important problems in neutrino astrophysics and nuclear astrophysics as well. The measured argon-37 production rate over a running period 1970-1983 (runs 18-80) is shown in Figure 1.. The combined argon-37 production rate over this period is 0.47 ± 0.04 argon-37-atoms per day in 615 tons of C_2Cl_4. Subtracting a background argon-37 production rate from cosmic ray muons and neutrinos of 0.08 argon-37-atoms per day, the argon-37 production rate that may be ascribed to solar neutrinos is 0.39 ± 0.04 argon-37-atoms per day (that value times 5.31 we get the above mentioned 2.1 ± 0.3 SNU, 1SNU = 10^{-36} neutrino captures per second per chlor-37-atom).

The idea to search for possible time variations in Davis' measurements of the argon-37 production rate in the solar neutrino experiment goes back to the papers of Sheldon (1969) and Sakurai (1979) which both try to find arguments in favour of an indication that certain averaged argon-37 production rate is correlated with the solar activity cycle. Therefore, we stated that not only the mean value of the neutrino capture rate of Davis' solar neutrino

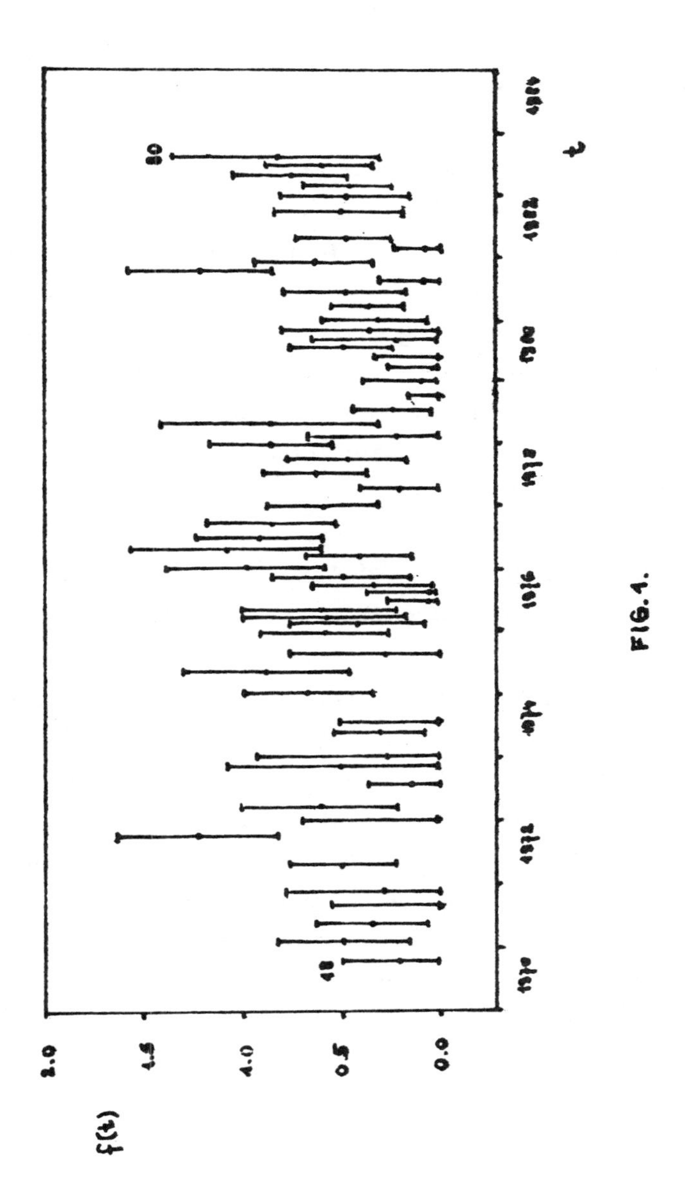

FIG. 1.

experiment as shown in Figure 1. is of interest for the interpretation in connexion with the standard solar model (Haubold and Gerth, 1983a). Performing the Fourier transformation of the argon-37 production rate for runs 18-69 we have shown that indeed there is the possibility that the solar neutrino flux may vary with time in periods of the order of years (Haubold and Gerth, 1983b,c).

It is the purpose of the present paper to succeed in the Fourier transformation of the argon-37 production rate for runs 18-80 as shown in Figure 1. and to explain our method of Fourier analysis with unequally-spaced data in some detail (Section 2.). In Section 3. we compare the present results of the harmonic analysis of runs 18-80 (1970-1983) with the recently published results for runs 18-69(1970-1981) and summarize arguments in favour of the examination of the solar neutrino data to find evidence for time variations.

2. Data and Analysis

The Fourier transformation of a stochastic function as we have in the case of the measurements of the argon-37 production rate shown in Figure 1. with arbitrary data spacing will have some special features in comparison with the Fourier transformation of equally spaced data. The observation function $\varphi(t)$ will be faded out by a so-called data window and really pointed out by the measurable function $f(t)$ that is the set of data reproduced in Figure 1.. In such a way, the spectral window $w(t)$ denotes a probability distribution function describing the flow of information between the observation function $\varphi(t)$ and the measurable function $f(t)$:

$$f(t) = w(t)\,\varphi(t). \tag{1}$$

In practice, the spectral window consists of an infinite series of factors,

$$w(t) = \prod_i w_i(t), \tag{2}$$

including aliasing and related effects. In every case one of the factors in (2) is the rectangular distribution for the time interval of the considered measurements. In the considered measurements of Davis and his associates the rectangular distribution is limited by the first run number 18(1970.279) and the last run number 80(1983.633). It is this rectangular distribution that provides the absolut convergence of the Fourier integral taken over the measurable function (1) :

$$F(\omega) = \int_{-\infty}^{+\infty} dt\, e^{-i\omega t}\, w(t)\,\varphi(t). \tag{3}$$

For the time interval $\left[t_a = t_0 - \frac{\Delta t}{2},\ t_b = t_0 + \frac{\Delta t}{2}\right]$ the Fourier transformation (3) leads to

$$F(\omega) = \frac{1}{\Delta t}\int_{t_0-\frac{\Delta t}{2}}^{t_0+\frac{\Delta t}{2}} dt\, e^{-i\omega t} = e^{-i\omega t_0}\ \frac{\sin\omega\frac{\Delta t}{2}}{\omega\frac{\Delta t}{2}}, \tag{4}$$

where the function $\operatorname{sinc} x = \frac{\sin x}{x}$ with $x = \omega\frac{\Delta t}{2}$ is the elementary function of the Fourier transformation of the spectral window as defined by Deeming (1975). It is of importance to notice here that in the limit $\Delta t \to 0$,

$$\lim_{\Delta t\to 0} \frac{\sin\omega\frac{\Delta t}{2}}{\omega\frac{\Delta t}{2}} = 1, \tag{5}$$

the spectral window (4) reflects the fundamental property of the Dirac delta-function,

$$\int_{-\infty}^{+\infty} dt\, \delta(t-t_0) = \lim_{\Delta t \to 0} \frac{1}{\Delta t} \int_{t_0-\frac{\Delta t}{2}}^{t_0+\frac{\Delta t}{2}} dt = 1, \tag{6}$$

which leads to the view to consider the definition of the spectral window (4) as a finite version of the Dirac delta-function. Only the normalization makes the difference. With (5) and (6) we can write for the Fourier transformation of the kth measurement f_k of Figure 1. :

$$F(\omega) = \int_{-\infty}^{+\infty} dt\, e^{-i\omega t} f_k\, \delta(t-t_k) = f_k\, e^{-i\omega t_k} , \tag{7}$$

or, more generally written for the series of measurements $f(t)$,

$$f_k = \delta(t-t_k)\, \varphi(t) \Rightarrow f(t) = \left(\sum_k \delta(t-t_k)\right) \varphi(t) , \tag{8}$$

we have,

$$F(\omega) = \frac{1}{n} \int_{-\infty}^{+\infty} dt\, e^{-i\omega t} \varphi(t)\, \delta(t-t_k) = \frac{1}{n} \sum_k f_k\, e^{-i\omega t_k} . \tag{9}$$

With equation (9) we obtained the basic formula for the Fourier transformation of stochastic distributed measurements with arbitrary data spacing normalized to the number of measurements. Hence, for the coefficients of the sine- and cosine-functions we can write

$$\begin{aligned} a(\omega) &= \frac{1}{n} \sum_k f_k \cos(\omega t_k), \\ b(\omega) &= -\frac{1}{n} \sum_k f_k \sin(\omega t_k). \end{aligned} \tag{10}$$

With this all informations concerning every frequency are given by its amplitude and phase of the harmonic waves contained in the spectrum. In the special case $\omega = 0$ we get

$$a(0) = \frac{1}{n} \sum_k f_k ,$$
$$b(0) = 0 , \tag{11}$$

where $a(0)$ represents the mean value of all measurements in Figure 1..

The power of a considered harmonic wave is defined by the half square of the belonging amplitude,

$$N = \frac{1}{2\pi} \int_0^{2\pi} dx \sin^2 x = \frac{1}{2} ,$$

and equivalently for the harmonic wave of frequency ω it holds

$$N(\omega) = \frac{1}{2} \left(a^2(\omega) + b^2(\omega) \right) = \frac{1}{2} F(\omega) F^*(\omega) , \tag{12}$$

where $F^*(\omega)$ denotes the complex conjugate of the Fourier transform (7). Again, for the special case of frequency $\omega = 0$ we obtain

$$N(0) = \frac{1}{n} \sum_k f_k^2 . \tag{13}$$

If we take the sum over the square of the differences between each single measurement f_k and the mean value of all measurements $a(0)$,

$$\sigma^2 = \frac{1}{n} \sum_k \left(f_k - a(0) \right)^2 , \tag{14}$$

we get the variance σ^2 of the deviation of the measurements from the mean value. With σ^2 we have not only a measure of the dispersion power N_s but also the value of the standard deviation σ . Remembering on the fundamental theorem of Parseval, we can state, that the total power of dispersion is equal to the sum of powers of each single harmonic wave of which the set of measurements will be represented as a trigonometric series of the observation function,

$$\varphi(t) = \sum_j \left[a(\omega_j) \cos(\omega_j t) + b(\omega_j) \sin(\omega_j t) \right], \tag{15}$$

that reads

$$\frac{1}{n} \sum_k (f_k - a(0))^2 = \frac{1}{2} \sum_j \left[a^2(\omega_j) + b^2(\omega_j) \right]. \tag{16}$$

Every harmonic wave with discrete frequency ω_j contributes a certain part of power

$$N(\omega_j) = \sigma^2(\omega_j)$$

to the total power of dispersion (14). The relative contribution,

$$Q = \frac{N(\omega_j)}{N_s} = \frac{\sigma^2(\omega_j)}{\sigma^2}, \tag{17}$$

is defined as the power quotient and can be taken into account as the criterion of significance for the jth harmonic wave contained in the power spectrum. The complementary power quotient,

$$1 - Q = \frac{\sigma^2 - \sigma^2(\omega_j)}{\sigma^2}, \tag{18}$$

denotes the remainder dispersion after reduction of the measurement by means of a sinusoidal wave of the respective frequency.

If we compare the representation of the measurable function $f(t)$ in equation (8) with that in equation (1) we have,

$$w(t) = \sum_k \delta(t - t_k), \tag{19}$$

for the spectral window with the property (6). In the series of probability factors defining the spectral window (2) which is by definition equal to (19) is included the time distribution of the data represented by Figure 1.. This is the fundamental meaning of the spectral window that it can be calculated from the data spacing alone and does not depend on the data themselves.

Performing the Fourier transformation of the measurable function (1) leads to

$$\int_{-\infty}^{+\infty} dt\, e^{-i\omega t}\, w(t)\varphi(t) = \int_{-\infty}^{+\infty} de\, W(\omega - e)\, \Phi(e), \tag{20}$$

where the convolution integral of the right-hand side of (20) contains the Fourier transforms

$$W(\omega) = \int_{-\infty}^{+\infty} dt\, e^{-i\omega t}\, w(t), \tag{21}$$

and

$$\Phi(\omega) = \int_{-\infty}^{+\infty} dt\, e^{-i\omega t}\, \varphi(t). \tag{22}$$

If the observation function $\varphi(t)$ consists of a sum of trigonometrical functions according to equation (15) then the convolution with the Fourier transform of the spectral window yields a shift of the spectral structure to the positive and negative frequencies $\pm\omega_j$. For the case of a single sinusoidal wave $\varphi(t) = \cos(\omega_j t)$ that means :

$$\int_{-\infty}^{+\infty} dt\, e^{-i\omega t}\, w(t)\cos(\omega_j t) = \frac{1}{2}\left[W(\omega - \omega_j) + W(\omega + \omega_j)\right]. \tag{23}$$

But, if there are several frequencies in the observation function $\varphi(t)$ the wings of the spectral window may overlap which can lead to interferences and difficulties arise in the interpretation of the extrema in the power spectrum. In our case the spectral window exhibits a simple structure because of the nearly equally distributed data in Figure 1. so that the power is concentrated in the main maximum as we can see in Figure 3. No interference effects appear.

Now we can summarize that we considered the measurable function $f(t)$ in (1) (Davis' measurements of the argon-37 production rate) as consisting of a product of two functions : the spectral window $w(t)$ (19) and the observation function $\varphi(t)$ (15). Then we defined with (9) the convolution of the true Fourier transform of the observation function $\varphi(t)$ with that spectral window. The problem of spacing of the Davis measurements, including aliasing and related effects (such as interferences) is completely contained in the spectral window. By means of the Fourier transformation of $\varphi(t)$ in (9) we obtain the significance criterion for the possible hidden periodicities in the power spectrum $N(\omega)$ in (12), that is the complementary power quotient (18). Based on the convolution theorem of the theory of Fourier transformation we pointed out that the Fourier transformation of the observation function $\varphi(t)$ with the spectral window (19) is equal to the convolution of the true Fourier transformation of $\varphi(t)$ that is $\Phi(\omega)$ with the Fourier transformation of the spectral window $w(t)$ that is $W(\omega)$. Beside the measurable function $f(t)$ reproduced in Figure 1. the normalized power spectrum $N(\nu)$ as derived in (12) is shown in Figure 2. and the Fourier transform $W(\omega)$ in (21) is given in Figure 3.. The discussion of both the numerical computated functions $N(\nu)$ and $W(\nu)$ will be given in Section 3..

3. Results and Discussion

Figure 2. shows the power spectrum $N(\nu)$ according to equation (12) which is connected with the Fourier transformation (9) of the measurable function $f(t)$. Clearly, there are sharp peaks well above the background which is arbitrarily fixed by $N(\nu) \leq 0.9500$. In the power spectrum $N(\nu)$ we plotted the

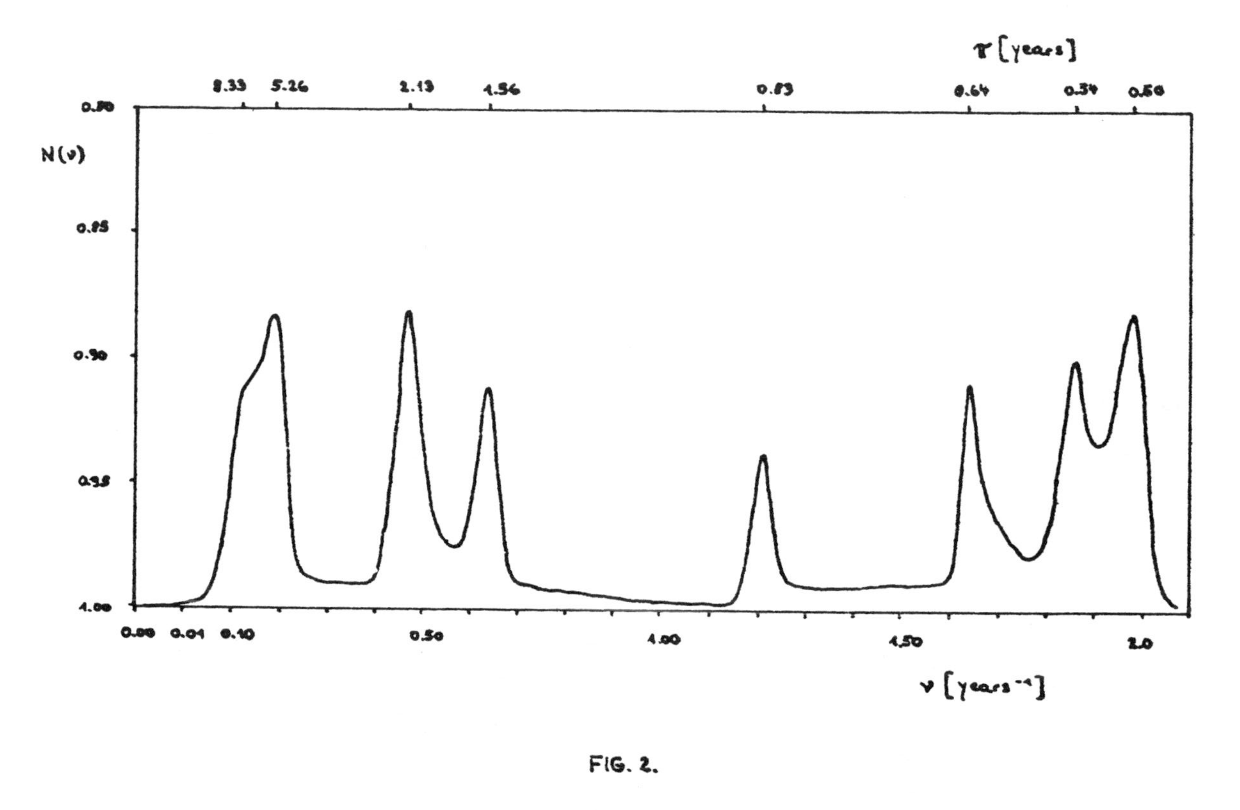
τ [years]
8.33
5.26
2.13
1.56
0.83
0.64
0.54
0.50
N(ν)
0.80
0.85
0.90
0.95
1.00
0.00
0.04
0.10
0.50
1.00
1.50
2.0
ν [years⁻¹]

FIG. 2.

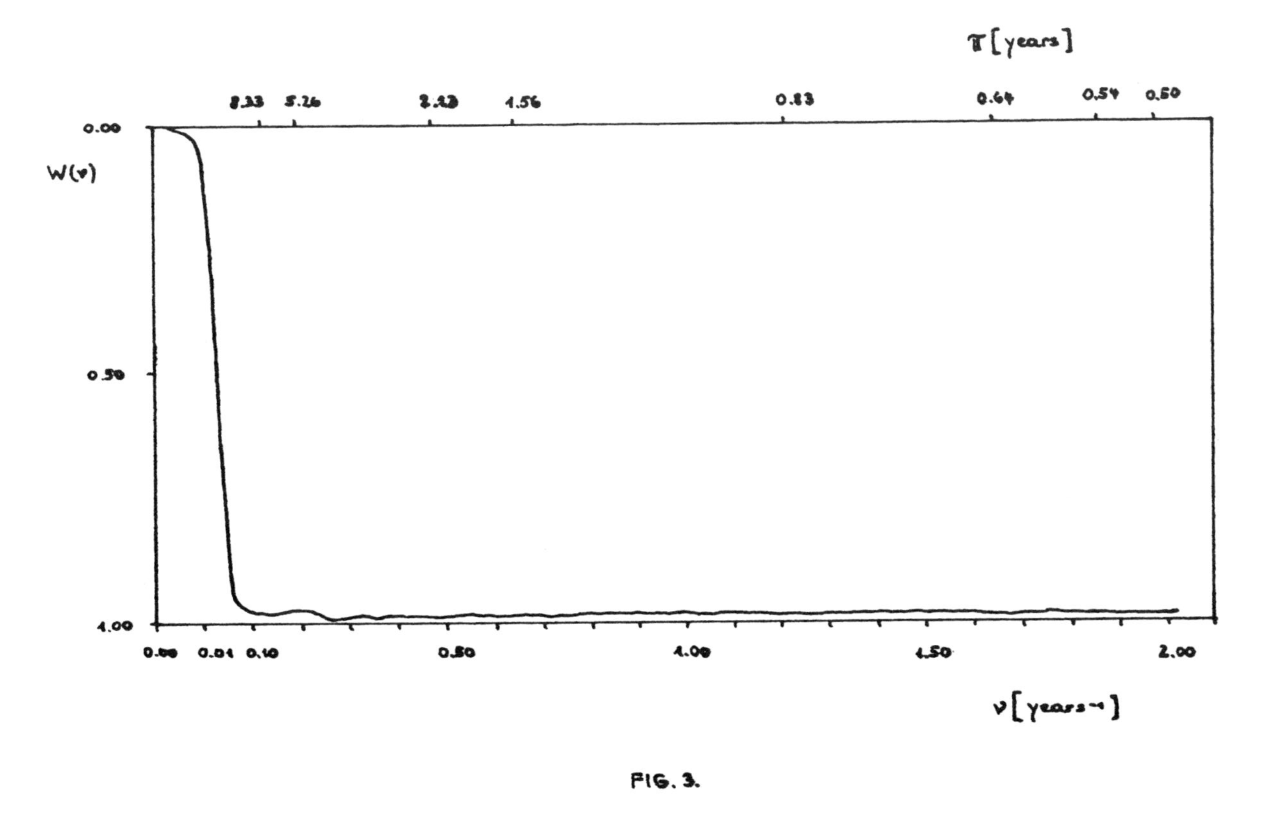
τ [years]
8.33
5.26
2.33
1.56
0.83
0.64
0.54
0.50
W(ν)
0.00
0.50
1.00
0.00
0.01
0.10
0.50
1.00
1.50
2.00
ν [years⁻¹]

FIG. 3.

complementary power coefficient $1-Q$ as given in (18) via the frequency of the harmonic waves present in the harmonic analysis of the argon-37 production rate data. With that one can conclude that the higher the power coefficient the greater is the significance of the respective period in the power spectrum. In the plot of the power spectrum $N(0)=1$ yields because of the fundamental wave at $\nu=0$ is suppressed by subtraction of the mean value of all data (cf. equations (13) and (14)). Table 1. summar zes the periods in the power spectrum of Figure 2. for runs 18-80 and mentioned all the periods localized in the Fourier analysis of the data of runs 18-69 (Haubold and Gerth, 1983b,c).

With Table 1. we find that the new periods in Figure 2. are really quite consistent with the old periods discovered in the Fourier analysis of runs 18-69. Therefore, one would have to conclude from all these data that in fact there is evidence for periodic variations of the argon-37 production rate in the solar neutrino experiment of Davis and his associates, even if this conclusion comes more from a mathematical point of view than from a physical one. In Figure 2. we observe in comparison with the old computations for runs 18-69 (Haubold and Gerth, 1983b,c), that the power of the periods π_1 and π_2 will be more concentrated in the shorter period of both and that in contradiction to that the period π_2 is splitted into two periods with rather equal power coefficients. The old period π_3 lies only a little bit over the background level and is quenched in the present calculations. At present time it is not easy to decide wether the old period π_6 is shifted to $\pi=$ 0.83 years or the new period $\pi_7=0.64$ years is identically with the old period $\pi_7=0.70$ years.

	π_1	π_2	π_3	π_4	π_5	π_6	π_7	π_8
runs 18-69	8.33	4.90	3.00	2.14	1.63	1.30	0.70	0,50
1-Q (18)	0.8885	0.8844	0.9475	0.8589	0.8358	0.9328	0.9470	0.8789
runs 18-80	8.33	5.26	%	2.13	1.56	0.83	0.64	0.54 0.60
1-Q (18)	0.9192	0.8830	%	0.8808	0.9122	0.9382	0.9094	0.8002 0.8820

TABLE 1.

Concerning the Fourier transformation of the spectral window as shown in Figure 3. according to equations (21) and (19) we do not find any evidence for correlations between peaks of $N(\nu)$ and $W(\nu)$. Therefore, there should be no influence of $N(\nu)$ through the finite lenght of data and no interferences from nearby frequencies which is usually observable in the spectral window. Our conclusion is, that all peaks of the power spectrum $N(\nu)$ in the region of low frequencies are significant. As expected the half width of the peak in the vicinity of $\nu = 0$ in Figure 3. is proportional to the total running time of the experiment of Davis and his associates.

Finally, we refer to the paper of Sakurai (1979) in which evidence for a quasi-biennial variation in the solar neutrino flux for equidistant four monthly mean values of the argon-37 production rate is reported. As one can see in Table 1. the same period with a high power quotient is clearly contained in the power spectrum analysis.

Acknowledgements

The authors are greatly indebted to Professor Dr. R.Davis, Jr. for his generosity in making his data of runs 18-80 available for our analysis and for suggestions.
One of us (H.J.H.) is grateful to Professor Dr. E.Bagge for stimulating discussions concerning the theory of solar neutrinos.

The authors' thanks go to Dr. V.N.Gavrin and Dr. E.A. Gavryuseva for discussions of the presented results.

References

Bahcall, J.N., Huebner, W.F., Lubow, S., Parker, P.D., and Ulrich, R.K., (1982), Rev.Mod.Phys. 54, 767.

Davis Jr., R., Cleveland, B.T., and Rowley, J.K., (1984), Report on solar neutrino experiment, presented at the conference in Steamboat Springs, Colorado, pp. 14.

Deeming, T.J., (1975), Astrophys.Sp.Sci. 36, 137.

Haubold, H.J., and Gerth, E., (1983a), Astron.Nachr. 304, 299.

Haubold, H.J., and Gerth, E., (1983b), Proc.18th Int.Cosmic Ray Conf., Bangalore, 10, 389.

Haubold, H.J., and Gerth, E., (1983c), Rapporto Interno, Instituto di Astrofisica Spaziale (CNR), C.P.67-00044 Frascati (Italia), No. 23, October 1983, pp. 22.

Sakurai, K., (1979), Nature 278, 146.

Sheldon, W.E., (1969), Nature, 221, 650.

THE EFFECTS OF Q-NUCLEI ON STELLAR BURNING

R.N. Boyd
Department of Physics, Department of Astronomy
The Ohio State University

R.E. Turner, B. Sur, and L. Rybarcyk
Department of Physics, The Ohio State University

and

C. Joseph
Department of Astronomy, The Ohio State University

ABSTRACT

The effects of anomalous nuclei, Q-nuclei, on stellar burning are examined. The baryon binding energies, beta-decay properties, and thermonuclear reaction rates for the Q-nuclei suggest they could catalyze a cycle in which four protons are combined to form a ^{4}He nucleus. The properties required of the Q-nuclei for them to solve the solar neutrino problem are determined. A solar modelling calculation was performed with Q-nuclei included, and several interesting results therefrom are compared to observations. Finally the solar neutrino detection rates for ^{71}Ga and ^{115}In detectors, in addition to that for ^{37}Cl, are estimated when Q-nuclei are included in the solar burning.

INTRODUCTION

Our present understanding of the Sun is based on the Standard Solar Model (SSM); it has done a remarkable job of combining basic physics and nuclear reaction rates to produce a detailed description of the stellar process of energy generation and nucleosynthesis. There are, however, several notable conflicts between its predictions and the corresponding observations, the best known of which is the Solar Neutrino (SN) problem.

The studies[1,2] discussed in the present paper have investigated the consequences on stellar burning and on the predictions of the SSM of a tiny abundance of abnormal nuclei, Q-nuclei, in the sun. Two essential features for these Q-nuclei have emerged from these studies. If they are to solve any of the problems of the SSM they must (1) have nucleons which are slightly more bound than the nucleons in normal nuclei, and (2) be compressed somewhat from normal nuclei. Those two requirements are not unrelated, as they could be satisfied simultaneously by a number of candidates for Q-nuclei which have been proposed by theorists.

Hadronic particles embedded in nuclei could form plausible candidates for Q-nuclei. Perhaps the best studied such case[3] is that of an embedded extra quark. The resulting Q-nuclei would indeed have more tightly bound nucleons than would normal nuclei[4], and the added attractive force could well produce the required density increase[2] of the Q-nuclei from that of normal nuclei. A similar situation could result if hadronic X-particles[5] or supersymmetric particles were embedded in nuclei. Another Q-nuclear candidate would be superdense nuclei[6], nuclei with roughly double the normal nuclear density. While this list is not exhaustive, it does show that there are several plausible candidates which could serve as the Q-nuclei hypothesized above.

Q-NUCLEAR PROPERTIES

How then would these Q-nuclei behave in a star? To study this question one needs to investigate (1) their baryon binding energies, (2) the beta-decay properties of Q-nuclei, (3) the thermonuclear reaction rates which would be associated with their interactions with other particles in the star, and (4) the abundance required for them to have an observable effect on stellar burning.

The baryon binding energies of Q-nuclei are crucial to their meaningful participation in stellar burning. The mass 5 nuclei, ^{5}He and ^{5}Li, are baryon unstable, resulting in the mass 5 gap in the periodic table. If, however, the hypothesized increase in binding energy of Q-nuclei from that of normal nuclei is more than 0.5 MeV per nucleon, the 5 baryon Q-nuclei will be baryon stable. This will allow them to catalyze the Q-nuclear cycle shown in Fig. 1 in which three proton radiative captures and two beta decays on $^{4}He^{Q}$ would form $^{7}Li^{Q}$, and a (p,α) reaction would then form a ^{4}He particle and return the $^{4}He^{Q}$ catalyst.

A requirement for the Q-nuclear cycle to operate in this way is that the beta decays of $^{5}Li^{Q}$ and $^{7}Be^{Q}$, either electron capture or positron decay, occur quickly. The energy available for these decays, ΔE_{β}, can be predicted using the semiempirical atomic mass formula, to be

$$\Delta E_{\beta} = K(Z - 1)/R - 0.78 \quad (1)$$

where R is the rms radius of the Q-nuclear charge distribution, Z is the charge of the initial Q-nucleus, K is a constant determined from fitting the value of ΔE_{β} for light nuclei, and 0.78 MeV is the neutron-hydrogen mass difference. The halflives are related to ΔE_{β} in a complicated way, but generally decrease with increasing ΔE_{β}. The halflife decreases much more rapidly[2] when ΔE_{β} surpasses the positron emission threshold.

The thermonuclear reaction rates were estimated using the statistical model prescription of Woosley et al.[7]. Since the dominant feature of any reaction between charged particles at stellar energies is the coulomb barrier between them, that prescription gives at least a qualitative idea of the thermonuclear reaction rates to be expected for Q-nuclei, even without basic information about their structure. The rate at which the Q-nuclear cycle shown in Fig. 1 will produce ^{4}He nuclei is determined by its slowest reaction rate, that (presumably) of the $^{6}Li^{Q}(p,\gamma)$ reaction. Comparison of the rates for ^{4}He production in the Q-nuclear cycle and for the SSM reactions show that those for the case in which the Q-nuclei result from an embedded up quark are greater[1] by about 15 orders of magnitude at 15 million degrees. Thus the required abundance of Q-nuclei would be expected to be very small, roughly one Q-nucleus per 10^{15} normal nuclei, for them to contribute appreciably to stellar burning.

An obvious consideration for any modification to the SSM is that of high-energy neutrino production, since the SN experiment[8] is one of the few reasons for considering any modifications to the SSM. As can be seen in Fig. 1,

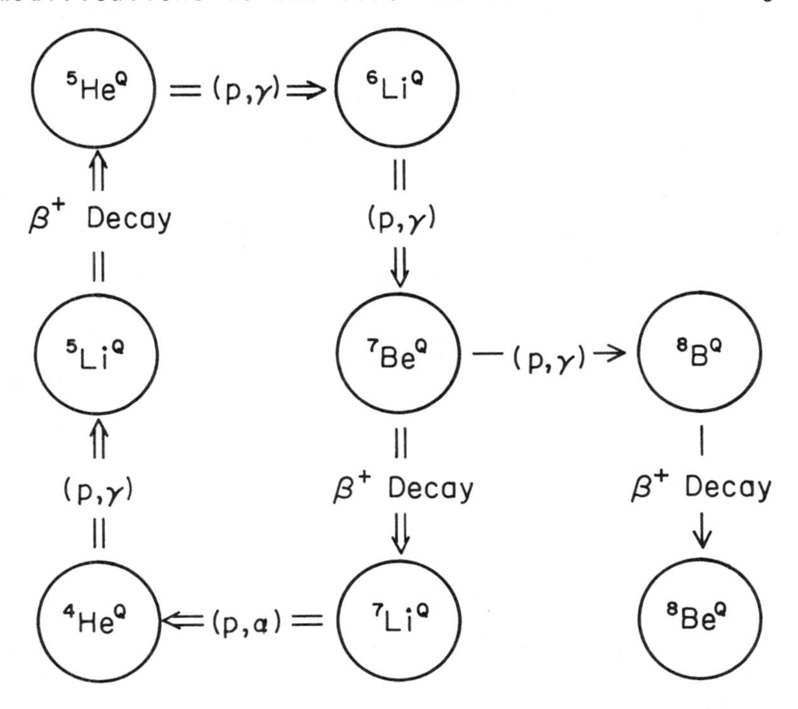

Figure 1. The Q-nuclear cycle (double lines), which converts four protons into an α-particle. The branch indicated by single lines can produce high-energy neutrinos.

proton radiative capture on ^7BeQ will occur if the half-life of ^7BeQ is too long. Then ^8BQ would beta decay with a neutrino similar in energy to that produced in the decay of ^8B. The comparison to the high-energy neutrino production in the SSM is useful; the only parameter once the abundance of Q-nuclei has been determined is the ^7BeQ halflife. It can thus be shown[1] that a halflife of about 1000 seconds is sufficiently short that very few high-energy neutrinos will be produced by the Q-nuclear cycle. This in turn requires that ΔE_β be about 2.0 MeV or more.

Is the inferred abundance of roughly one Q-nucleus in 10^{15} nuclei in conflict with limits for the abundances of the various candidates for Q-nuclei? The situation for searches for fractionally charged entities has recently been reviewed by Lyons[9]. The conclusion is that, while their existence in specific materials has been determined with extraordinary accuracy, their average abundance in nature certainly has not been determined at that level. Limits estimated from cosmological considerations[10] suggest that free quarks could have been left from the big bang at a level of about one in 10^{14} baryons; this also allows the required Q-nuclear abundance, assuming the residual quarks would have ultimately been captured by nuclei either during the big bang or in first generation stars. Similar conclusions apply to the abundances of the other Q-nuclear candidates.

RESULTS OF STELLAR MODELLING CALCULATIONS

Thus stellar modelling calculations were performed, using a modified version of the code of Paczynski[11] to determine the effect of Q-nuclei on the evolution of the Sun. Q-nuclear abundances ranging from zero to 2×10^{-15} per normal nucleus were assumed, and the convection parameter and zero age ^4He abundance were varied to achieve the Sun's luminosity and radius after 4.7 billion years. Several remarkable conclusions[12] result from these calculations. (1) If it is assumed that no observable neutrinos (for the ^{37}Cl detector) are added by the Q-nuclear cycle, it is found that the predicted SN rate changes from 6.9 SNUs at a Q-nuclear abundance of zero to 2.0 SNUs at an abundance of 2×10^{-15} per normal nucleus[2]. At that Q-nuclear abundance, the Q-nuclear cycle produces about two-thirds of the ^4He particles, hence of the Sun's energy. (2) The depth of the Sun's convective envelope increases by about 20% with respect to the SSM result[12]. This is germane to the five-minute oscillations of the Sun, discussed by Ulrich and Rhodes[13]. Observations suggest that the depth of the convective envelope is roughly 30% greater than that predicted by the SSM[14]. (3) The temperature at the bottom of the convective envelope in the SSM is 2.0 million degrees. When the Q-nuclei are added at an abundance of $2x10^{-15}$ per normal nucleus, that temperature increases to 2.5 million degrees[12]. This is of importance to the question of Li abundance in the Sun; it is observed to be much lower than would be expected from the temperature at the bottom of the convective envelope in the SSM. It has been noted[14] that a temperature of around 2.5 million degrees at the bottom of the convective envelope would solve the problem. (4) The zero-age ^4He abundance required when Q-nuclei are abundant at $2x10^{-15}$ per normal nucleus is 17% higher than that required by the SSM. The value[16] for the SSM, 0.25 ± 0.01, is in good agreement with the primordial value[17]. Since the Sun is thought to be a second- or subsequent-generation star, it might be thought that the zero-age ^4He abundance should be above the primordial value. Observations[18] suggest that the ^4He abundance is near 0.30, in good agreement with that of other stars having nearly solar abundances of heavy elements[17], and with the results of the Q-nuclear model. (5) Recent estimates of the ages[19] of the oldest stars range from about 16 to 20 billion years. The age of the universe[20], predicted from the Hubble constant, ranges from 9 to 12 billion years, producing an irreconcilable inconsistency. However the ages of stars having masses of about one solar mass are reduced from those of the SSM by about 20% when Q-nuclei are included, greatly alleviating, and possibly eliminating, the conflict between stellar astrophysics and cosmology.

Thus all the above conflicts between the SSM and reality are removed or greatly mitigated by a stellar Q-nuclear abundance of $2x10^{-15}$ per normal nucleus. Several other problems with the SSM have also been discussed[12]; they also appear to be reduced by the inclusion of a tiny abundance of Q-nuclei.

NEUTRINO DETECTION RATE ESTIMATES

Recently we have improved the estimate[21] of the SN detection rate when Q-nuclei are included in stars. The neutrino flux is determined by the fact that two neutrinos are produced per ^4He particle in any hydrogen burning cycle. What determines the detectable SN rate, then, is the energy of the neutrinos produced. Those from the Q-nuclear cycle will come from the decays of $^5Li^Q$ and $^7Be^Q$. If the halflives for these Q-nuclei are not short, then $^6Be^Q$ and $^8B^Q$ could be formed from proton radiative capture; both would produce neutrinos of sufficient energy to increase greatly the ^{37}Cl detection rate, thus exacerbating the SN problem.

To estimate the SN detection rates we have assumed ΔE_β for $^5Li^Q$ and $^7Be^Q$ to be the same; this is expected[21] to be a reasonable estimate. The cross sections for various neutrino detectors given by Bahcall[22] were then used to calculate the detectable neutrino flux for detectors of ^{37}Cl, ^{71}Ga and ^{115}In. Our stellar modelling calculations provided the fraction of energy generated, hence of the ^{4}He particles produced by the Q-nuclear cycle, for each Q-nuclear abundance, as well as the high-energy neutrino flux (from decay of normal ^{8}B). The major results from these calculations are threefold. (1) The value of ΔE_β for $^{5}Li^Q$ and $^{7}Be^Q$ must be between 1.8 and 2.1 MeV, i.e., the neutrino end point energy must be from 0.8 to 1.1 MeV. This requires some compression of the Q-nuclear charge distribution from that of normal nuclei (see eq. 1). As noted above the lower limit is required for the beta decays of $^{5}Li^Q$ and $^{7}Be^Q$ to be fast. The upper limit is established from the fact that, for neutrino end point energies above 1.1 MeV, the ^{37}Cl detector rate actually increases above the SSM rate. (2) For ΔE_β in that range, the detectable ^{37}Cl rate is decreased, while those for both ^{71}Ga and ^{115}In are increased by from 50% to more than 100%. (3) The fraction of the Sun's energy generated by the Q-nuclear cycle must be greater than about two-thirds if Q-nuclei are to solve the SN problem.

CONCLUSIONS

Our investigations have shown that the Q-nuclear model solves several longstanding problems with the SSM. That circumstantial evidence for the existence of Q-nuclei, together with the fact that none of the Q-nuclear candidates can be ruled out by present abundance limits, strongly suggests that improved searches for such entities be conducted. In addition, since no other explanation of the SN problem predicts an increase in the neutrino detection rates for the ^{71}Ga or ^{115}In detectors from those of the SSM, they could provide a crucial test of the existence of Q-nuclei.

This work was supported in part by the National Science Foundation Grant 82-03699.

REFERENCES

1. R.N. Boyd, R.E. Turner, M. Wiescher and L. Rybarcyk, Phys. Rev. Letters 51, 609 (1983).
2. R.N. Boyd, R.E. Turner, L. Rybarcyk and C. Joseph, to be published in Ap. J., 1985.
3. A. deRujula, R.C. Giles and R.L. Jaffe, Phys. Rev. D17, 285 (1978).
4. G.F. Chapline, Phys. Rev. D25, 991 (1982).
5. R.N. Cahn and S.L. Glashow, Science 213, 607 (1981).
6. See, e.g., T. Ohnishi, Nucl. Phys. A362, 480 (1981).
7. S. Woosley, W.A. Fowler, J.A. Holmes and B.A. Zimmer-

man, At. Data Nucl. Data Tables 22, 371 (1978).
8. R. Davis, Jr., Science Underground, ed. by M.M. Nieto et al., AIP Conf. Proc. No. 96 (Am. Inst. Phys., N.Y.), p. 2.
9. L. Lyons, Oxford Univ. Report No. 80-77, 1982, unpublished.
10. R.V. Wagoner and G. Steigman, Phys. Rev. D20, 825 (1979).
11. B. Paczynski, Acta Astr. 20, 47 (1970).
12. C. Joseph, to be published in Nature, 1984.
13. R.K. Ulrich and E.J. Rhodes, Ap. J. 265, 551 (1983).
14. R. Scuflaire, M. Gabriel and A. Noels, Astron. Astrop. 99, 39 (1981).
15. R. Weymann and R.L. Sears, Ap. J. 142, 174 (1965); A.M. Boesgaard, Pub. Astr. Soc. Pacific 88, 353 (1976).
16. J.N. Bahcall, W.F. Heubner, S.H. Lubow, P.D. Parker and R.K. Ulrich, Rev. Mod. Phys. 54, 767 (1982).
17. J. Yang, M.S. Turner, G. Steigman, D.N. Schramm and K.A. Olive, Ap. J. 281, 493 (1984).
18. J.N. Heasley and R.W. Milkey, Ap. J. 221, 667 (1978).
19. K. Janes and P. Demarque, Ap. J. 264, 206 (1983); A. Sandage, Astron. J. 88, 1159 (1983); R.D. Cannon, Highlights of Astron. 6, 109, ed. by R.M. West (1983); B.W. Carney, Ap. J. Suppl. Ser. 42, 481 (1980).
20. G. de Vaucouleurs and G. Bollinger, Ap. J. 233, 433 (1979).
21. B. Sur and R.N. Boyd, submitted to Phys. Rev. Lett., 1984.
22. J.N. Bahcall, Rev. Mod. Phys. 50, 881 (1978).

A PROPOSED SOLAR NEUTRINO EXPERIMENT USING $^{81}Br(\nu,e^-)^{81}Kr$

G. S. Hurst
Oak Ridge National Laboratory, Oak Ridge, Tenn. 37831
and University of Tennessee, Knoxville, Tenn. 37996

C. H. Chen, S. D. Kramer, and S. L. Allman
Oak Ridge National Laboratory, Oak Ridge, Tenn. 37831

ABSTRACT

It has now been shown that it is feasible to measure the ^7Be neutrino source in the sun by using the reaction $^{81}Br(\nu,e^-)^{81}Kr$ in a radiochemical experiment. Such an experiment would be quite similar to the Davis, Cleveland, and Rowley method for measuring the ^8B neutrino using $^{37}Cl(\nu,e^-)^{37}Ar$ except that the resonance ionization spectroscopy (RIS) method (instead of decay counting) would be employed to count the 2 x 10^5-yr ^{81}Kr atoms.

INTRODUCTION

In this paper we can now describe a feasible method for measuring the flux of ^7Be neutrinos on the earth by using the interaction $^{81}Br(\nu,e^-)^{81}Kr$. It has recently been shown[1] that ^{81}Kr atoms can be counted at the few hundred atom level by using the method of resonance ionization spectroscopy (RIS).[2] Thus, the bromine experiment initially considered by Scott[3] can now be performed as a Davis-type radiochemical experiment and is proposed[4] as a natural sequel to the chlorine experiment--the only solar neutrino experiment to this date.

In this paper we first discuss the scientific justification of the bromine experiment. The method of counting ^{81}Kr atoms depends on RIS which will briefly summarized. An interesting new way of counting noble gas atoms based on RIS is quite reminiscent of Maxwell's sorting demon. This method, to be published in greater detail elsewhere,[5] will be outlined here for completeness. Finally, the bromine experiment itself will be described and is similar to the existing chlorine experiment at Homestake.

SCIENTIFIC JUSTIFICATION FOR THE BROMINE EXPERIMENT

Several speakers at this symposium have already described the models of the stellar interior and predictions of the neutrino flux associated with the sun. These are based on the "standard stellar model" or variations of it,[6] and Fig. 1 is a compact way to present the results.[7] Labels have been added to the diagram of Burks to point out the experiment in progress and the two which are now proposed. Of course, only the chlorine experiment has been done, and it continues as the only active experiment to monitor solar neutrino flux. Addition of the gallium experiment to measure the important p-p source and bromine to measure the ^7Be flux would complement the chlorine experiment, and the sum of the three would provide a radiochemical method for solar neutrino spectroscopy.

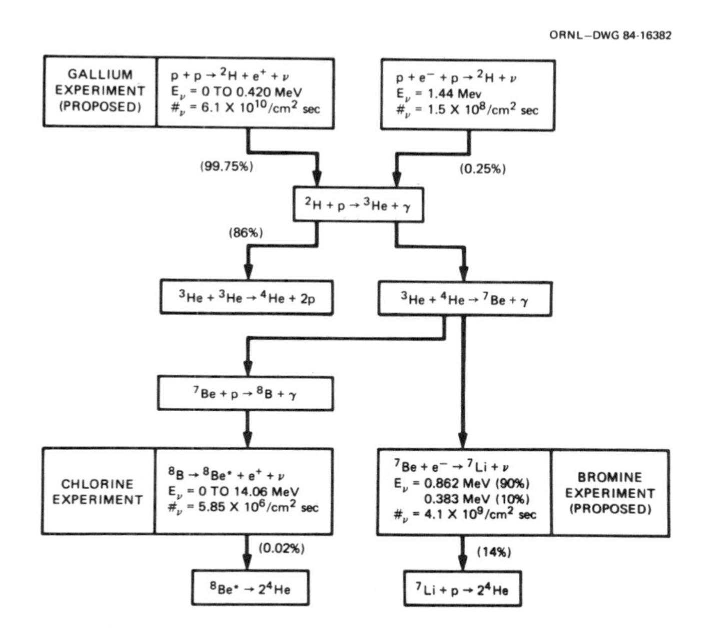

Fig. 1. Solar neutrino sources, energy, and fluxes on the earth according to the standard stellar model.

Table I elaborates on the response of the three detectors to the various neutrino sources. Note that the ^{71}Ga detector responds appreciably to the ^{7}Be source according to both the estimates of Bahcall and to Ohihara et al. Likewise, but to a smaller degree, the ^{81}Br detector mixes some of the ^{8}B source with ^{7}Be, and with ^{37}Cl a small component of ^{7}Be is mixed with the ^{8}B source. However, each detector responds primarily to one of the important neutrino sources; to first order, each would measure a prominent feature of the sun, and in combination they would provide essentially the neutrino spectrum.

The above conclusion on the value of a bromine experiment has long been recognized. However, no method has been available for counting a few hundred atoms of ^{81}Kr because of its long half-life for decay. We now divert our discussion into RIS which provides the basis for a laser method of counting ^{81}Kr.

Table I. Capture rates in SNU

Target (source)	$^{71}Ga^a$	$^{71}Ga^b$ *	$^{81}Br^a$	$^{81}Br^c$ **	***	$^{37}Cl^a$
p-p	67.2	67.5	0	0	0	0
pep	2.4	5.4	1.2	1.2	0.8	0.23
^7Be	28.5	58.7	10.6	10.1	4.6	1.02
^8B	1.7	4.4	2.2	1.0	0.9	6.05
^{13}N	2.7	5.4	1.0	0.9	0.3	0.08
^{15}O	3.8	8.2	1.6	1.5	0.9	0.26
Total	106.3	149.6	16.6	14.7	7.5	7.6

*p,n experiments
**($3^-/2 \rightarrow 1^-/2$)
***($3^-/2 \rightarrow 5^-/2$)

[a]J. N. Bahcall, W. F. Huebner, S. H. Lubow, P. D. Parker, and R. K. Ulrich, Rev. Mod. Phys. 54, 767 (1982).
[b]H. Orihara, C. D. Zifiratos, S. Nishihara, K. Furukawa, M. Kabasawa, K. Maeda, K. Miura, and H. Ohnuma, Phys. Rev. Lett. 51, 1328 (1983).
[c]F. K. Liu and F. Gabbard, Phys. Rev. C 27, 93 (1983).

RESONANCE IONIZATION SPECTROSCOPY

Resonance ionization spectroscopy was developed initially for absolute measurements of the population of He(2^1S) states produced by proton excitation of helium gas.[8] From the beginning, RIS has been conceived as a quantum state selective method of ionizing atoms (or molecules). Further, this highly selective method for ionizing atoms of a particular Z is also a very efficient means of ionization. In fact, a pulsed laser can be tuned to ionize nearly 100% of the atoms of a selected Z while not ionizing any atoms of another Z. For those accustomed to nonselective ionization for the detection of nuclear decays, the laser is a selective detector that can be deployed to pick out a single atom from a multitude of other types, ionize the selected atom, and count it with nearly 100% efficiency.

Figure 2 illustrates two methods of RIS. In the most elementary scheme, to the left, photons of one wavelength excite atoms from the ground state to a bound state; thus, atoms are

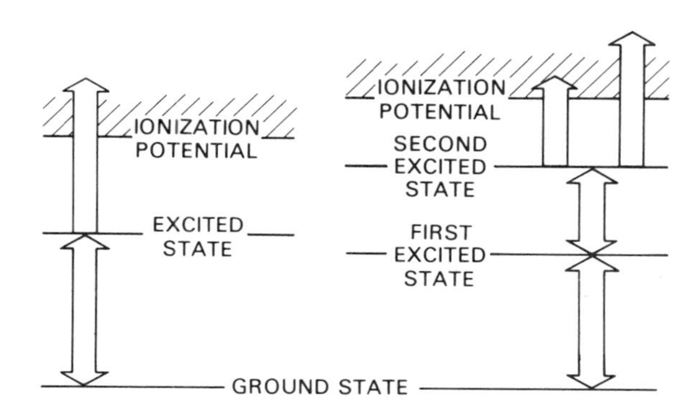

Fig. 2. Two schemes for RIS.

chosen from their known spectroscopy. Following the bound-bound transition, another photon of the same wavelength creates a bound-continuum process (or ionization). Of course, more than one laser can be used by overlapping beams of different wavelengths in both space and time, as illustrated in Fig. 2. In contrast to the nonselective ionization assoociated with radioactive sources or accelerated particles, RIS is resonance ionization. Nuclear physicists have taught us how to measure one electron with a Geiger-Mueller counter or a proportional counter and one positive ion with an evacuated detector. Thus, with RIS we can selectively count single atoms.[9] Figure 3 shows that the method can now be generalized to every atom in the periodic table except perhaps helium and neon.[2]

At this conference it is tempting to show various types of applications that are possible with RIS in particle physics. Table II shows that RIS can be used as three basic methods, i.e., (a) to count stable (or radioactive) noble gas atoms, (b) to count lone atoms in solids [low levels of impurities in materials using sputter-initiated RIS (SIRIS)[10]], and (c) to detect atoms or rare events as they happen (live-time methods).

Table II. Methods of RIS analysis in particle physics

1. Maxwell's demon for counting noble gas atoms
   * $^{81}Br(\nu,e^-)^{81}Kr$ solar neutrino experiment
   * $^{82}Se(\beta\beta)^{82}Kr$
   * $^{128}Te(\beta\beta)^{128}Xe$ and $^{130}Te(\beta\beta)^{130}Xe$
2. SIRIS FOR LONE ATOMS IN SOLIDS
   * $Mo(\nu,e^-)Te$ Solar Neutrino Experiment
   * Fractional Charge
   * Superheavy Atoms
3. LIVE-TIME METHODS
   * Passage of Magnetic Monopole
   * Rare Atom of Nuclear Process

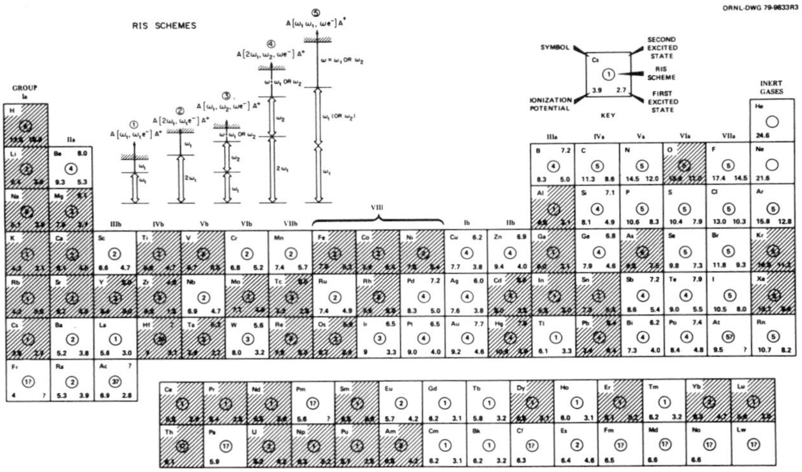

Fig. 3. RIS schemes applicable to the periodic table.

For further information on these methods, the interested reader may consult the various papers in the proceedings of a symposium on RIS.[11] For example, the SIRIS technique is described by Parks et al.,[10] while its use for fractional charge and superheavy atom searches is discussed by Fairbank et al.[12] The possibility of using RIS in monopole searches was discussed by Hurst[13] following a paper by Kroll and Ganapathi[14] on the theory of monopole interactions with helium to produce He(2^3S).

MAXWELL'S DEMON

The work of Ray Davis and his colleagues (Cleveland, Rowley, and others) inspired the realization of a Maxwell demon[15] which could look into an enclosure and sort out desired isotopes of an atom. For example, a few hundred atoms of ^{81}Kr can be sorted out from an enormous number of other isotopes of krypton, and these ^{81}Kr atoms can be counted one by one as they are stored in a solid. This selection and counting process can be continued until all of the ^{81}Kr atoms have been counted. Such has now been done[1], and a result is shown in Fig. 4 for 1000 atoms of ^{81}Kr which was prepared for us at the National Bureau of Standards to simulate a possible solar neutrino exposure of a tank like the one at Homestake but filled with a bromine compound.

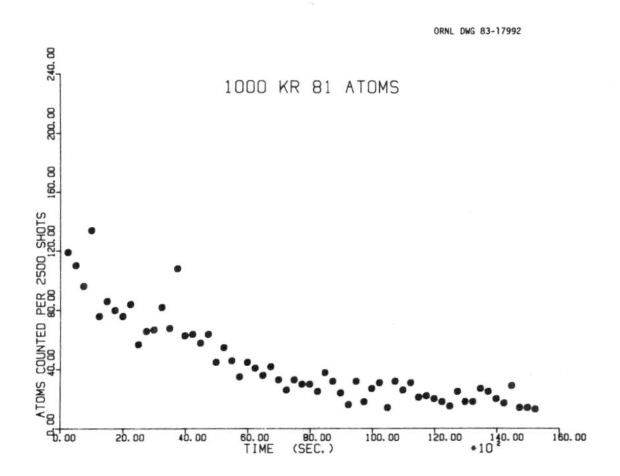

Fig. 4. Demonstration of the use of RIS to count approximately 1000 atoms of ^{81}Kr.

The interested reader may consult the article by Chen et al.[1] or a much more detailed review by Hurst et al.[5] to learn how this type of atom counting is done. Basically, a laser system is used to do RIS on krypton atoms isolated in a small vacuum system of high quality. The system contains a quadrupole mass spectrometer to add A selection. After Z and A selection, the ^{81}Kr atoms are accelerated into a silicon target where they are implanted (the demon's door!) and are individually counted by recording a burst of electrons emitted by the target on impact of each ion. While all of the ^{81}Kr atoms are being implanted in silicon, a small fraction (10^{-4}) of the ^{82}Kr atoms are also implanted due to the limited abundance sensitivity of the small quadrupole. But, after evacuating the isolated system to clear out the remaining krypton background, laser annealing of the silicon can be done to release the implanted atoms for another isotopic enrichment and counting cycle. Thus, even two cycles give rejection of ^{82}Kr by a factor of 10^8. For the neutrino experiment, this should be adequate even if some air leaks cannot be avoided.

The laser RIS process[16] uses the 116.5-nm resonance transition in krypton, and it efficiently ionizes krypton in a volume which is small compared to the total volume of the chamber. To overcome this limit, an atom buncher was developed so that atoms can be made to sit on a cold finger adjacent to the path of the RIS beams and be prompted to desorb from the finger into the RIS beam at the time it is pulsed on. Naturally, another laser is pulsed onto the cold tip to prompt the atom evaporation. With this arrangement, the ^{81}Kr atoms in the 2-liter volume appear to have a half-life of less than 1 hr, as shown by the nearly exponential decay in Fig. 4. Of course, nothing is decaying--the true half-life is 2 x 10^5 yr, and the atoms counted by RIS can be recovered by laser annealing of the target and counted again!

THE PROPOSED BROMINE EXPERIMENT

Even if ^{81}Kr atoms can be counted in small numbers, there are other problems which must be dealt with to make a $^{81}Br(\nu,e^-)^{81}Kr$ detector work. First, the cross section for neutrino capture must be known. In Fig. 5, some of the relevant nuclear levels are shown in context with the ^7Be neutrino energy. Shown also are some log(ft) values for the (ν,e) process to the $^{81}Kr(7/2^+)$ state and to the $^{81}Kr(1/2^-)$ state. The value of 4.88 was measured by Bennett[17] by looking at the decay rate of the inverse process from the $^{81}Kr(1/2^-)$, 13-sec state. These cross section estimates may be no better than +20% and are further complicated by possible capture into $^{81}Kr(5/2^-)$ (see a discussion by Liu and Gabbard[18]). Even this complication does not change the main argument that the bromine detector would sense mainly the ^7Be neutrino, see Table I. Furthermore, the elegant (p,n) experiments performed at Indiana University (see Charles Goodman's paper at this symposium) should help to resolve the contributions of each channel and to improve on the total cross section for the ^7Be neutrino.

Davis and Cleveland have already demonstrated that krypton atoms can be recovered from the perchloroethylene solution in the Homestake tank, using their helium recirculation system.

Finally, background questions must be carefully considered in all experiments of this type. Those due to cosmic rays, α particles and neutrons have been estimated[19] just as they were for

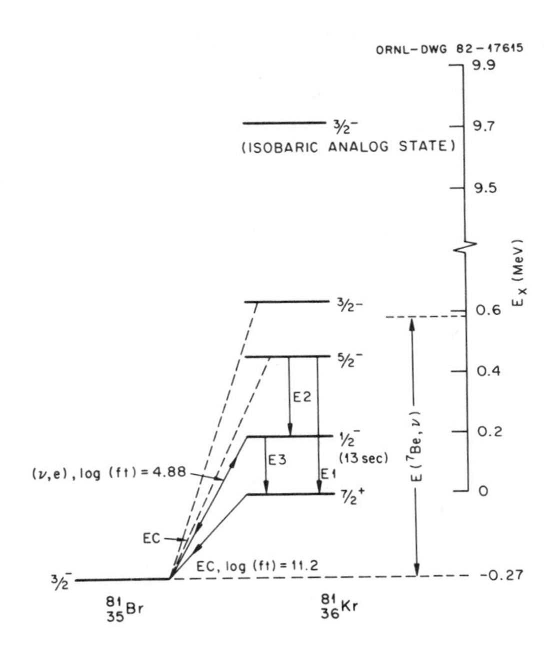

Fig. 5. Energy levels involved in the interaction of $^{81}Br(\nu,e^-)^{81}Kr$.

the chlorine experiment and are satisfactory. There is one special background to consider due to the fact that excessive laser ionization of stable krypton atoms could interfere with the selection of $^{81}Kr^+$ due to space charge. However, a stage of isotopic enrichment before admitting the sample to the demon can be done by using a quadrupole mass spectrometer operated with a special electron gun to reduce memory effects.[20]

As the summary (Table III) shows, we believe that all aspects of the bromine experiment are feasible. With an experiment the size of the chlorine experiment (380 m^3), from two to five atoms of ^{81}Kr should be created per day; thus, with a fraction of the patience of Ray Davis, we could see the bromine experiment--a natural companion to the famous chlorine experiment--in successful operation in a few years.

Table III. Summary of proposed bromine solar neutrino experiment

1. Reaction: $^{81}Br(\nu,e^-)^{81}Kr$, primarily ^{7}Be

2. Facilities: Davis radiochemical, like Cl except that atom counting requires RIS method

3. Compound: 380 m^3 (1000 tons) $CHBr_3$

4. Signal Levels (Atoms ^{81}Kr per day):

   * Consistent Model -------2

   * Standard Solar Model ---5

5. Noise Levels:

   * Cosmic rays, alpha particles, and neutrons

   * Atmospheric krypton - ^{82}Kr can interfere with counting ^{81}Kr

   * Negligible ^{81}Kr in atmosphere

The authors appreciate the continued interest and support of our work by Wick Haxton and George Cowan at LANL and the interest of W. M. Bugg, J. O. Thomson, and Paul Huray of the University of Tennessee. Marvin Payne of the Oak Ridge National Laboratory contributed to many of the technical aspects of this work.

This research was sponsored in part by the Office of Health and Environmental Research, U.S. Department of Energy under contract DE-AC05-84OR21400 with Martin Marietta Energy Systems, Inc.

REFERENCES

1. C. H. Chen, S. D. Kramer, S. L. Allman, and G. S. Hurst, Appl. Phys. Lett. 44, 640 (1984).
2. G. S. Hurst, M. G. Payne, S. D. Kramer, and J. P. Young, Rev. Mod. Phys. 51, 767 (1979).
3. R. D. Scott, Nature 264, 729 (1976).

4. G. S. Hurst, C. H. Chen, S. D. Kramer, B. T. Cleveland, R. Davis, Jr., R. K. Rowley, Fletcher Gabbard, and F. J. Schima, Phys. Rev. Lett. 53, 1116 (1984).
5. G. S. Hurst, M. G. Payne, S. D. Kramer, C. H. Chen, R. C. Phillips, S. L. Allman, G. D. Alton, J.W.T. Dabbs, R. D. Willis, and B. E. Lehmann, Reports on Progress in Physics (to be published).
6. J. N. Bahcall, W. F. Huebner, S. H. Lubow, P. D. Parker, and R. K. Ulrich, Rev. Mod. Phys. 54, 767 (1982).
7. Barry Lee Burks, PhD Thesis, University of North Carolina (1983).
8. G. S. Hurst, M. G. Payne, M. H. Nayfeh, J. P. Judish, and E. B. Wagner, Phys. Rev. Lett. 35, 82 (1975); M. G. Payne, G. S. Hurst, M. H. Nayfeh, J. P. Judish, C. H. Chen, E. B. Wagner, and J. P. Young, Phys. Rev. Lett. 35. 1154 (1975).
9. G. S. Hurst, M. G. Payne, S. D. Kramer, and C. H. Chen, Phys. Today 33(9), 24 (1980).
10. J. E. Parks, H. W. Schmitt, G. S. Hurst, and W. M. Fairbank, Jr., in Resonance Ionization Spectroscopy 1984, edited by G. S. Hurst and M. G. Payne (The Institute of Physics, Bristol, 1984), pp 167-174.
11. G. S. Hurst and M. G. Payne, eds., Resonance Ionization Spectroscopy 1984 (The Institute of Physics, Bristol, 1984).
12. W. M. Fairbank, Jr., G. S. Hurst, J. E. Parks, and C. Paice, in Resonance Ionization Spectroscopy 1984, edited by G. S. Hurst and M. G. Payne (The Institute of Physics, Bristol, 1984), pp 287-296.
13. G. S. Hurst, in Resonance Ionization Spectroscopy 1984, edited by G. S. Hurst and M. G. Payne (The Institute of Physics, Bristol, 1984), pp 309-318.
14. N. M. Kroll and V. Ganapathi in Resonance Ionization Spectroscopy 1984, edited by G. S. Hurst and M. G. Payne (The Institute of Physics, Bristol, 1984), pp 297-308.
15. G. S. Hurst in Resonance Ionization Spectroscopy 1984, edited by G. S. Hurst and M. G. Payne (The Institute of Physics, Bristol, 1984), pp 7-18.
16. S. D. Kramer in Resonance Ionization Spectroscopy 1984, edited by G. S. Hurst and M. G. Payne (The Institute of Physics, Bristol, 1984), pp 205-212.
17. C. L. Bennett, M. M. Lowry, R. A. Naumann, F. Loeser, and W. H. Moore, Phys. Rev. C 22, 2245 (1980).
18. F. K. Liu and F. Gabbard, Phys. Rev. C 27, 93 (1983).
19. J. K. Rowley, B. T. Cleveland, R. Davis, Jr., W. Hampel, and T. Kirsten, Geochim. Cosmochim. Acta, Suppl. 13, 45 (1980).
20. R. D. Willis, S. L. Allman, C. H. Chen, G. D. Alton, and G. S. Hurst, J. Vac. Sci. Technol. A 2(1), 57 (1984).

THE GALLIUM SOLAR NEUTRINO DETECTOR

W. Hampel
Max-Planck-Institut für Kernphysik
P.O. Box 103980, 69 Heidelberg 1, F.R. Germany

ABSTRACT

This paper summarizes the efforts to perform a radiochemical gallium solar neutrino experiment based on $GaCl_3$ solution. The motivation and the general experimental procedure are illustrated first. We next focus on the results of a pilot experiment. It follows a discussion on the ^{71}Ga neutrino capture cross section. We then describe the present status of the project and mention future plans. A new collaboration has been formed in Western Europe with the intention to perform a full scale experiment with 30 tons of gallium in the Gran Sasso Underground Laboratory in Italy. Finally, we shall briefly discuss the interpretation of different possible results which the Ga detector in principle could give.

1. INTRODUCTION

It is well known that experiments designed to detect the neutrinos generated in the fusion processes in the sun provide the only direct experimental possibility to test the theories of nuclear energy generation in stars. There is also a strong motivation coming from particle physics: solar neutrino experiments may be the only way to gain information on neutrino mixing (oscillations) occuring over large (astronomical) distances (see section 6), a question of great importance in the context of Grand Unified Theories. So far, the only experimental attempt to detect solar neutrinos is the Brookhaven Chlorine Experiment[1], a radiochemical detector based upon the $^{37}Cl(\nu_e, e^-)^{37}Ar$ neutrino capture reaction. The signal measured with this detector, 2.1 ± 0.3 SNU (1 SNU = 1 neutrino capture per sec in 10^{36} target nuclei), is more than a factor of three lower than the prediction from the Standard Solar Model (SSM), 6.9 SNU[2,3]. However, because of the 814 keV threshold of the Cl detector, more than 75% of the signal expected according to the SSM is due to the high-energy (0 - 14 MeV) ^8B neutrinos, generated in a very rare and strongly temperature-dependent side branch of the proton-proton reaction chain (see Fig. 1). On the other hand, this detector is not sensitive to the bulk of solar neutrinos, the low-energy (0 - 420 keV) pp neutrinos from the primary fusion reaction in the sun. It has long been recognized that an experiment capable to measure these pp neutrinos should be able to provide the answer to the solar neutrino problem posed by the Cl experiment. The only experiment of this type which is in an advanced stage is the Gallium Experiment.

0094-243X/85/1260162-13 $3.00

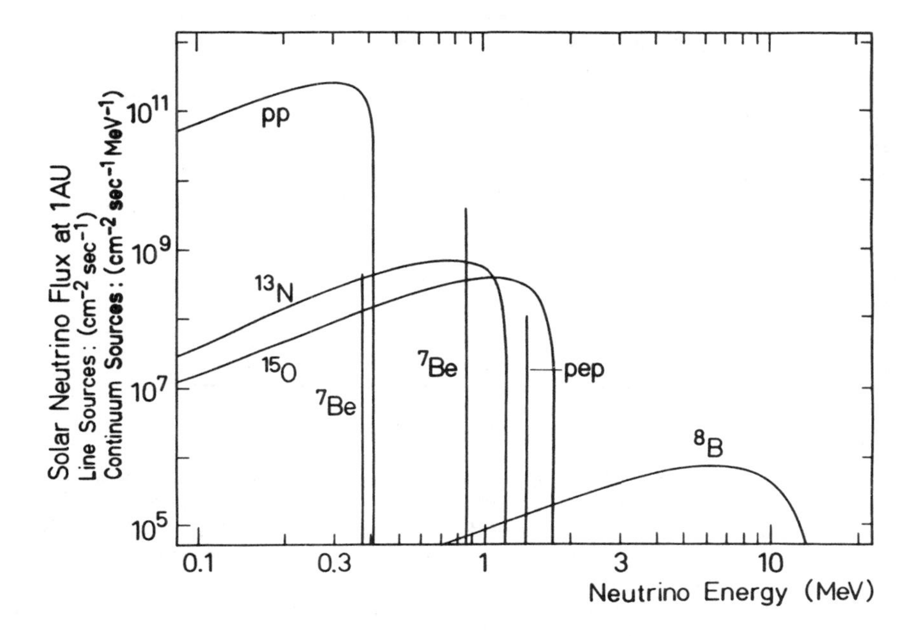

Fig 1: Solar neutrino energy spectrum (Standard Solar Model[2,3])

2. BASIC PRINCIPLES OF THE GALLIUM DETECTOR

The gallium detector has first been suggested by Kuzmin[4]. It is based on the neutrino capture reaction $^{71}Ga(\nu_e,e^-)^{71}Ge$ (see Fig. 2). The energy threshold (233.2 keV[5]) is well below the maximum energy (420 keV) of the pp neutrinos. ^{71}Ge decays back to ^{71}Ga by electron capture with a halflife of 11.43 days[6]. Table I lists the neutrino fluxes for the different neutrino sources (according to the SSM[2,3]), the ^{71}Ga neutrino capture cross sections[5] (averaged over the energy spectrum for the continuum sources) (see also section 4) and the resulting capture rates. The total rate of 113 SNU is dominated by 74 SNU from the pp and pep neutrinos.

The experimental procedure is as follows. 30 tons of gallium in form of a concentrated $GaCl_3$-HCl solution will be exposed to solar neutrinos. In such a medium, the neutrino induced ^{71}Ge atoms (as well as the inactive Ge carrier atoms added at the beginning of a run) form the volatile $GeCl_4$, which at the end of an exposure is simply swept out of the solution by bubbling He, Ar or even air through it. The gas stream is passed through two gas scrubbers where the $GeCl_4$ is absorbed in water. The $GeCl_4$ is then extracted into CCl_4, back-extracted into tritium-free water and finally reduced to the gas GeH_4 by means of KBH_4. GeH_4, together with xenon, is introduced into a small proportional counter, where the number of ^{71}Ge atoms is determined by observing their radioactive decay.

Table I: Solar neutrino fluxes[2,3], cross sections[5] and capture rates for the gallium detector.

Neutrino source and energy [MeV]	Flux on earth $[10^{10}cm^{-2}sec^{-1}]$	Cross section $[10^{-46}cm^2]$	Capture rate [SNU]
pp (0-0.42)	6.05	11.7	71.0
pep (1.44)	0.0149	167	2.5
^{7}Be (0.38,0.86)	0.457	68.2	31.3
^{8}B (0-14.06)	0.00049	2800	1.4
^{13}N (0-1.20)	0.05	57	2.9
^{15}O (0-1.73)	0.04	99	4.0
		Total	113.0

3. THE PILOT EXPERIMENT

A pilot experiment with 4.6 tons of $GaCl_3$ solution (equivalent to 1.26 tons of Ga) has been performed by an international collaboration (Brookhaven Nat. Lab., MPI Kernphysik Heidelberg, WIS Rehovot, IAS Princeton)[7]. This pilot experiment has been completed in 1983. We shall summarize the results in the following.

The germanium extraction from $GaCl_3$ solution and the conversion of the extracted $GeCl_4$ to GeH_4 has been studied extensively by the BNL group. More than 30 runs with the pilot tank have been performed. These include runs where small amounts (a few mg down to 100 µg) of stable Ge carrier were introduced into the pilot tank, as well as runs in which ^{71}Ge was produced in the $GaCl_3$ solution either by cosmic rays, by a neutron source, or by in situ decay of ^{71}As.

A typical example is shown in Fig. 3. More than 99% of the 2 mg Ge added to the $GaCl_3$ solution before the sweep have been extracted

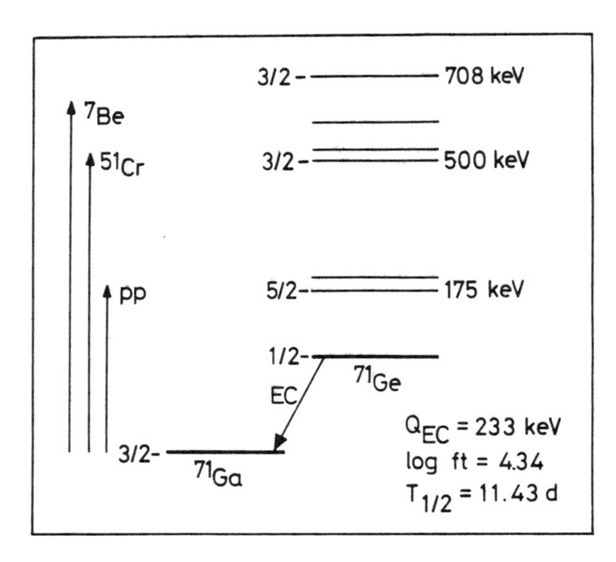

Fig. 2: ^{71}Ga-^{71}Ge energy levels.

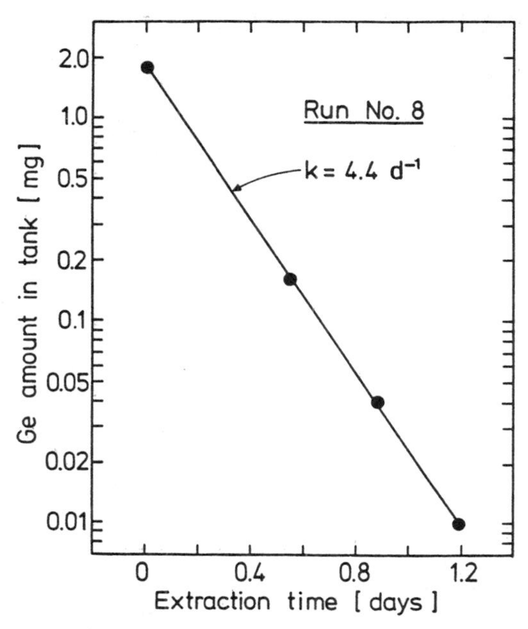

Fig. 3:

Example of a germanium extraction from the pilot tank.

in a 28 hour run. It may be concluded that the entire chemical procedure (extraction and the subsequent conversion to GeH_4) can be carried out with > 95% overall yield.

The ^{71}Ge decay rates expected for a full scale solar neutrino experiment are below 1 per day. The development of a counting system capable to measure such low decay rates was the main responsibility of the Heidelberg group. This is an extreme low-level counting problem and can be achieved only by counting ^{71}Ge as GeH_4 in a miniaturized proportional counter. The energy deposition from Auger electrons and X rays emitted in the ^{71}Ge electron capture decay results in a spectrum with 2 peaks: an L peak at 1.2 keV and a K peak at 10.4 keV. All provisions typical for low-level counting have to be applied in order to reach the extremely low background rates reqired: use of ultrapure materials for counter construction; anticoincidence shield with NaI and plastic scintillation detectors; heavy passive shielding with lead and iron. The remaining background in the L and K peak energy windows is of the order of 1 cpd (cpd = counts per day). It is mainly caused by three sources: (1) beta particles coming from natural radioactivity in the construction materials; (2) Compton electrons caused by external gamma rays; and (3) electronic noise pulses. Fortunately, all these events produce in most cases pulse shapes different from those of ^{71}Ge decays. Our computer-controlled counting system is therefore designed to record the whole shape of each proportional counter pulse by means of a transient digitizer. All counting data including the pulse shape are finally stored on magnetic tape.

The two most important properties for a counter are the counting efficiency ε and the background rate b. The values achieved with

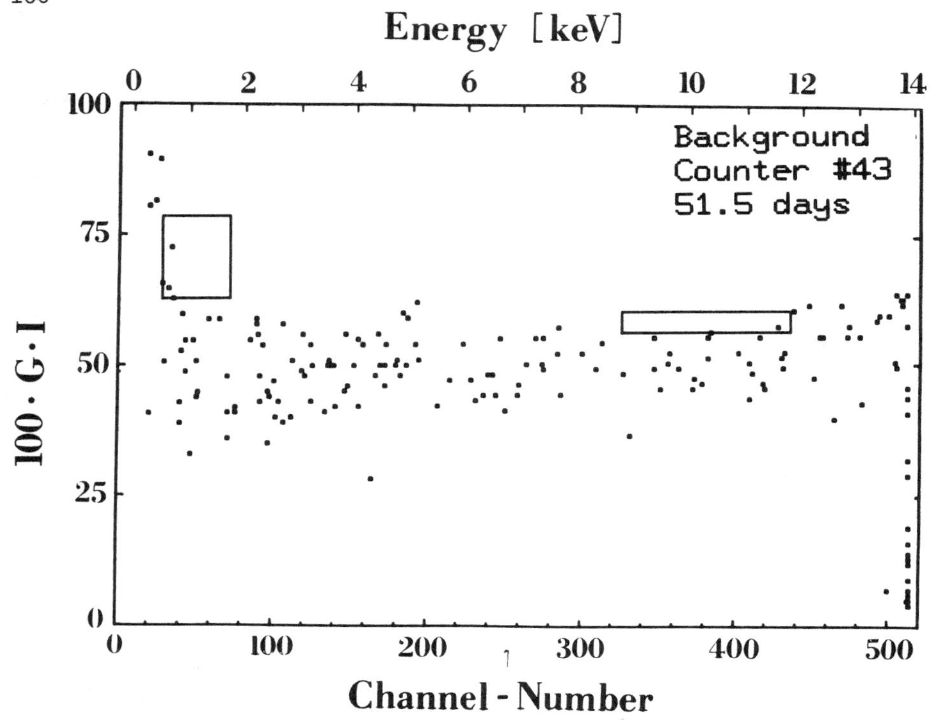

Fig. 4: Proportional counter background measurement (see text).

many counters (L peak: ε = 25%, b = 0.2 cpd; K peak: ε = 38%, b = 0.08 cpd) would allow to determine a solar neutrino rate of 90 SNU with a standard error of 14%. However, since the pilot experiment has been completed, improvements have been made concerning the dead volume of the counters, resulting in increased efficiency values of 28% for the L peak and 43% for the K peak. Also, background rates measured with some counters are now about a factor of two lower than the values listed above. Fig. 4 presents the results of a 51.5 day background measurement obtained with one of the best counters. G*I, a quantity characterizing the pulse shape, is plotted versus the energy for each count. The boxes represent the regions where 95% of all ^{71}Ge events with energies in the L or K peak windows will be located. The background obtained with this counter is 0.08 cpd in the L peak and 0.04 cpd in the K peak.

Radiochemical solar neutrino detectors provide no direct way to distinguish the neutrino signal from the signal due to side reactions. It is therefore essential to examine carefully the production of ^{71}Ge in the $GaCl_3$ solution by sources other than solar neutrinos. The most serious side reaction is $^{71}Ga(p,n)$, the protons being generated in the $GaCl_3$ solution as secondaries from (α,p) and (n,p) reactions and from cosmic ray muon interactions. In the course of the

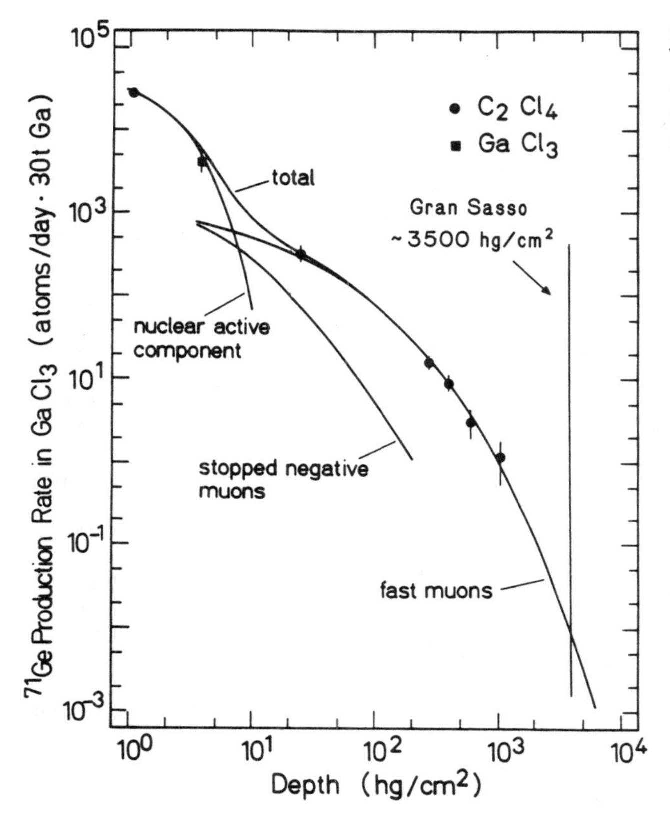

Fig. 5:

Cosmic ray production of ^{71}Ge in $GaCl_3$ versus shielding depth.

pilot experiment, the tolerable levels of uranium, thorium and radium in the $GaCl_3$ solution have been determined and it has been demonstrated that ton quantities of $GaCl_3$ solution that meet these specifications are commercially available. The ^{71}Ge production by fast neutrons entering the solution from outside has been studied with a Pu-Be neutron source. This effect sets limits on the fast neutron flux at the detector site, which (if necessary) can be met by use of a water shield around the detector.

The depth dependence of the ^{71}Ge production in $GaCl_3$ by cosmic ray muons has been derived from measurements and calculations on the same effect for the Cl detector[8] and from the measured cross section ratio (^{71}Ge from $GaCl_3$)/(^{37}Ar from C_2Cl_4) for 225 GeV muons. The result is shown in Fig. 5, where the rate for ^{71}Ge production by cosmic ray muons in 30 tons of Ga is plotted as a function of shielding depth (1 hg/cm^2 = 100 g/cm^2 = 1 m.w.e). For 3500 hg/cm^2 (this is roughly the shielding depth of the Gran Sasso Underground Laboratory, see section 5) the ^{71}Ge production rate is 0.01 atoms per day in 30 tons of gallium. This is of the order of 1% of the rate expected from solar neutrinos (assuming 90 SNU).

4. THE ^{71}Ga NEUTRINO CAPTURE CROSS SECTION

For solar neutrinos, the cross section for the ^{71}Ga(ν_e,e^-) reaction is dominated by transitions to the ^{71}Ge ground state. This main contribution to the total cross section can be calculated reliably. The most important experimental input data needed in this calculation are the halflife and the Q_{EC} value for the ^{71}Ge electron capture decay. We have redetermined both the halflife (11.43+0.03 days)[6] and the Q_{EC} value (233.2±0.5 keV)[5] and have recalculated the ground state cross sections for the different solar neutrino sources[5]. They are presented in Table I.

There are two excited states in ^{71}Ge at 175 and 500 keV excitation energy which can be populated in allowed Gamow-Teller transitions in particular by the ^{7}Be neutrinos (see Fig. 2). In order to calculate the cross section contribution from these two states one needs the corresponding beta-decay ft values, which cannot be measured directly. Bahcall[9] has estimated these values from beta-decay systematics. He obtained a 8 SNU increase of the solar neutrino rate due to the contribution of these two excited states.

Another way to gain the necessary information is a measurement of the (p,n) forward scattering cross section (see for instance Rapaport et al.[10]). The (p,n) reaction connects the same states in the target and product nuclei as neutrino capture or (in inverse direction) beta-decay. Orihara et al.[11] have measured the ^{71}Ga(p,n) reaction at 35 MeV proton energy and have extracted ft values for the 175 and 500 keV states. Normalizing their data to the new ground state ft value (2.179×10^4 sec, obtained with the ^{71}Ge halflife and Q_{EC} value mentioned above) yields an excited state contribution of 33 SNU (total 146 SNU). This is quite in contrast to the results obtained by Bahcall. However, it has recently been shown by Baltz et al.[12] that the forward scattering differential cross section at 35 MeV proton energy is not directly proportional to the Gamow-Teller strength. There are contributions which do not simply correspond to Gamow-Teller transitions. Consequently, the neutrino capture cross section extracted from the measurements by Orihara et al. will be overestimated. Baltz et al. point out that these problems in the interpretation of the (p,n) data in terms of Gamow-Teller strength disappear at higher proton energies. Thus there is hope that a new measurement of the (p,n) cross section on ^{71}Ga with 120 MeV protons presently carried out at the Indiana University Cyclotron[13] will settle this question.

The most direct way to obtain information about the ^{71}Ga neutrino capture cross section is a calibration experiment using an artificial ^{51}Cr neutrino source (see also section 5). The 746 keV neutrinos from the ^{51}Cr decay can populate both the 175 and 500 keV states in ^{71}Ge (see Fig. 2). Thus the ^{51}Cr signal measured in excess of the expected ground state contribution can be attributed to these two excited states. However, when translating the rate measured with the ^{51}Cr source into a cross section for solar neutrinos, there arises an uncertainty from the fact that the individual contributions from both states are not known. In order to evaluate this effect, a

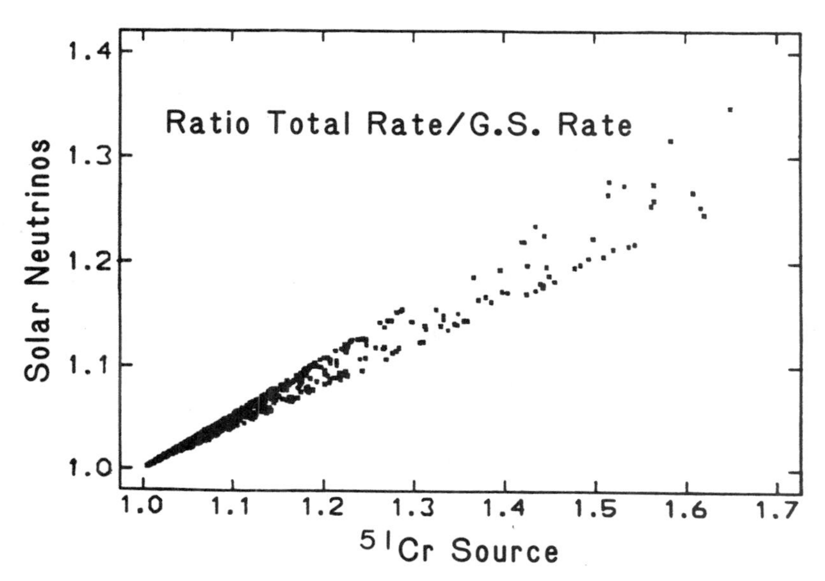

Fig. 6: Monte-Carlo calculation: solar neutrino rate versus ^{51}Cr rate assuming different ft values for the 175 and 500 keV excited states in ^{71}Ge (see text).

Monte-Carlo procedure has been performed in which different ft values were assigned to both states and the corresponding rates for both solar neutrinos and the ^{51}Cr were calculated. The results are plotted in Fig. 6 (all rates are normalized to those for the ground state transition). It turns out that this effect is small. For instance, if the ^{51}Cr experiment yields a rate increased by a factor of 1.3 over the g.s. rate with an error of 12%, this translates into a factor of 1.15 for the solar neutrino rate, with 8% error.

Finally, it should be noted that there is a state in ^{71}Ge at 708 keV excitation energy which could slightly contribute to the cross sections for the pep, ^{13}N and ^{15}O neutrinos, with a negligible effect on the total rate. On the other hand, the Gamow-Teller strength connected with ^{71}Ge states at higher excitation energies up to the neutron separation energy (7.4 MeV) could give a non-negligible contribution to the cross section for ^{8}B neutrinos. Apart from theoretical calculations based on different nuclear models[14,15], information on this question will be obtained from the (p,n) experiments mentioned above.

5. STATUS OF THE PROJECT AND FUTURE PLANS

After it was demonstrated by the outcome of the pilot experiment that a full scale Ga detector is feasible, it would have been natural to continue the existing collaboration in order to perform the solar neutrino experiment. Unfortunately, due to funding problems in the United States, Brookhaven Nat. Lab. was forced to terminate work on the Ga experiment at the end of 1983. thus also terminating the collaboration. In the meantime, a new collaboration[16] has been formed in Western Europe with scientists from West Germany (MPI Heidelberg, KFZ Karlsruhe, TU München), France (Saclay, Collège de France Paris, University of Nice), Italy (Universities of Milano and Rome) and Israel (WIS Rehovot). It is envisioned to perform the experiment with 30 tons of gallium as $GaCl_3$ solution in the Gran Sasso Underground Laboratory in Italy. Gallium funding is requested through German sources.

The calibration of the Ga detector by means of an artificial ^{51}Cr neutrino source provides the most direct test of the whole experimental procedure and would yield also information on the ^{71}Ga neutrino capture cross section (as discussed in the previous section). ^{51}Cr (halflife 27.7 days) can be produced by neutron activation of natural chromium ($^{50}Cr(n,\gamma)^{51}Cr$). It decays by electron capture and emits monoenergetic neutrinos of 426 keV (10%) and 746 keV (90%) (see also Fig. 2). A source strength of roughly 1 Megacurie is necessary in order to perform a sensible experiment. In Western Europe there are only two possibilities to produce such a source:

(1) A test irradiation with 16 kg of Cr powder has been performed at the ESSOR reactor at Ispra (Italy), operated by EURATOM. It was concluded that a 1 MCi source can be produced at this reactor[17].

(2) An investigation at the SILOE reactor in Grenoble[18] has revealed that the capacity of this reactor combined with that of the OSIRIS reactor at Saclay will allow the production of a 0.8 MCi source of ^{51}Cr. A test irradiation is under way in order to verify this number.

A source of 0.8 MCi, surrounded by 13 cm of lead shield and placed into the center of a standard Pfaudler tank with 30 tons of gallium (as $GaCl_3$ solution), would produce 5.5 ^{71}Ge atoms per day. Taking into account the decay of ^{51}Cr during the experiment, this value has to be compared with the unavoidable "background" resulting from solar neutrinos, 0.8 atoms per day (= 90 SNU) (both rates correspond to the ground state transition only, any contribution from excited states would increase these numbers, see section 4). Under these circumstances, the ^{51}Cr signal can be determined with a 12% statistical error in a series of measurements with 4 sources (4 extractions per source), provided that the solar neutrino rate (to be subtracted) is known to $\pm$ 0.25 atoms per day.

6. DISCUSSION OF POSSIBLE GALLIUM DETECTOR RESULTS

Finally, we shall briefly discuss the response of the Ga detector to different suggestions which have been made in order to solve the problem posed by the Cl experiment. In Fig. 7 the expected rate for the Cl detector is plotted versus the corresponding Ga rate for a variety of so-called non-standard solar models which have been suggested over the last 15 years. They include models with departures from the standard model itself (e.g. rotating core, internal magnetic fields, non-maxwellian particle velocity distribution) as well as changes of the input data to the model (S-factors, chemical

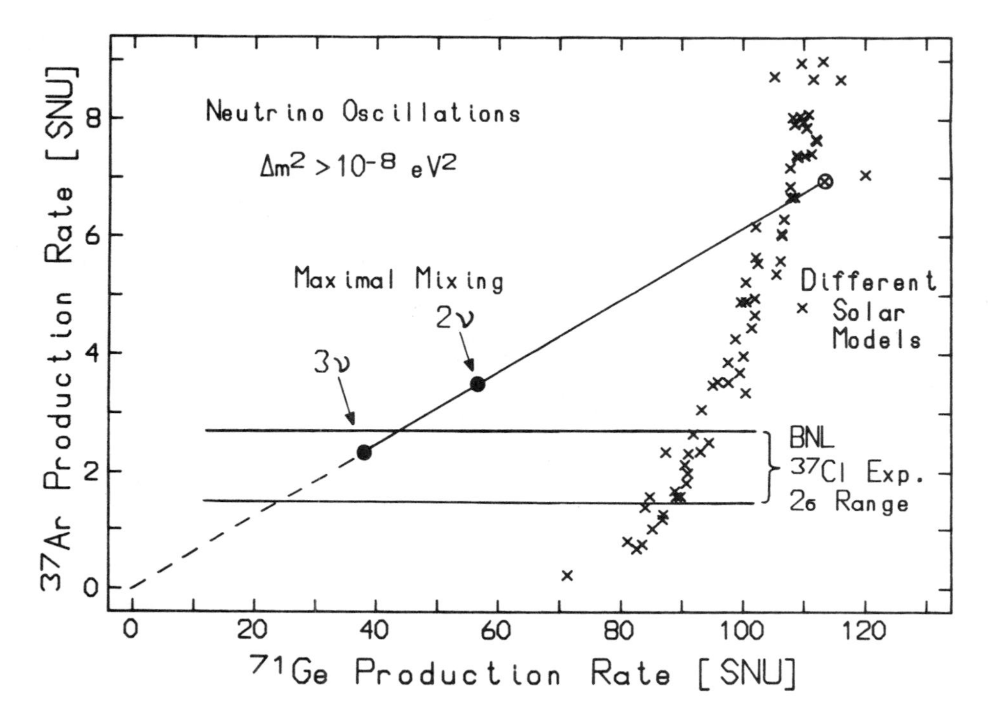

Fig. 7: Response of the signals expected from the Cl and Ga detectors to neutrino oscillations ($\Delta m^2 > 10^{-8}$ eV^2) and to non-standard solar model assumptions.

composition. opacities)[19]. In most cases these models were invented in order to reduce the strongly temperature-dependent ^8B neutrino flux to a level consistent with the Cl observation. However, all models plotted in Fig. 7 still assume that hydrogen burning is the only energy source of the sun and that energy generation is in secular equilibrium with the observed luminosity. It turns out that models in agreement with the measured 2 sigma range for the Cl experiment predict a rate around 90 SNU for the Ga experiment.

The other possible explanation seriously discussed at present is a reduced electron neutrino flux incident at the detector due to neutrino oscillations[20]. This implies that neutrinos are superpositions of different mass eigenstates m_j:

$$\nu_i = U_{ij}\, m_j, \quad i=e,\mu,\tau,\ldots, \quad j=1,2,3,\ldots,$$

where U_{ij} is the orthogonal mixing matrix. The oscillation length is given by:

$$L_{ij}\ [\mathrm{m}] = 2.48\ E_\nu\ [\mathrm{MeV}]\ /\Delta m_{ij}^2\ [\mathrm{eV}^2],$$

where $\Delta m_{ij}^2 = |m_i^2 - m_j^2|$ is the difference between the squared masses of 2 mass eigenstates. Oscillations may occur simultaneously

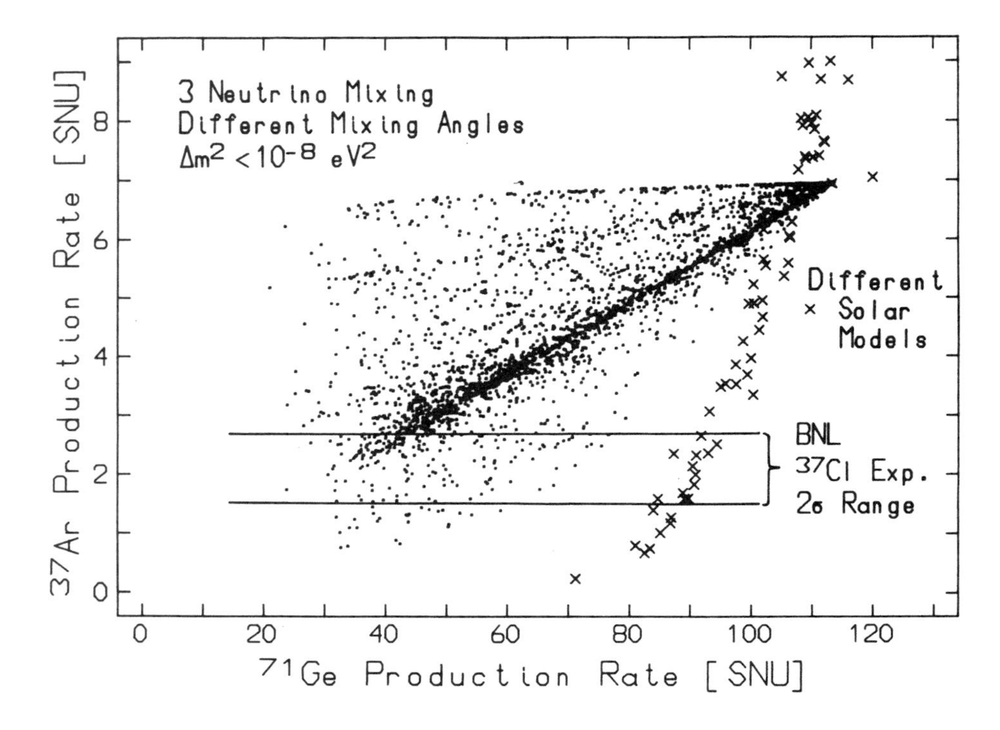

Fig. 8: Cl detector signal versus Ga detector signal: The result of a MC calculation for oscillations with $\Delta m^2 < 10^{-8}$ eV^2 is compared to non-standard solar model predictions and the experimental Cl detector rate.

between 2, 3 or (if they exist) even more different neutrino flavors. If all Δm^2 involved in the oscillations are $> 10^{-8}$ eV^2, then the electron neutrino flux incident at the detector is decreased by a constant factor independent of the neutrino energy. Thus the Cl and Ga detector rates are reduced by the same factor. This is usually the only case considered when the Cl discrepancy is discussed in terms of neutrino oscillations. Almost maximal mixing of 3 neutrino flavors is required in order to reduce the Cl rate to the experimental 2 sigma range. In that case the Ga detector would measure around 40 SNU (see Fig. 7).

On the other hand, if at least one of the Δm^2 values is $< 10^{-8}$ eV^2, then the decrease of the electron neutrino flux at the detector site is a function of the neutrino energy, therefore the reduction factors for the Cl and Ga detector rates may be quite different[20]. Fig. 8 shows a plot similar to that of Fig. 7, now for neutrino oscillations with $\Delta m^2 < 10^{-8}$ eV^2. It presents the results of 3000 simulations where different Δm^2 values and matrix elements U_{ij} have been selected by a Monte-Carlo method and the corresponding Cl and

Ga rates have been calculated. Again, the Cl experimental result and the predictions from different non-standard solar models have been included for comparison. It follows that for Cl rates consistent with the experiment the Ga detector can now measure rates up to 75 SNU, quite in contrast to the case in Fig. 7, but still distinct from the non-standard solar model predictions (85 - 95 SNU). We therefore conclude that the Ga experiment is in principle able to settle the question whether the Cl discrepnacy is due to problems with solar models or to neutrino oscillations.

Acknowledgement. The Heidelberg group would like to thank all members of the pilot experiment collaboration for the stimulating cooperation and the excellent work they have accomplished in the frame of the pilot experiment.

7. REFERENCES

1. B. Cleveland, R. Davis Jr., and J.K. Rowley, Inst. Phys. Conf. Ser. No. 71, 241 (1984).
2. J.N. Bahcall, W.F. Huebner, S.H. Lubow, P.D. Parker, and R.K. Ulrich, Rev. Mod. Phys. 54, 767 (1982).
3. W.A. Fowler, Science Underground, AIP Conf. Proc. 96, 80 (1983).
4. V.A. Kuzmin, Sov. Phys. JETP 22, 1051 (1966).
5. W. Hampel and R. Schlotz, Paper presented at 7th Int. Conf. on Atomic Masses and Fundamental Constants (AMCO-7), Darmstadt-Seeheim, F.R. Germany, 3-7 Sept. 1984. To be published in the Conference Proceedings.
6. W. Hampel and L.P. Remsberg, submitted to Phys. Rev. C (1984).
7. R. Davis Jr., B. Cleveland, G. Friedlander, S. Katcoff, L.P. Remsberg, J.K. Rowley, J. Weneser (Brookhaven Nat. Lab.); T. Kirsten, W. Hampel, G. Heusser, M. Hübner, J. Kiko, E. Pernicka, R. Schlotz (MPI Kernphysik Heidelberg); I. Dostrovsky (WIS Rehovot); J.N. Bahcall (IAS Princeton).
8. A.W. Wolfendale, E.C.M. Young, and R. Davis Jr., Nature Phys. Sci. 238, 130 (1972).
9. J.N. Bahcall, Rev. Mod. Phys. 54, 881 (1978).
10. J. Rapaport, T. Taddeucci, P. Welch, C. Gaarde, J. Larsen, C. Goodman, C.C. Foster, C.A. Goulding, D. Horen, E. Sugarbaker and T. Masterson, Phys. Rev. Lett. 47, 1518 (1981).
11. H. Orihara, C.D. Zafiratos, S. Nishihara, K. Furukawa, M. Kabasawa, K. Maeda, K. Miura, and H. Ohnuma, Phys. Rev. Lett. 51, 1328 (1983).
12. A.J. Baltz, J. Weneser, B.A. Brown and J. Rapaport, Phys. Rev. Lett. 53, 2078 (1984).

13. J. Rapaport, priv. communication (1984).
14. N. Itoh and Y. Kohyama, Ap. J. 246, 989 (1981).
15. K. Grotz, H.V. Klapdor, and J. Metzinger, Paper presented at the Fifth Internat. Symposium on Capture Gamma Ray Spectroscopy and Related Topics, Knoxville, Tennessee (USA), 10-14 Sept. 1984.
16. GALLEX Collaboration: T. Kirsten, B. Povh, H. Völk, W. Hampel et al. (MPI Kernphysik Heidelberg), K. Ebert, E. Henrich et al. (KFZ Karlsruhe), R. Mössbauer et al. (TU München), M. Spiro et al. (CEN Saclay), A. de Bellefon et al. (Collège de France Paris), E. Schatzman (Univ. of Nice), E. Fiorini, E. Bellotti et al. (Univ. of Milano), L. Paoluzi et al. (Univ. of Rome), I. Dostrovsky (WIS Rehovot).
17. J.L. Bourdon, R. Richena, M. Aglietti-Zanon, EURATOM Ispra (Italy), priv. communication (1984).
18. G. Dupont, CEN Grenoble, priv. communication (1984).
19. For references see: W. Hampel, Proc. Second Workshop on Nuclear Astrophysics, Ringberg Castle, MPI f. Physik und Astrophysik, Munchen, MPA 90, 11 (1983).
20. W. Hampel, Paper presented at the 11th Int. Conf. on Neutrino Physics and Astrophysics, Nordkirchen, F.R. Germany, June 11-16, 1984. To be published in the Conference Proceedings.

PILOT INSTALLATION OF THE GALLIUM-GERMANIUM SOLAR NEUTRINO TELESCOPE

I. R. Barabanov, E. P. Veretenkin, V. N. Gavrin, S. N. Danshin,
L. A. Eroshkina, G. T. Zatsepin, Yu. I. Zakharov, S. A. Klimova,
Yu. B. Klimov, T. V. Knodel, A. V. Kopylov, I. V. Orekhov,
A. A. Tikhonov, M. I. Churmaeva
Institute for Nuclear Research of the AS USSR

ABSTRACT

The pilot gallium-germanium installation with 7 t of Ga metal is described. Preliminary results of the yields of ^{71}Ge and ^{69}Ge from cosmic rays are given. The possibility of conducting a calibrating experiment using a metal target and a Cr neutrino source produced from enriched chromium is considered.

INTRODUCTION

For the gallium solar neutrino experiment several attractive possibilities exist for the separation of germanium from the gallium target. Two schemes are discussed now extensively, one based on using a metal gallium target and the other based on using an acid gallium chloride solution.[1,2,3] The first method has been chosen at the Institute for Nuclear Research of the AS USSR. This method has the following advantages:

(1) The Ga metal target is less sensitive to the background reaction produced by radioactive impurities.

(2) A metal target has a smaller volume in comparison with a gallium chloride solution target.

(3) Gallium metal is advantageous in an experiment with a neutrino source.

On the other hand an experiment based on using a metallic gallium target has a number of disadvantages. The main one is that each time the germanium is separated it is necessary to add fresh reagents. Severe control of germanium impurities in necessary because of this fact.

In this report the pilot installation, the chemical procedure for germanium extraction from seven tons of metallic gallium and the counting system are described. The preliminary results are presented. In addition, the possibility is considered of conducting a neutrino source experiment.

THE PILOT INSTALLATION AND CHEMICAL PROCEDURE

The design of the pilot installation is presented in Figure 1. Seven tons of gallium are placed in a reactor 1, made of PTFE. The gallium extraction procedure is as follows: 70 kg of a solution with 6.9 mass % H_2O_2 and 1.7 mass % HCl are added. Then after intensive mixing during 15 min, 50 kg of 22% HCl solution are added. The solution obtained is separated from gallium and then is reduced two times in volume by means of evaporation in a vacuum circulating glass

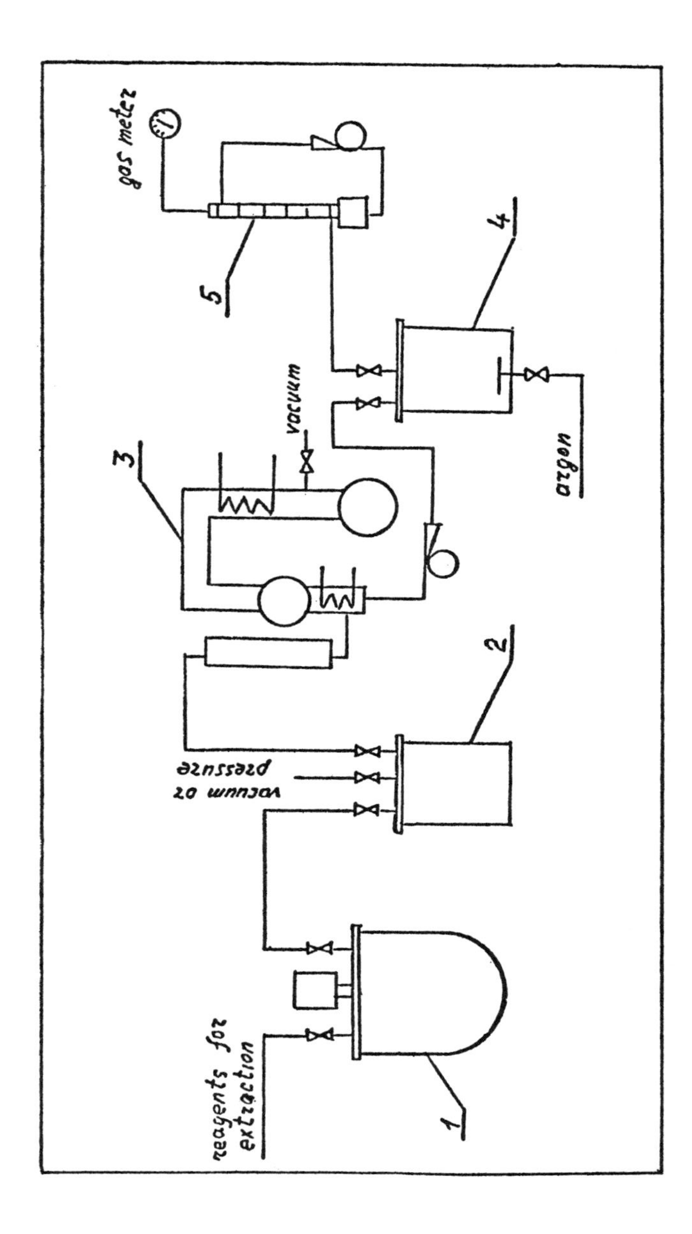

Fig. 1. Schematic of the pilot installation.

apparatus. The solution obtained is pumped to tank 4 where 100 ℓ of concentrated HCl are added. Then germanium is swept out at 60 °C by a stream of argon. The total volume of gas passing through tank 4 is ≈ 4 m^3. The germanium carried out is caught by two liters of H_2O in the absorption column 5. The results presented in Table I illustrate the extraction efficiency.

Table I

No. of extraction	Mass of carrier (mg)	Extraction coefficient (%)
16	0.924	95 ± 4
19	0.841	84 ± 5
21	0.793	88 ± 8

After the solution in the trap 5 is adjusted to 9 Mol/l of HCl, $GeCl_4$ is swept to a small trap containing 50 ml of 0.5 N LiOH. After addition of HCl, germanium is extracted by CCl_4 and reextracted by 100 ml H_2O. Then GeH_4 is synthesized from the resultant solution on the installation for GeH_4 synthesis. After purification by gas chromatography, GeH_4 is placed in a proportional counter. The total procedure takes approximately 20 hours. Figure 2 shows a schematic of the Ga-Ge telescope developed on the basis of experience obtained using the pilot installation and work with small scale laboratory apparatus.

COUNTING SYSTEM

The detection of germanium decays was conducted using the system presented in Fig. 3. The proportional counter is placed in the well of a NaI(Tl) crystal of diameter 150 and length 150 mm. The NaI(Tl) crystal, in turn, is placed in a plastic scintillator used as an anticoincidence shield. The total assembly is placed in a passive shield of 8 cm Pb and 3 cm W. The background of the NaI(Tl) detector is 4 c/sec in the 0.2-3.0 MeV energy range using an anti-coincidence plastic shield. The system makes possible the separation of the ^{69}Ge events from ^{71}Ge decays in the proportional counter. Here the ^{69}Ge events are selected by the presence of a pulse in the NaI(Tl) detector accompanying the pulse in proportional counter corresponding to K or L peak of germanium. The efficiency of ^{69}Ge counting is 50 % of the efficiency of ^{71}Ge.

The electronics of the counting system consists of pulse shape discrimination channels using the ADP-method[4] (ADC-1, ADC-2), a channel for the measurement of the amplitude of the pulse from the NaI(Tl) crystal (ADC-3), a pulse shape recording channel on the basis of storage oscilloscope C8-12 and an anticoincidence unit in the channel of external triggering of the oscilloscope. All information about the pulse (energy-E, ADP, amplitude U from NaI(Tl), time of the event and number of counter) is recorded by a printer immediately

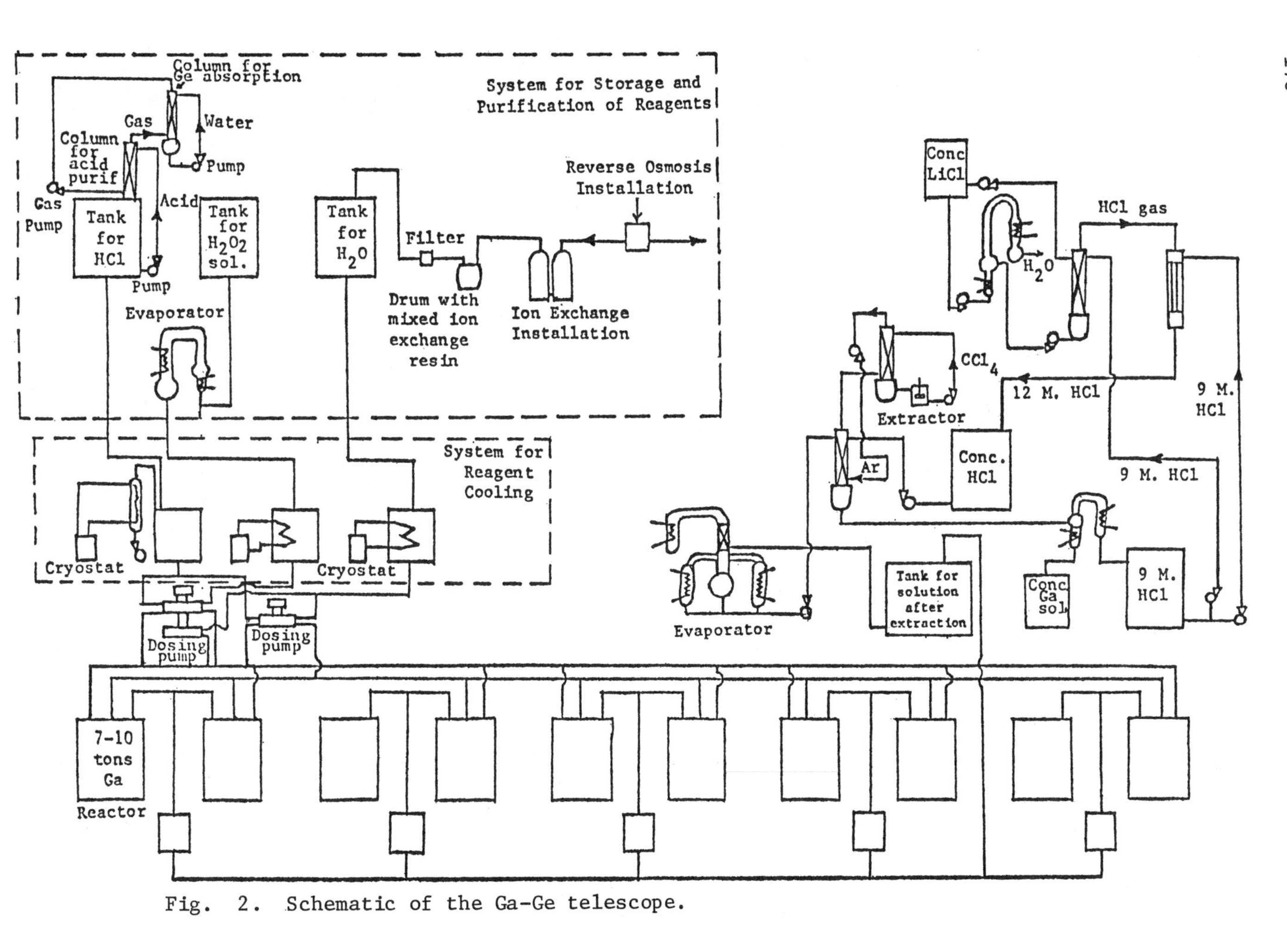

Fig. 2. Schematic of the Ga-Ge telescope.

after the pulse. The pulse shape is monitored on a TV screen fed by a TV camera placed before the screen of the oscilloscope. The pulse shape can be recorded on a video tape recorder used in the system.

Figure 5 presents the time dependence of the count rate in the K-peak of germanium for a sample of GeH_4 obtained from the pilot installation. In Figure 5 is presented also the count rate of the proportional counter in the K-peak region in coincidence with the pulse from the NaI(Tl) crystal for the 50-1550 KeV energy region (^{69}Ge events).

Figure 4 shows the spectrum of pulses from the NaI(Tl) crystal in coincidence with the proportional counter pulses from the K or L peaks of Ge. The picture definitely shows the presence of γ-lines, corresponding to the decays of ^{69}Ge.

The data obtained were treated by a least squares method, it was assumed that there are two decaying components, ^{69}Ge and ^{71}Ge, and the background. The results obtained were corrected for the efficiency of the counting system, the extraction coefficient, and the coefficient due to limited time of exposure. The corresponding production rates of ^{69}Ge and ^{71}Ge are respectively, 240 at/hour and 120 at/hour for 50 days counting time.

The expected production rate of ^{71}Ge in the pilot gallium-germanium installation from nuclear-active component of cosmic rays obtained from a comparison with the production rate of ^{37}Ar[5] in C_2Cl_4 is 150 at/hour.

POSSIBILITY OF CONDUCTING A NEUTRINO SOURCE EXPERIMENT[6]

4.1. Choice of parameters of the gallium target and of the neutrino source.

The most suitable isotopes for calibrating a gallium-germanium neutrino telescope are ^{51} Cr, proposed by Raghavan[7] and ^{65}Zn, proposed by Alvarez.[8] These isotopes can be obtained by neutron irradiation of the corresponding isotopes of chromium and zinc.

Let us consider what is the influence of the geometry of the proposed experiment on the production rate of ^{71}Ge. Apparently the optimal geometry is the target in the form of sphere with a pointlike source at the center of it.

If the dimensions of the source are much less than the dimension of the target then the production rate of ^{71}Ge could be calculated with good accuracy by substituting for the real source a pointlike source with the same activity. Then we'll get

$$N = B \cdot G \quad {}^{71}\mathrm{Ge/day} \qquad (1)$$

where:
$B = \sigma n S$ - coefficient describing the chosen target and source (σ - cross section of ν - interaction with nucleus ^{71}Ga: n - number of atoms of ^{71}Ga in 1 cm^3 of target, S - activity of the source in ν per sec), $G = (1/4\pi)\int 1/r^2 dv$ - a factor, describing the geometry of the target.

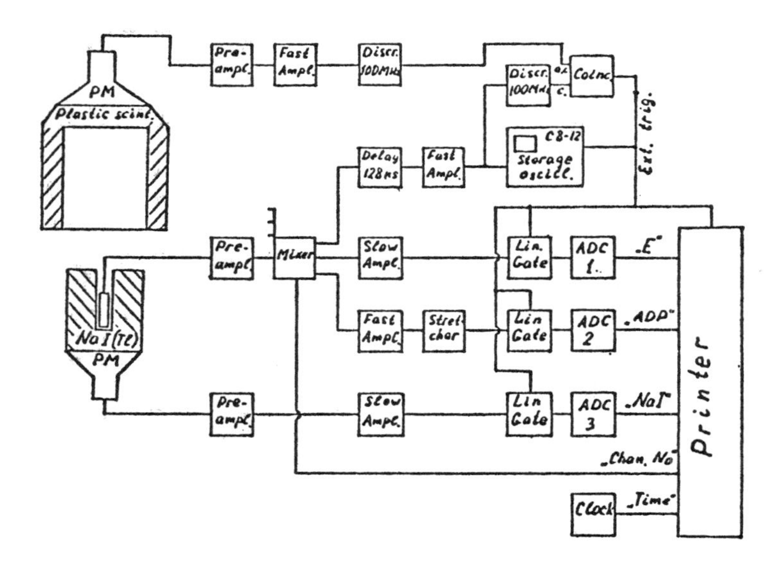

Fig.3. Scheme of the counting system.

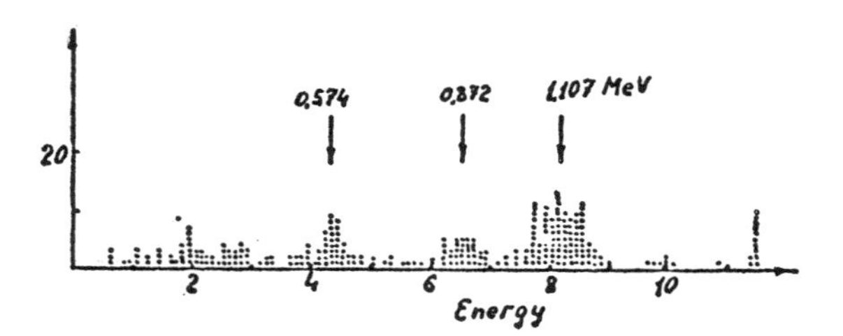

Fig.4. Spectrum of the pulses from NaI(Tl) crystal in coincidence with K or L peak pulses of Ge from proportional counter.

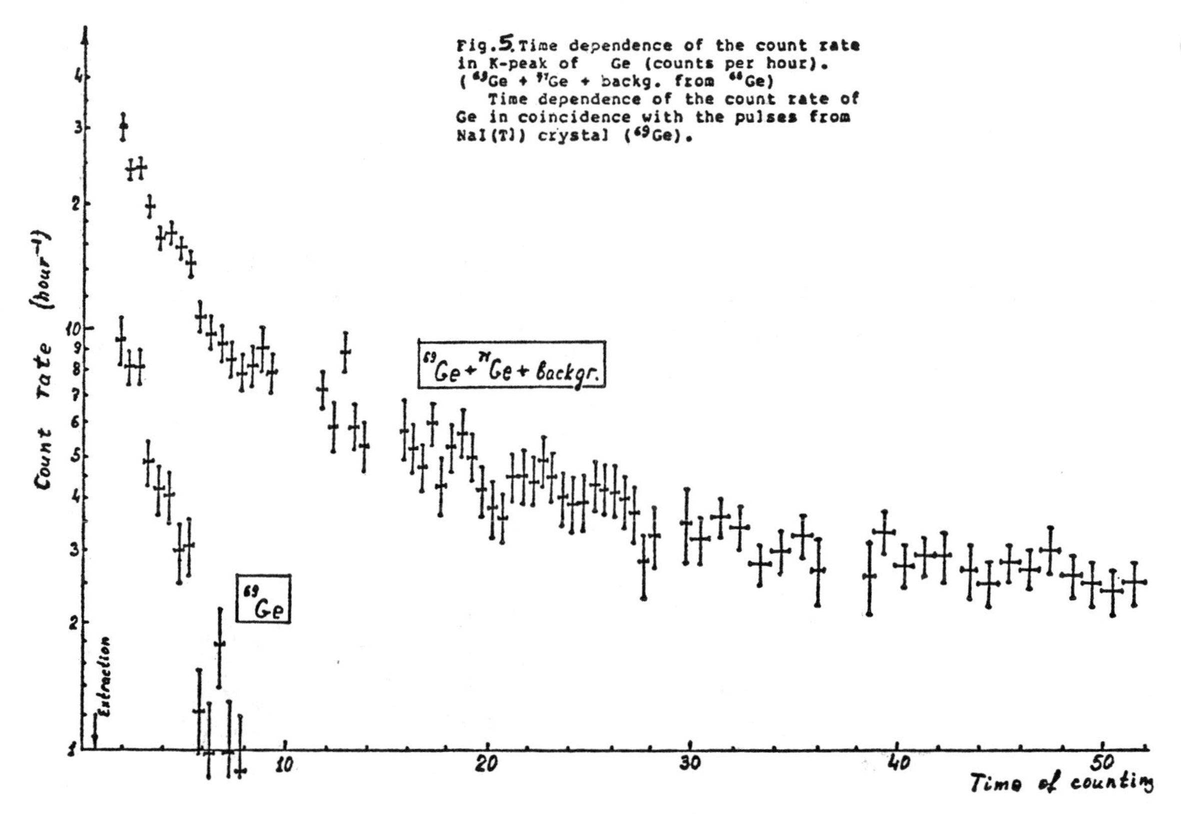

Fig. 5. Time dependence of the count rate in K-peak of Ge (counts per hour). (^{69}Ge + ^{71}Ge + backg. from ^{68}Ge)
Time dependence of the count rate of Ge in coincidence with the pulses from NaI(Tl) crystal (^{69}Ge).

For a spherical target with radius R: G = (T - R_s).

Taking into account that $n = N_A \rho \eta / A$, where N_A - number of Avogadro, A - atomic number of Ga, ρ - density of Ga in the target (g/cm^3), η - natural abundance of ^{71}Ga, R_s - radius of the source.

$$N = \frac{\sigma S N_A \eta \rho}{A} (R - R_s) \qquad (2)$$

One can see that for a spherical target of radius R and activity S the production rate of ^{71}Ge is proportional to the density of Ga in the target and decreases with the increase of the source dimension. For a point-like source $N = \sigma n S r$, i.e., $N \sim m^{1/3}$ and for a given mass of Ga $N \sim \rho^{2/3}$. Thus the production rate of Ge has a weak increase with the increase of Ga mass and is very sensitive to the density changes of Ga in the target. So from the analysis of (1/G)(dG(m)/dm) - one can see that for a spherical target of Ga metal with a mass greater than 3 t, the additional increase of the production rate from every subsequent layer of target with a mass 1 t is less than 10 %.

The inferences made for a spherical target can be applied with good accuracy to a target with a form near the spherical one; this means that they are valid for Ga metal placed in the chemical reactor of our pilot installation, which is the same as one module of the neutrino telescope.

Calculation shows that for equal activities the use of an enriched (87.7 %) chromium source reduces the volume of the source by 20 times in comparison with a source from natural chromium; for zinc this ratio is equal to 88.

The use of highly enriched chromium allows more efficient use of thermal neutron flux, by increasing the transparency of the sample used in irradiation. The use of chomium has some advantages in comparison with zince. The first is that the experiment could be performed in much less time and need much less radiation shielding (because of the smaller energy of γ-rays emitted).

4.2. Main parameters of neutron activation of the enriched chromium.

It is planned to use a sample of 190 g of Cr (87.7 % ^{50}Cr enrichment) and the research reactor CM-2 of the Institute of Atomic Reactors.[10] The main parameters of the reactor:

(1) Active zone dimensions 420 x 420 x 350 mm

(2) Central neutron trap dimensions 140 x 140 x 300 mm

(3) Number of vertical channels of diameter 10 mm, placed in neutron trap - 17

(4) Flux of thermal neutrons in the neutron trap - 3.2×10^{15} ncm^{-2} s^{-1}

In the calculation made the flux of thermal neutron was taken to be 2.9×10^{15} ncm^{-2} s^{-1} because of the absorption of the neutrons in the sample. It is planned to place in every channel 16 chromium rods 0.3 mm in diameter and 75 mm long. Put together, these rods make a cylinder 75 mm high and 25 mm in diameter. This geometry ensures a decrease of the thermal neutron flux due to screening in the sample of not more than 11 %.

From an analysis of the dependence of the activity of the sample on the irradiation time (from the point of view of obtaining sufficient activity - 0.9 MCi and the effective using of ^{50}Cr) it was obtained that the optimal time is ~ 20 days.

It seems to be impossible to make a correct calculation of the activity of the neutrino source because of some ambiguities in the thermal neutron flux, the activation cross section etc. Because of this it is necessary to measure the activity of the sample. There are two says to do this: by means of calorimetry and by measuring the concentration of ^{51}V after ^{51}Cr decay, which is estimated to be 7 %.

4.3. Main parameters of the radiation shield.

A calculation of the radiation shield required was made for 190 g chromium sample (166.67 g of ^{50}Cr), which after 20 days of irradiation and 3 days past the end of irradiation had an activity of 0.84 MCi.

The results obtained show that to decrease the radiation level to 100 mR/hour a shield should be used of 21.9 kg, 20.6 kg, 4.9 kg of Pb, W, and U respectively. The source is planned to be placed in the gallium together with the shield. Because of this the uranium shield is preferable.

4.4. The statistical accuracy of the experiment.

It is planned to conduct 4 exposures, each of 12 days duration followed by a 1 day interval for ^{71}Ge extraction. The counting of ^{71}Ge begins 1 day after the end of exposure.

The total efficiency of the ^{71}Ge detection is determined by:

(1) The coefficient of extraction of ^{71}Ge from gallium, which was taken to be 85 % - that achieved at the present time and 95 % - that which is planned to be achieved.

(2) The efficiency of the counting system is 72 %, achieved at the present time (using the two-chambered proportional counter[11]) and 80 % - which is planned to be achieved.

(3) Efficiency determined by the time of counting which is planned to be $5T_{1/2}$ (57 days).

Thus the total achieved efficiency is 59 % and the planned one is 83 %.

The cross section for the interaction of the neutrinos from ^{51}Cr with ^{71}Ga was taken from ref. 12. The new values of log(ft) were taken for the excited states of ^{71}Ge.[13] Thus the total cross section of the neutrino interaction with ^{71}Ga was taken to be 8.8×10^{-21} b.

The results obtained did not take into account the background from solar neutrinos which is equal to about 1.5 %. The background from cosmic rays is much less because the experiment is planned to be performed in the special underground chamber of the Baksan Neutrino Observatory of INR AS USSR. This chamber is situated at a depth of ~ 4000 m.w.e.

The results of the calculations are presented in Table II.

Table II

Mass of target	Total efficiency of detection	Number of detected pulses				Total number of pulses	σ%
		run1	run2	run3	run4		
10 t	0.59	126	92	66	48	332	5.5
	0.83	179	129	94	66	468	4.6

ACKNOWLEDGEMENTS

In conclusion we would like to express our sincere gratitude to A. E. Chudakov, G. V. Domogatskij, R. A. Eramgjan, A. V. Klinov, A. C. Kochenov, V. A. Legasov, M. A. Markov, V. M. Pontecorvo, A. N. Tavkhelidze, V. A. Tsikanov for stimulating attention and interest to this problem.

REFERENCES

1. J. K. Rowley, B. T. Cleveland, R. Davis, Jr. and J. C. Evans, Neutrino-77, Baksan Valley, Vol. 1, p. 15.
2. I. R. Barabanov, A. I. Egorov, V. N. Gavrin, Yu. S. Kopysov and G. T. Zatsepin, Neutrino-77, Baksan Valley, Vol. 1, p. 20.
3. R. Davis, Jr., B. T. Cleveland, J. K. Rowley, S. Katcoff, L. P. Remsberg, G. Friedlander and J. Weneser, Proposal for a Fundamental Test of the Theory of Nuclear Fusion in the Sun with a Gallium Solar Neutrino Detector, 1981.
4. R. Davis, Jr., J. Evans, V. Radeka, L. Rogers, Neutrino-72, Balatonfured, Vol. 1, p. 23.
5. I. R. Barabanov, E. P. Veretenkin, V. N. Gavrin, G. I. Gordeeva, G. T. Zatsepin, A. V. Kopylov, I. V. Orekhov, Neutrino-77, Baksan Valley, Vol. 1, p. 62.
6. V. N. Gavrin, S. N. Danshin, G. T. Zatsepin, A. V. Kopylov, Preprint INR AS USSR P-0335, 1984.
7. R. S. Raghavan, Neutrino-78, Brookhaven.
8. L. W. Alvarez, Physics Notes, Lawrence Radiation Lab., 1973.
9. A. M. Petrosjanz, Problemi atomnoi nauki i techniki, Moscow, Atomizdat, 1979.
10. C. M. Feinberg, et al., Third Int. Conf. UNO, 1969.
11. I. R. Barabanov, V. N. Gavrin, A. A. Gogin, Yu. I. Zakharov, Preprint INR AS USSR, P-0318, 1983.
12. J. N. Bahcall, Rev. Mod. Phys., 50, 881 (1978).
13. H. Orihara, et al., Phys. Rev. 51, 1328 (1983).

THE INTEGRAL METHOD OF TREATMENT OF EXPERIMENTAL DATA FROM RADIOCHEMICAL SOLAR NEUTRINO DETECTORS

V. N. Gavrin, A. V. Kopylov, A. V. Streltsov
Institute for Nuclear Research of the AS USSR

ABSTRACT

An analysis is made of the statistical errors in solar neutrino detection by radiochemical detectors at different times of exposure. It is shown that short exposures (τ_e = one-half to one half-life) give minimal one-year error. The possibility is considered of the detection of the solar neutrino flux variation due to annual changes of the Earth-Sun distance. The integral method of treatment of the experimental data is described. Results are given of the statistical treatment of computer simulated data.

INTRODUCTION

Nearly forty years have passed since the time when B. Pontecorvo proposed the chlorine-argon detector for solar neutrino detection using a radiochemical technique.[1] In 1967 this detector was built by R. Davis and his colleagues and since 1970 till nowadays, i.e., during 15 years, this detector has measured the flux of solar neutrinos, being the only telescope looking into the interior of the sun.[2,3] It is well known that the flux of solar neutrinos measured by this detector is less than the theory predicts.[4] This is what is known as the solar neutrino problem.

At the present time the group of G. Zatsepin from INR of the Academy of Sciences of the USSR is doing extensive work building a gallium-germanium detector with 60 t of Ga metal and a chlorine-argon telescope with 3000 t of C_2Cl_4. The results obtained on these detectors will allow us to get answers to many exciting questions of solar physics and the physics of elementary particles.

A specific feature of radiochemical detectors of solar neutrinos is the extremely small number of events received during the long time of measurements. For example, in Davis' experiment the mean production rate of ^{37}Ar Q = 0.42 atoms per day in 610 t of C_2Cl_4. For an exposure time τ_e = 70 days, extraction efficiency ε_e = 0.935, efficiency of the counting system ε_c = 0.44, there will be 6.5 events from argon decay for 300 days of counting. In addition to this there will be approximately 6 counts as the background of the counting system. If we take the total muon intensity at the depth of Davis' detector to be $I_\mu = 3.9 \cdot 10^{-9}$ μcm^{-2} s^{-1}, then 1.2 events out of 6.5 will be from muons.[5] (The intensity of muons for Davis' detector was measured[6] and according to this data there will be approximately 1.5 events from muons.)

The calculated intensity of muons for the chlorine-argon detector with 3000 t of C_2Cl_4 is I_μ = (1.6 to 1.8)$\cdot 10^{-9}$ $\mu \cdot cm^{-2}$ s^{-1}. If we take $I_\mu = 1.8 \cdot 10^{-9}$, then the ^{37}Ar

0094-243X/85/1260185-11 $3.00

production rate from muons will be ≈ 0.19 SNU which corresponds for this detector to 0.18 atoms of ^{37}Ar per day and the production rate from neutrinos scaled from Davis' detector will be 1.70 atoms per day. If we take the same parameters as before, i.e., $\tau_e = 70$ days, $\varepsilon_e = 0.935$, $\varepsilon_c = 0.44$, for 300 days of measurement there will be 29.1 events, 26.3 from neutrinos and 2.8 from muons. The background of the counting system should not depend on the mass of the detector essentially, so we can expect in this case approximately 6 background pulses provided the counting system works as well as in Davis' experiment.

Because of the small number of events the statistical analysis of the data was the subject of special investigation. Methods of statistical treatment of this data were described by Aurela[7] and by Cleveland.[8] In this paper we analyze the statistical errors for gallium-germanium and chlorine-argon detectors for different times of exposure and for different counter backgrounds. We describe also the integral method of statistical treatment of the experimental data.

THE ANALYSIS OF THE STATISTICAL ERRORS

Let us consider the most simple procedure which is often used in low activity measurement. Suppose the extracted sample is measured during the time t_o, then after some interval to allow the sample to decay, the background is measured during the time t_1. Then the number of decays, detected by the counter is

$$N = n(t_o) - (t_o/t_1)\cdot N(t_1) \qquad (1)$$

where $N(t_o)$, $N(t_1)$ are the number of counts for t_o, t_1. If the sample is counted right after the extraction, the rate of the counts at the beginning of the counting is

$$a = Q\cdot(1-\exp(-\lambda\cdot\tau_e)\cdot\varepsilon_e\cdot\varepsilon_c \text{ atoms per day}$$

If b is the number of background pulses per day then for the mean values we get

$$N(t_o) = a\cdot\Delta(t_o)/\lambda + bt_o$$

$$N(t_1) = bt_1$$

where λ-decay constant, $\Delta(t_o) = 1-\exp(-\lambda\cdot t_o)$. If the background doesn't change in time, then the values of $N(t_o)$ and $N(t_1)$ have a Poisson distribution and a dispersion, D, of N equal to

$$D = a\cdot\Delta(\tau_o)/\lambda + bt_o(1 + t_o/t_1) \qquad (2)$$

This expression can be easily rewritten to be suitable for any detector:

$$D = A \cdot \Delta(T_o) + BT_o \cdot (1 + T_o/T_i) \tag{3}$$

where $A = a/\lambda$, $B = bT_{1/2}$, $\Delta(T_o) = 1-0.5^{T_o}$, $T_o = t_o/T_{1/2}$, $T_1 = t_1/T_{1/2}$ and $T_{1/2}$ - the half-life of the detected isotope. If $T_i \gg T_o$, i.e., the accuracy of the background measurement is much higher than the accuracy of the $N(t_o)$ measurement,

$$D = A \cdot \Delta(T_o) + BT_o \tag{4}$$

The standard deviation of N is

$$\sigma = 100 \quad \% \cdot \sqrt{A \cdot \Delta(T_o) + BT_o}/(A \cdot \Delta(T_o)) \tag{5}$$

From Eq. 5 it follows that there is some optimal time T_o which gives the minimal error.

Table 1 shows the errors for the chlorine-argon detector with 3000 tons of C_2Cl_4 for $\varepsilon_e = 0.935$, $\varepsilon_c = 0.44$, Q = 1.88 at/day for different times of exposure: $3T_{1/2}$, $2T_{1/2}$, $T_{1/2}$, $0.5T_{1/2}$ for the background B = 0.7, i.e., equal to the background in Davis' experiment, and also for B = 0.35 and B = 1.4. The results show that the error for one measurement is increased from 18.2% for $\tau_e = 3T_{1/2}$ to 34.2% for $\tau_e = 0.5T_{1/2}$ if we take B = 0.7. But the number of measurements is increased from 3.4 per year for $\tau_e = 3T_{1/2}$ to 20 for $\tau_e = 0.5T_{1/2}$, so the one-year error is decreased for $\tau_e = 0.5T_{1/2}$ in comparison with $\tau_e = 3T_{1/2}$. In addition to statistical error there is a systematic error of about 10% per every measurement. If $\tau_e = 0.5T_{1/2}$ then 20 measurements per year could be done and the increase of the one-year error due to the systematic error will be less than for longer times of exposure. This an another reason to prefer shorter exposures. As a result the total one-year error, including both statistical and systematic errors, is expected to be 11.2%, 9.9%, 8.5%, 8.0% for the exposure times $3T_{1/2}$, $2T_{1/2}$, $T_{1/2}$, $0.5T_{1/2}$. For $\tau_e = 0.5T_{1/2}$ and 10 years of experiment the production rate can be measured with an accuracy of about 2.5%. Then the accuracy of the neutrino flux measurement will be approximately 2.7% if we take into account that the accuracy of the muon production rate evaluation is about 10% [5] and the muon's part in the total production rate is about 10% as was shown earlier.

Table 1 shows also that the optimal time of measurement is $t_o = 4T_{1/2}$ for $\tau_o = 0.5T_{1/2}$ and B = 0.7. To reduce the statistical error the condition should be fulfilled $t_1 > t_o$. Therefore, it is reasonable to accept the total time of measurement $t_o + t_1 = 10\ T_{1/2}$. The increase of this time inevitably results in an increase in the number of counting channels. For the chlorine-argon detector, if we take the time of exposure plus the time of extraction to be 18 days and the total time of counting $t_o + t_1 = 360$ days, then we will need 20 counting channels. This is the price for high accuracy. Here it is worth mentioning that short exposures result in high time resolution, which is important for the chlorine-argon detector especially because, as was shown by Bahcall,[4] this detector is the most sensitive among radiochemical detectors to the neutrinos from collapsed stars.

Table 1 shows also the influence of the background on the error. If the background is two times lower (higher), the statistical error for $\tau_e = 0.5T_{1/2}$ will decrease (increase) from 34.2% to 32.3% (37.5%). For 10 years of experiment this will result in a total reduction (increase) of this time of 1.1 (2.0) years. Thus the background of B = 0.7 is not negligible for this experiment. Eventually, the value of the background determines the minimal time of exposure because the shorter time of the exposure the greater is the influence of the background on the statistical error.

Table 1

Statistical errors for chlorine-argon detector with 3000 t of C_2Cl_4, see equation (5). Q=1.88 at/day, $\mathcal{E}_e$=0.935, $\mathcal{E}_c$=0.44.

$\tau_e=3T_{1/2}$, A=34.179, B=.35										
T_o:	0.5	1.0	1.5	2.0	2.5	3.0	3.5	4.0	4.5	5.0
δ%:	31.8	24.4	21.5	20.0	19.1	18.6	18.2	18.0	17.9	17.8
$\tau_e=3T_{1/2}$, A=34.179, B=.70										
T_o:	0.5	1.0	1.5	2.0	2.5	3.0	3.5	4.0	4.5	5.0
δ%:	32.1	24.6	21.7	20.2	19.4	18.9	18.6	18.4	18.3	18.2
$\tau_e=3T_{1/2}$, A=34.179, B=1.4										
T_o:	0.5	1.0	1.5	2.0	2.5	3.0	3.5	4.0	4.5	5.0
δ%:	32.6	25.1	22.2	20.8	19.9	19.5	19.2	19.1	19.1	19.1
$\tau_e=2T_{1/2}$, A=29.296, B=.35										
T_o:	0.5	1.0	1.5	2.0	2.5	3.0	3.5	4.0	4.5	5.0
δ%:	34.4	26.4	23.2	21.6	20.7	20.1	19.7	19.5	19.4	19.3
$\tau_e=2T_{1/2}$, A=29.296, B=.70										
T_o:	0.5	1.0	1.5	2.0	2.5	3.0	3.5	4.0	4.5	5.0
δ%:	34.8	26.7	23.6	22.0	21.0	20.5	20.2	20.0	19.9	19.8
$\tau_e=2T_{1/2}$, A=29.296, B=1.4										
T_o:	0.5	1.0	1.5	2.0	2.5	3.0	3.5	4.0	4.5	5.0
δ%:	35.5	27.3	24.2	22.6	21.7	21.3	21.0	20.9	20.9	20.9
$\tau_e=1T_{1/2}$, A=19.531, B=.35										
T_o:	0.5	1.0	1.5	2.0	2.5	3.0	3.5	4.0	4.5	5.0
δ%:	42.4	32.5	28.7	26.7	25.6	24.9	24.5	24.2	24.1	24.0
$\tau_e=1T_{1/2}$, A=19.531, B=.70										
T_o:	0.5	1.0	1.5	2.0	2.5	3.0	3.5	4.0	4.5	5.0
δ%:	43.0	33.1	29.2	27.3	26.2	25.6	25.2	25.0	25.0	25.0
$\tau_e=1T_{1/2}$, A=19.531, B=1.4										
T_o:	0.5	1.0	1.5	2.0	2.5	3.0	3.5	4.0	4.5	5.0
δ%:	44.2	34.2	30.3	28.5	27.5	26.9	26.7	26.7	26.7	26.9
$\tau_e=0.5T_{1/2}$, A=11.441, B=.35										
T_o:	0.5	1.0	1.5	2.0	2.5	3.0	3.5	4.0	4.5	5.0
δ%:	56.0	43.0	38.0	35.5	34.0	33.2	32.7	32.4	32.3	32.3
$\tau_e=0.5T_{1/2}$, A=11.441, B=.70										
T_o:	0.5	1.0	1.5	2.0	2.5	3.0	3.5	4.0	4.5	5.0
δ%:	57.4	44.2	39.2	36.8	35.4	34.7	34.4	34.2	34.3	34.4
$\tau_e=0.5T_{1/2}$, A=11.441, B=1.4										
T_o:	0.5	1.0	1.5	2.0	2.5	3.0	3.5	4.0	4.5	5.0
δ%:	60.0	46.6	41.6	39.3	38.1	37.6	37.5	37.6	37.9	38.3

Table 2 shows the values, obtained for a gallium-germanium detector with 60 t of Ga metal for the parameters: $\varepsilon_e = 0.95$, $\varepsilon_c = 0.80$. The production rate was taken as Q = 135 SNU or 2.4 atoms of ^{71}Ge per day according to the standard model [4] and the new values of log(ft) for the excited states of ^{71}Ge, obtained in [9]. The calculation was made using [5] with the background B equal to 0.15, 0.30, and 0.60. The background expected in the experiment was taken to be 0.30, i.e., equal to 1 count per month with a counting efficiency $\varepsilon_c = 0.80$. The errors for one measurement (one run) are 20.3%, 22.1%, 27.5%, 37.1% for τ_e equal to $3T_{1/2}$, $2T_{1/2}$,

Table 2

Statistical errors for gallium-germanium detector with 60 t of Ga metal. Q=2.4 at/day, ε_e=0.95, ε_c=0.80

$\tau_e=3T_{1/2}$, A=26.337, B=.15										
T_o:	0.5	1.0	1.5	2.0	2.5	3.0	3.5	4.0	4.5	5.0
δ%:	36.1	27.7	24.3	22.6	21.6	21.0	20.6	20.3	20.1	20.0
$\tau_e=3T_{1/2}$, A=26.337, B=.30										
T_o:	0.5	1.0	1.5	2.0	2.5	3.0	3.5	4.0	4.5	5.0
δ%:	36.3	27.8	24.5	22.8	21.8	21.2	20.8	20.6	20.4	20.3
$\tau_e=3T_{1/2}$, A=26.337, B=.60										
T_o:	0.5	1.0	1.5	2.0	2.5	3.0	3.5	4.0	4.5	5.0
δ%:	36.6	28.1	24.8	23.1	22.2	21.6	21.2	21.0	20.9	20.9
$\tau_e=2T_{1/2}$, A=22.574, B=.15										
T_o:	0.5	1.0	1.5	2.0	2.5	3.0	3.5	4.0	4.5	5.0
δ%:	39.1	29.9	26.3	24.5	23.4	22.7	22.3	22.0	21.8	21.7
$\tau_e=2T_{1/2}$, A=22.574, B=.30										
T_o:	0.5	1.0	1.5	2.0	2.5	3.0	3.5	4.0	4.5	5.0
δ%:	39.3	30.1	26.5	24.7	23.6	23.0	22.5	22.3	22.1	22.1
$\tau_e=2T_{1/2}$, A=22 574, B=.60										
T_o:	0.5	1.0	1.5	2.0	2.5	3.0	3.5	4.0	4.5	5.0
δ%:	39.7	30.5	25.9	25.1	24.1	23.5	23.1	22.9	22.8	22.8
$\tau_e=1T_{1/2}$, A=15.049, B=.15										
T_o:	0.5	1.0	1.5	2.0	2.5	3.0	3.5	4.0	4.5	5.0
δ%:	48.0	36.8	32.4	30.1	28.8	28.0	27.5	27.1	26.9	26.8
$\tau_e=1T_{1/2}$, A=15.049, B=.30										
T_o:	0.5	1.0	1.5	2.0	2.5	3.0	3.5	4.0	4.5	5.0
δ%:	48.4	37.1	32.7	30.5	29.2	28.4	28.0	27.7	27.5	27.5
$\tau_e=1T_{1/2}$, A=15.049, B=.60										
T_o:	0.5	1.0	1.5	2.0	2.5	3.0	3.5	4.0	4.5	5.0
δ%:	49.2	37.8	33.5	31.3	30.0	29.3	28.9	28.7	28.7	28.7
$\tau_e=0.5T_{1/2}$, A=8.816, B=.15										
T_o:	0.5	1.0	1.5	2.0	2.5	3.0	3.5	4.0	4.5	5.0
δ%:	63.1	48.4	42.7	39.7	38.0	37.0	36.4	36.0	35.8	35.6
$\tau_e=0.5T_{1/2}$, A=8.816, B=.30										
T_o:	0.5	1.0	1.5	2.0	2.5	3.0	3.5	4.0	4.5	5.0
δ%:	64.0	49.2	43.5	40.6	38.9	38.0	37.5	37.2	37.1	37.1
$\tau_e=0.5T_{1/2}$, A=8.816, B=.60										
T_o:	0.5	1.0	1.5	2.0	2.5	3.0	3.5	4.0	4.5	5.0
δ%:	65.7	50.7	45.0	42.2	40.7	39.9	39.6	39.5	39.5	39.7

$T_{1/2}$, $0.5T_{1/2}$, respectively. The numbers of runs per year for these times are 10, 15, 29 and 54. The one-year errors including a systematic error of about 10% are 7.1%, 6.3%, 5.4%, 5.2%. Taking into account that the difference between the last two figures is small (= 0.2%) the optimal τ_e was taken to be $1T_{1/2}$. The total time of counting $t_o + t_1$ for this case is equal to 120 days and then we need 10 counting channels. The accuracy of the neutrino flux measurement for 10 years of experiment is 1.7%.

For a chlorine-argon detector the figures obtained in this way are really very close to what could be found in the experiment, because these figures were obtained by scaling the results of Davis's experiment. For a gallium-germanium detector the figures may be far from the ones obtained in the experiment, because at the present time we don't know what is reason for the persistent discrepancy between theory and experiment. The experiment can show the production rate to be 135 SNU if the flux of ^{8}B neutrinos is small due to lower temperature and there are no oscillations, but if neutrinos oscillate, the production rate for the gallium-germanium detector could be as much as 6 times lower, as was discussed earlier.[10] So if we take Q = 45 SNU, ε_e = 0.95, ε_c = 0.80 and make a similar calculation, then the optimal time τ_e will be $1T_{1/2}$ for the background B = 0.3, the error for one run including a 10% systematic error will be 52.6% and the one-year error - 9.8%. Then for 10 years of experiment σ = 3.1%.

Here is is worth mentioning that the results of Davis' experiment were examined for possible time variations due to the neutrino oscillation effect correlated with the annual variation in Earth-Sun distance[11] following the suggestion made by Pomeranchuk.[12] The short time of exposures which are considered here will make it easier to do this kind of examination also, because, firstly, there are more runs per year in this case and therefore one can detect shorter oscillation lengths, and, secondly, by long times of exposures an isotope born at the beginning of the exposure has a high probability to decay by the end of it, effectively decreasing the live time of the detector.

CONSIDERATION OF THE POSSIBILITY TO PROVE THAT NEUTRINOS ARE FROM THE SUN

Let us consider the possibility for the chlorine-argon telescope with 3000 t of C_2Cl_4 to detect the solar neutrino flux variation due to the annual changes in Earth-Sun distance. We assume here that there's no neutrino oscillation but that the neutrino flux just varies proportionally with $1/R^2(t)$, where $R(t)$ - Earth-Sun distance. In this case the production rate is described by

$$q(\phi) = 0.18 + 1.7(1+2e\cos\phi) \text{ at/day} \tag{6}$$

where $e = 0.01675$ - the eccentricity of the Earth's orbit, ϕ - the angle measured relative to the direction which corresponds to the shortest $R(t)$. The value 0.18 corresponds to the background from muons and the value 1.7 describes the mean production rate of ^{37}Ar by neutrinos, as was discussed earlier. If we take two integrals of (6), one corresponding to $2\Delta t$-interval around the direction of the "shortest" sun and the other around the direction to the "farthest" sun and then take the ratio of these values we get

$$R = \int_{-\Delta t}^{\Delta t} q(2\pi t/T)dt \Big/ \int_{T/2-\Delta t}^{T/2+\Delta t} q(2\pi t/T)dt = 1 + \delta \tag{7}$$

where T is a one-year period and δ is a quantity which depends on Δt. In the case considered here $\delta = 0.062$ if $\Delta\tau \to 0$. (If there's no background from muons the value $\delta = 0.069$ if $\Delta t \to 0$). If the duration of one run including the extraction time is t_r and the error of one measurement is σ then the error for each integral in (7) for t_m years years of experiment is

$$\sigma_i = \sigma / \sqrt{2\Delta t \cdot t_m / t_r}$$

Then the error for R, taking into account that the ratio R is close to 1 is

$$\sigma_R \simeq \sqrt{2}\, R\sigma_i = R\sigma / \sqrt{\Delta t \cdot t_m / t_r} \tag{8}$$

The optimal Δt corresponds to maximal ratio δ/σ_R. This ratio was calculated for different Δt for the chlorine-argon detector using the following input data: $\sigma = 0.356(35.6\%)$, $t_r = 18$ days, $t_m =$ 10 years. It was found that the maximal value of δ/σ_R is equal to 0.798 with $2\Delta t = 135$ days. If we take R = 1 in (8) then the ratio $\delta/\sigma_r = 0.837$. So if R = 1, i.e., there's no variation of the solar neutrino flux due to changes in Earth-Sun distance, then the probability that the experimental value $R < 1 + 0.837 \cdot \sigma_R$ is

$$P(R < 1 + 0.837 \cdot \sigma_R) = 80\%$$

if we assume that the value R has a normal distribution. This means that if, as a result of measurement, the obtained value of R turned out to be higher than the expected one of 1.049 (the chance of which is 50%) at $2\Delta t = 135$ days, then the statement could be made at the 80% confidence level that the detected neutrinos are from the sun. If the obtained value turned out to be lower than 1.049 then the collection of the statistics should be continued. (If the obtained value turns out to be less than 1.0 then at the 79% confidence level the statement could be made that neutrinos oscillate provided that the effect is from solar neutrinos).

A similar calculation was made for a gallium-germanium detector with 60 t of Ga metal. If we take as the input parameters Q = 135 SNU, $\varepsilon_e = 0.95$, $\varepsilon_c = 0.60$, $\tau_e = T_{1/2}$, $t_r = 12.4$ days, $\sigma = 0.337(33.7\%)$ then we get $\delta/\sigma_R = 1.12$ at the optimal value $2\Delta t =$ 135 days. If we take in (8) R = 1, then $\delta/\sigma_R = 1.18$. The corresponding confidence levels are 87% and 88%.

4. The integral method of treatment of the experimental data.

The result of a one-run measurement in the radiochemical solar neutrino experiment is a sequence of times of count detection. Because the number of counts is very small it's impossible to group the counts to get the characteristic picture of a decaying isotope plus some constant level of a background. If this were possible, then the subtraction of the background could be done by the usual fitting procedure with two parameters to be found: effect - a, and background - b. To restore the shape of the time distribution of the counts we used the integral representation in a way similar to the one in Kolmogorov's method. If the intensity of the counts is described by $b + a\exp(-\lambda \cdot t)$, then the integral of this value at the time t will be

$$F(t) = bt + a \cdot (1-\exp(-\lambda t))/\lambda \tag{9}$$

where b and a are, respectively, the number of background counts and the number of decays per day. The empirical integral intensity (the total number of counts at the time t) is a histogram and now the parameters b and a may be found by the fitting of this histogram by the curve (9). This fitting could be done by several ways, for example: by minimizing the maximal deviation of the curve from the corresponding point of the histogram, by minimizing the sum of the deviations of the curve from the histogram in the breakpoints of this histogram, etc. To get a balance between the total number of events for the whole experiment and the value $\sum_{i=1}^{n} F_i(T_i)$, where T_i - the time of counting, n - the number of runs, it was demanded that the condition

$$b_i T_i + a_i(1-\exp(-\lambda \cdot T_i))/\lambda = l_i \qquad (10)$$

be fulfilled for every run, where l_i - the number of counts.

To test the method simulation was made of times of count detection for several hundreds of runs for different value a and b. The data obtained in this way were analyzed using the integral method of treatment. The results of this analysis for every run are values a_i and b_i. Then the mean values and dispersions of a and b were calculated. A comparison was made of the obtained mean values and the input values in the simulation. The frequencies obtained for different approximations were compared with each other and with Poisson distribution for a and b used in the simulation.

The result of the measurement in a given run is the sequence of times of count detection. The aim of an analysis of this sequence is not only the values a_i and b_i, but also the corresponding errors. To solve this problem we did the following. If the result of the statistical treatment of a given run are the values a_i and b_i, a simulation is made, this time for a=a and b=b of several hundreds of runs. Then the calculated dispersions for a and b obtained as a result of a statistical treatment of the simulated data give the errors of a_i and b_i.

It was shown in ref. 13 that the frequency distribution for the number of ^{37}Ar-events should be broader than the corresponding Poisson distribution because of the influence of the background. This should be taken into account in an analysis of the experimental data. For example, in Davis's experiment there are five points with the effect equal to zero, while for the corresponding Poisson distribution for the total number of runs ≃ 50 the number of such points is practically zero. But this is not the case for a theoretical distribution which could be obtained as a result of a statistical treatment.

Fig. 1a shows: a) the theoretical distribution obtained by the method of maximum likelihood for the mean number of background events equal to 6 and ^{37}Ar-events equal to 6.53 which corresponds to the chlorine-argon detector with 610 t of C_2Cl_4, 300 days of counting and the parameters: Q = 0.42 at/day, $\tau_e = 2T_{1/2}$, $\varepsilon_c = 0.935$, $\varepsilon_c = 0.44$, B = 0.7 and b) the Poisson distribution for the mean value equal to 6.53. One can see in this figure that the theoretical distribution is broader, especially marked is the difference between these distributions for small number. For example the probability for a zero number of ^{37}Ar-events is equal to 0.034, i.e. from 50 runs approximately two runs may give zero ^{37}Ar-events. So from this point of view the presence of 5 zero points in Davis' experiment is quite possible.

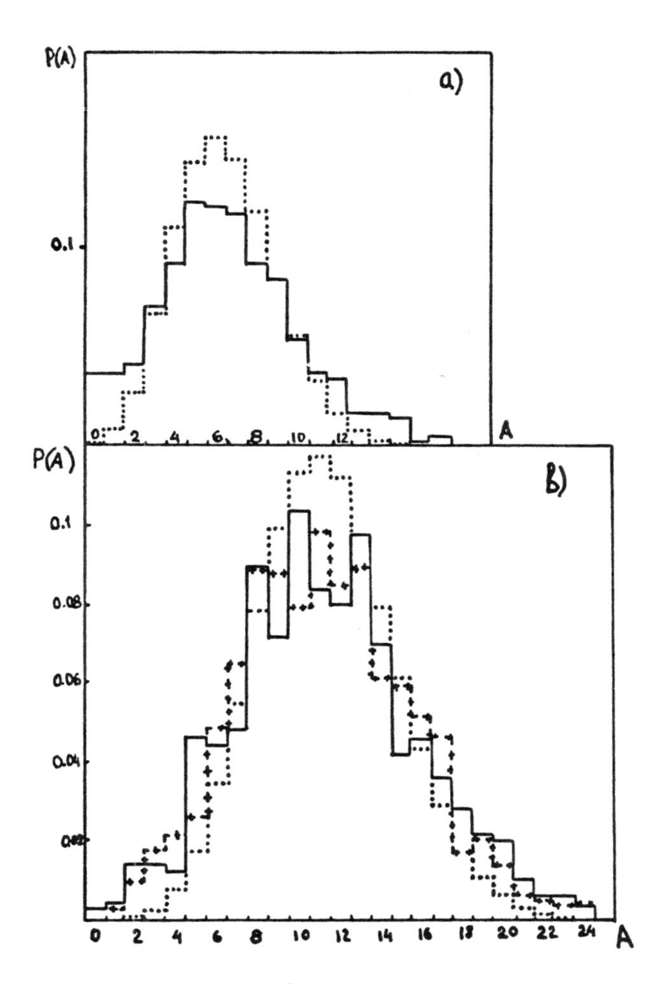

Fig. 1
Frequency distributions and corresponding Poisson distributions.
1a) points - Poisson distr., solid line - frequency distr.
1b) points - Poisson distr., solid line - frequency distr. for min(max Dev), crosses - frequency distr. for min(Dev).

Fig. 1b shows the results of a treatment of the computer simulated data for 3000 t of C_2Cl_4 for the following parameters: $Q = 1.88$ at/day, $\tau_e = 0.5\ T_{1/2}$, $\varepsilon_e = 0.935$, $\varepsilon_c = 0.44$, $B = 0.7$. Here we used the integral method of treatment. The approximation was made both by minimizing the maximal deviation (a) and by minimizing the sum of modules of deviations (b). A comparison is made with the corresponding Poisson distribution. One can see from this figure that in this case the theoretical distributions are broader than the Poisson distribution also, but because of the better statistics (the number of ^{37}Ar-events for 3000 t and $\tau_e = 0.5\ T_{1/2}$ is more than for 610 t of C_2Cl_4 and $T_e = 2T_{1/2}$) there is no plateau at a small number of ^{37}Ar-events. By simulation of the "experimental" data we used a = 0.226, b = 0.02. The mean values obtained as a result of treatment are a = 0.225, b = 0.02 for the approximation by min(max Dev) and a = 0.224, b = 0.02 for the approximation by $\min(\sum|Dev|)$. The corresponding errors are $\sigma_a = 38.8\%$, $\sigma_b = 64.7\%$, $\sigma_a = 38.2\%$, $\sigma_b = 62.8\%$. Approximately the same result was obtained using approximation by $\min(\sum(Dev)^2)$. It was shown in Table I that if $t_1 \gg t_o$ then the error $\sigma_a = 34.2\%$.

Fig. 2 shows an example of the integral representation of one run. Here one can see the empirical histogram and two curves, one

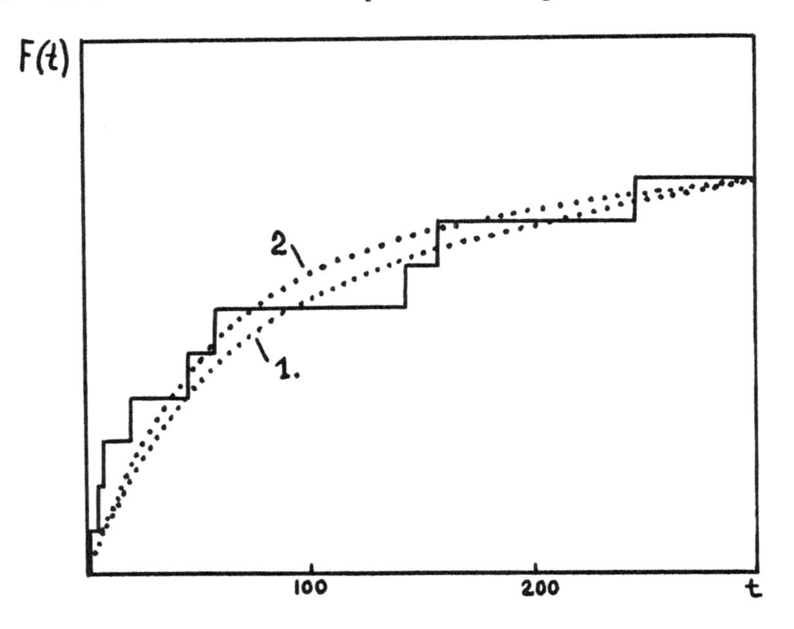

Fig. 2
Integral representation of data of one run.
1 - method of maximum likelihood
2 - integral method
F(t) - total number of events by the time t(days).

obtained by the method of maximum likelihood, the other by the integral method. Not for every run are these curves different. One can see from (9) that the slope of the curve at $t \gg T_{1/2}$ determines the background. If we approximate the curve at large t by a straight line and continue this line to the axis Y, then it will cross the axis at the point at which the ordinate Y_i is equal to the total number of ^{37}Ar-events, while the difference $l_i - Y_i$ is equal to the total number of background events. The general run of the histogram relative to the curve gives some indication of a likelihood of the obtained sequence. So this mode of representation of one-run-data is rather illustrative.

ACKNOWLEDGMENTS

The authors are grateful to G. T. Zatsepin, A. E. Chudakov and also to I. R. Barabanov for fruitful discussions and interest in this work.

REFERENCES

1. B. Pontecorvo, Chalk River Report P.D. - 205, 1946.
2. R. Davis Jr., D. S. Harmer, and K. C. Hoffman, Phys. Rev. Lett. 20, 1205 (1968).
3. R. Davis Jr., B. Cleveland, and J. K. Rowley, in "Workshop on Science Underground", AIP Conference Proc. No. 96, Los Alamos, Sept. 1982, published by AIP, New York, 1983.
4. J. N. Bahcall, Rev. Mod. Phys., 50, 881 (1978).
5. G. T. Zatsepin, A. V. Kopylov, and E. K. Shirokova, Sov. J. Nucl. Phys. 33(2), 200 (1981).
6. E. V. Fenyves, M. Cherry, et al., Proc. of 17th ICCR, Paris, Vol. 10, p. 317, 1981.
7. A. M. Aurela, Comput. Phys. Commun. 13, 281 (1977); A. M. Aurela, Comput. Phys. Commun. 17, 301 (1979).
8. B. T. Cleveland, Nucl. Instrum. and Meth. 214, 451 (1983).
9. H. Orihara, et al., Phys. Rev. 51, 1328 (1983).
10. G. T. Zatsepin, in Proc. of 9th Intern. Workshop on Weak Interactions and Neutrinos, Sept. 1982, Javea, Spain, Ed., A. Morales.
11. R. Ehrlich, Phys. Rev. D, 25, 2282 (1982).
12. V. Gribov and B. Pontecorvo, Phys. Lett. 28B, 493, 1969.
13. V. N. Gavrin and A. M. Kopylov, Pis'ma v Astronom. J. 10, 154, (1984).

THE MOLYBDENUM SOLAR NEUTRINO EXPERIMENT

K. Wolfsberg, G. A. Cowan, E. A. Bryant, K. S. Daniels,
S. W. Downey, W. C. Haxton, *V. G. Niesen, N. S. Nogar,
C. M. Miller, and D. J. Rokop

Los Alamos National Laboratory
Los Alamos, New Mexico 87544

The goal of the molybdenum solar neutrino experiment[1] is to deduce the ^{8}B solar neutrino flux, averaged over the past several million years, from the concentration of ^{98}Tc in a deeply buried molybdenum deposit. The experiment is important to an understanding of stellar processes because it will shed light on the reason for the discrepancy between theory and observation of the chlorine solar neutrino experiment. Possible reasons for the discrepancy may lie in the properties of neutrinos (neutrino oscillations or massive neutrinos) or in deficiencies of the standard solar model. The chlorine experiment only measures the ^{8}B neutrino flux in current times and does not address possible temporal variations in the interior of the sun,[2] which are also not considered in the standard model. In the molybenum experiment, we plan to measure ^{98}Tc (4.2 Myr), also produced by ^{8}B neutrinos, and possibly ^{97}Tc (2.6 Myr), produced by lower energy neutrinos.

We anticipate (see below) a capability of measuring 10^{7} atoms of ^{98}Tc by a mass spectrometric method. We will require about 2600 tons of ore, from a deeply buried deposit, which contain 13 tons of molybdenite, to obtain this number of atoms.[1] A problem, then, is how to chemically isolate a trace quantity of technetium for mass spectrometry from such a large quantity of material. Fortunately, some of the work is done for us in the commercial processing of molybdenum by the AMAX Corporation in the production of molybdenum

---

*Currently at University of Washington, Seattle; consultant, Los Alamos National Laboratory.

0094-243X/85/1260196-07 $3.00

oxide. Flotation separation at the Henderson molebdenum mine in Colorado produces a concentrate that is 90 to 99% molybedenite. The concentrate is shipped to the AMAX roasting plant at Ft. Madison, Iowa, for conversion of the MoS_2 to MoO_3. There are two roasters at the plant, each with a capacity of roasting as up to 50 tons of concentrate per day. So, our experiment requires only part of a day's production. A schematic of the plant is presented in Fig. 1. The MoS_2 concentrate has a residency time of about 8 hr in the roaster at temperatures from 740 to 615°C. The exhaust gas, containing principally SO_2 and SO_3, is treated for dust removal, cooled, scrubbed, and converted to sulfuric acid. The scrub operation removes fluorides, chlorides, some of the sulfur gases, selenium, and rhenium from the gas stream before it enters the sulfuric acid plant. In the scrub stream, the last traces of molybdenum dust are filtered out, selenium is precipitated with H_2S (probably as the metal) for environmental reasons, and lime is added to precipitate sulfates and sulfites.

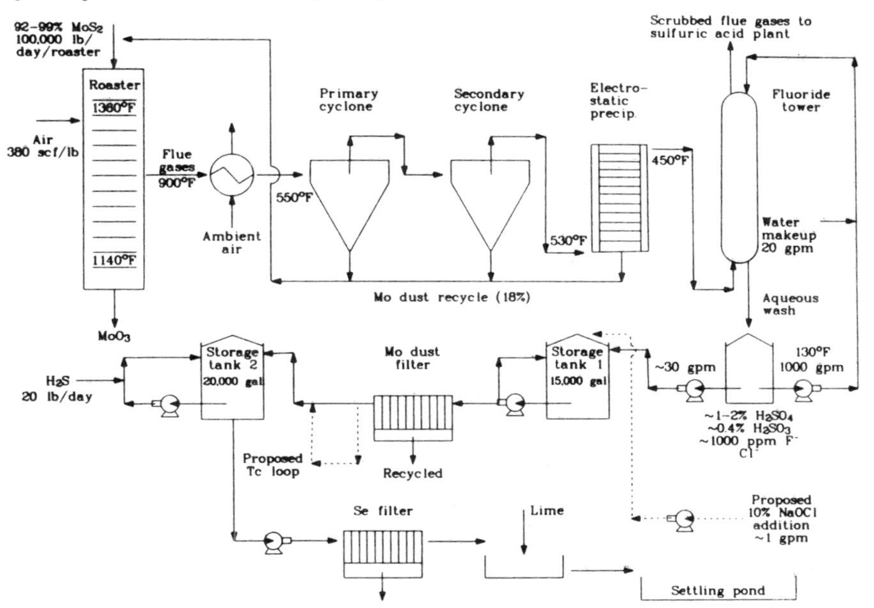

Fig. 1. Schematic of AMAX molybdenum roasting plant. Dashed lines are modifications for the technetium experiment.

To infer the fate of technetium in the process, we are currently relying largely on our measurements of its chemical homolog rhenium, which has similar properties. The boiling points of Re_2O_7 and Tc_2O_7 are 360 and 311°C, respectively; the redox potentials for the ReO_4^-/ReO_2 and TcO_4^-/TcO_2 couples are 0.510 and 0.738 volts, respectively. The MoS_2 concentrate fed into the roaster typically contains 7 ppm rhenium, whereas the MoO_3 product contains <0.1 ppm rhenium, indicating that rhenium is volatilized in the roaster. Molybdenum dust recycled from the precipitators contains about 6 ppm rhenium, indicating that there is no sink for that element in the hot gas exhaust line prior to the acid scrub (fluoride tower). Rhenium is condensed in the acid scrub and is found at concentrations of about 6 ppm both before and after the molybdenum dust filter and the selenium filter. Only about 10% of the rhenium is in the selenium filter; Re_2S_7 is not precipitated quantitatively because H_2S cannot be added in excess on account of of the large amount of H_2SO_3 in the scrub stream. A process had to be developed for removing technetium and rhenium from this acid scrub stream, preferably ahead of the H_2S addition. It was also desirable that this process cause minimum disruption in the commercial operation of the plant.

Ion exchange appeared to be a good choice. Commercial columns of the type used for deionizing or softening water are rated for flow rates of up to 30 ℓ/min. Perrhenate and pertechnetate are strongly absorbed on anion-exchange resin, and our experiments showed that many commercial resins work well. However, technetium does not sorb well out of H_2SO_3 solution, probably because it is reduced to a neutral or cationic species; rhenium still absorbs strongly. We solved the problem by oxidizing the solution with NaOCl, the addition of which is acceptable to the AMAX plant management.

Exploratory experiments were conducted with columns of Duolite A-162 resin, 0.7 cm diameter by 95 cm long, with flow rates of 25 mℓ/min; the linear flow rate is equivalent to one of 26 ℓ/min for a commercial 23-cm diameter column. Rhenium breakthrough curves and distribution on the column were measured. Technetium should be absorbed more strongly than rhenium. Laboratory experiments with H_2SO_4-H_2SO_3 solutions oxidized with NaOCl, indicated that 98% of the rhenium could be retained for a half-day run. We did a similar experiment at the plant, using actual plant-stream solution and found that 87% of the rhenium was captured on the column. The poorer, but acceptable, performance of the column under real conditions is caused by competition of molybdate for sorption sites and elution by molybdate. We have now designed the modifications at the plant required to remove technetium and rhenium from the acid scrub stream. These are indicated by the dashed lines in Fig. 1 for the addition of NaOCl to a storage tank already in the line and a by-pass loop to be constructed for the ion-exchange columns. The plan for this loop, shown in Fig. 2, allows for the interception of all or part of the stream through anion-exchange columns.

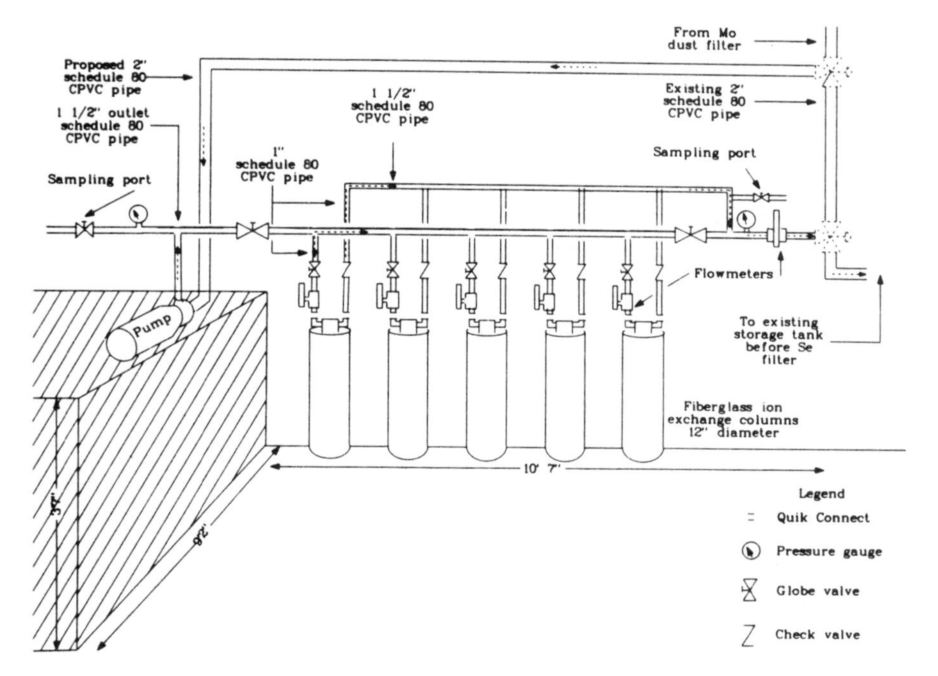

Fig. 2. Proposed technetium bypass loop.

Our plan for a large-scale experiment is to return to Los Alamos with these columns to continue purification of the technetium. The next step is to recover the technetium from the ~100 ℓ of resin in a relatively small volume. Pertechnetate has a very high distribution coefficient on anion-exchange resin, and we invesigated methods for stripping technetium that involved reduction to non-sorbing species. It appears that once TcO_4^- sorbs on the resin it is very difficult to reduce, and none of the reducing agents we investigated was successful. Nitric acid is effective as a stripping agent, but it would be unacceptably hazardous on a large scale. At the present time we propose to ash the resin at 420°C for ~1 day. Tests have shown only ~10% loss of technetium, and the residue is soluble. Additional purification steps prior to mass-spectrometric measurement are being developed. We will still have to develop methods to separate the technetium, with high recovery, from rhenium (~100 g), molybdenum (~1 kg) (An overall decontamination factor of 10^{21} is required!), and other impurities such as arsenic.

We are pursuing the investigation of both thermal-ionization and laser resonance-ionization methods for mass spectrometric measurement of the technetium at the levels required. The principal problems are to achieve high ionization efficiencies and acceptably low levels of interference from molybdenum isobars.

For thermal ionization, we start with filaments made from zone-refined rhenium metal and heat them at 2300°C for 30 to 50 min. Filaments that exhibit low molybdenum signals at 1900°C are hand picked from these. To date the resin-bead method has given the best results both for high ionization efficiency and for low isobaric interference from the sample itself. In the last chemical manipulation, technetium is loaded from NH_4OH onto a single, ~200-μm bead of anion-exchange resin bead that is subsequently washed with HCl. The bead is then put onto the rhenium canoe filament for mass spectroscopy in a tandem-magnet mass spectrometer. At the present time we can typically achieve ionization efficiencies of 8×10^{-5}

for samples containing 50 pg (3×10^{11} atoms) of ^{99}Tc, although efficiencies as high as 7×10^{-2} have been achieved. To do the experiment, we will have to be able to routinely achieve an efficiency of about 5×10^{-4} to 10^{-3}; this will give a count rate of 10 to 20 counts per second above a similar background. We are investigating the bead loading and filament parameters involved in this process and also schemes such as negative thermal ionization, which will produce TcO_4^- but no similar molybdenum ion.

Resonance ionization is a laser-based ionization technique that allows discrimination against unwanted atoms in the sample.[3] For technetium, a variety of two-color, three-photon resonance ionization schemes have been explored that greatly reduce the problems associated with molybdenum contamination.[4] To date, however, insufficient efficiency has been demonstrated for the solar neutrino measurements; efforts in this phase of the work are also presently centered on increasing the ion yields.

Technetium-99 will be present at several orders of magnitude greater than ^{98}Tc as a spontaneous fission product of the ^{238}U in the molybdenite. Demonstration of secular equilibrium is a requirement of the experiment. The ^{99}Tc will be traced for overall yield and mass spectrometry with artificially produced ^{97}Tc. The known concentration of ^{99}Tc can then be used as a tracer for ^{98}Tc in the large experiments.

ACKNOWLEDGEMENT

This work is performed under the auspices of the U.S. Department of Energy.

The management of the AMAX Molybdenum Division of the Climax Molybdenum Company have been most gracious in allowing us to pursue the possibility of this experiment. In particular we are indebted to T. Kearns, R. H. Cale, and their staff at the Ft. Madison, Iowa plant for much help and advise.

REFERENCES

1. G. A. Cowan and W. C. Haxton, "Solar Neutrino Production of Tecnetium-97 and Technetium-98," Science 216, 51-54 (1982).

2. G. A. Cowan and W. C. Haxton, "Solar Variability, Glacial Epochs, and Solar Neutrinos," Los Alamos Science 3, No. 2, 46-57 (1982).

3. C. M. Miller, W. R. Shields, D. J. Rokop, A. J. Gancarz, and N. S. Nogar, "Laser-Based Resonance-Ionization Mass Spectrometry," in "Isotope and Nuclear Chemistry Division Annual Report FY1982," Los Alamos National Laboratory report LA-9797-PR, 19-21 (June 1982).

4. S. W. Downey, N. S. Nogar, and C. M. Miller, "Resonance Ionization Mass Spectrometry of Technetium," Int. J. Mass Spec. and Ion Physics (in press, 1984).

THE ^{205}Tl EXPERIMENT

W. Henning and W. Kutschera
Argonne National Laboratory, Argonne, IL 60439, USA

H. Ernst, G. Korschinek, P. Kubik, W. Mayer, H. Morinaga,
E. Nolte and U. Ratzinger
Physik Department, Technische Universität München,
8046 Garching, W. Germany

M. Müller and D. Schüll
GSI Darmstadt, 6100 Darmstadt, W. Germany

ABSTRACT

^{205}Tl has been previously proposed as a geological detector for solar neutrinos, making use of the reaction ^{205}Tl(ν,e$^-$)^{205}Pb with a neutrino threshold of only ≈43 keV. We report on an experiment performed to study the feasibility of detecting radioactive ^{205}Pb nuclei ($T_{1/2}$ = 15 million years) at very low concentrations using the recently developed technique of accelerator mass spectrometry. Employing the high-energy ion beams of good quality from the UNILAC heavy-ion accelerator at the GSI Darmstadt, we are able to demonstrate a suppression of neighbouring isotopes to better than 1 in 10^{16} and of neighbouring elements to about 1 in 10^3. While these results are very encouraging, the minimum number of atoms detectable is still severely limited by the efficiency of producing multiply-charged ions from present ion sources. Future improvements in ion-source performance are briefly discussed.

INTRODUCTION

^{205}Tl was proposed as a geological solar-neutrino detector eight years ago by M. Freedman and his collaborators at Argonne.[1] Its attraction comes from an extremely low neutrino threshold (≈43 keV) for the ^{205}Tl(ν,e$^-$)^{205}Pb reaction,[2] and from the long half-life of the ^{205}Pb radioisotope ($T_{1/2}$ = 15 million years). A partial level scheme is shown in Figure 1. The low neutrino threshold provides sensitivity to the low-energy part of the solar neutrino spectrum which is dominated by the well-known hydrogen burning reaction $p + p \rightarrow d + e^+ + \nu$ + 430 keV. This leads to a ν-capture rate up to a 100 times higher than that from the ^{37}Cl(ν,e$^-$)^{37}Ar reaction.[3] The long half-life has two attractive consequences: i) collecting radioactive ^{205}Pb nuclei over times comparable to their half-life allows accumulation of a large number of ^{205}Pb atoms which can be more easily detected. ii) The long collection time averages over possible short-time fluctuations in the solar neutrino flux.

These properties make ^{205}Tl a potentially interesting solar neutrino detector with features complementary to ^{37}Cl and ^{71}Ga. In particular, if the ^{71}Ga experiment were to reveal a solar neutrino

0094-243X/85/1260203-09 $3.00

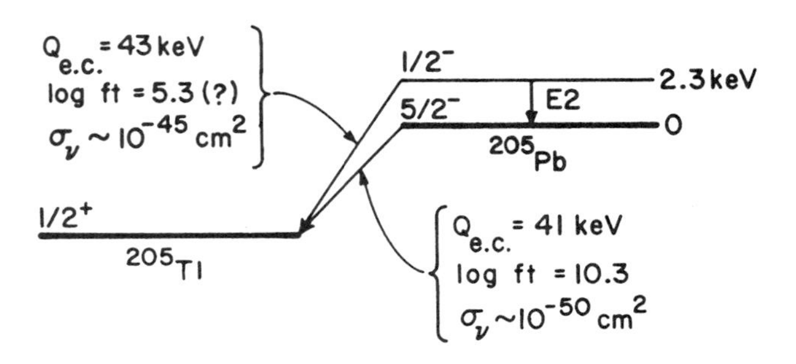

Fig. 1. Partial level scheme[4] for ^{205}Tl and ^{205}Pb with electron-capture Q-values from reference 2.

deficit similar to the ^{37}Cl experiment, ^{205}Tl might help to distinguish between the possible causes of either short-term fluctuations in the solar neutrino flux or neutrino oscillations and decay.

However, a ^{205}Tl experiment has several severe problems which need to be solved before it can be useful. These regard i) questions of a suitable geological site with known history, ii) possible background nuclear reactions producing ^{205}Pb at such a site, iii) the unknown log ft value for the first excited state at 2.3 keV excitation energy in ^{205}Pb which, with high probability, dominates the neutrino absorption reaction, and iv) the difficulties in determining the ^{205}Pb content in a geological sample.

The first two, in particular, have been investigated in detail by M. Freedman.[4] He has proposed what seems to be a suitable geological site, a mine in Macedonia, where the Tl mineral lorandite is rather abundantly found with low (few ppm) Pb contamination. For this particular location, background reactions were calculated to contribute less than 20% to the ^{205}Pb yield. Considerably more uncertainty arises from the unknown log ft value. Of some promise is the recent success in relating forward angle (p,n) cross sections at high incident energies to Gamow-Teller decay strengths,[5] although for a $\Delta\ell = 1$, first-forbidden transition such a relation has not yet been established. The present experiment was directed at the last question, the detection and identification of the ^{205}Pb atoms. Our recent involvement in the technique of accelerator mass spectrometry (AMS) which allows the detection of radioisotopes at extremely low concentrations suggested to us the application of this technique to ^{205}Pb detection.

ACCELERATOR MASS SPECTROMETRY OF ^{205}Pb

In a geological ^{205}Tl experiment rather involved physical and chemical procedures will be used to separate the Tl mineral lorandite (~10-100 kg) from the mining ore (a few tons), and then to extract the Pb fraction, which, anticipating a few ppm Pb contamination in the lorandite, will result finally in a sample of up to 1 gm Pb containing 10^6- 10^7 ^{205}Pb atoms. Thus, we expect a Pb sample with a $^{205}Pb/Pb$ concentration of the order 10^{-13} to 10^{-14} and possibly a $^{205}Pb/^{205}Tl$ concentration of 10^{-3} to 10^{-5}. The latter is expected if chemistry can reduce Tl to ppb levels in the Pb sample. The use of 99.9% enriched ^{203}Tl as a diluent in a second stage of chemical processing should help in reducing the ^{205}Tl isotope concentration to the required low level.

The recently developed technique of AMS is now routinely used to measure very low concentrations ($\leq 10^{-14}$) of light radioisotopes in various applications e.g., using ^{10}Be, ^{14}C and ^{36}Cl.[6] However, for the heaviest radioisotopes this is much more difficult and has not yet been achieved. Very briefly, in AMS a high-energy heavy ion accelerator is used to accelerate the radioisotope to energies of the order of MeV/nucleon at which point ions can be unambiguously identified and individually counted by mass and, in particular, by nuclear charge Z. The identification is greatly facilitated by using standard instrumentation and established techniques from nuclear physics heavy ion research or by expanding them. The field of AMS has grown rapidly over the past few years. An overview can be obtained from the proceedings of the three international symposia held on this topic.[6]

To attempt identification of ^{205}Pb atoms by AMS, a large heavy-ion accelerator is necessary in order for the ion to be accelerated to sufficiently high energies where nuclear charge determination via differential energy loss becomes possible. The only such accelerator presently in operation and capable of accelerating the heaviest

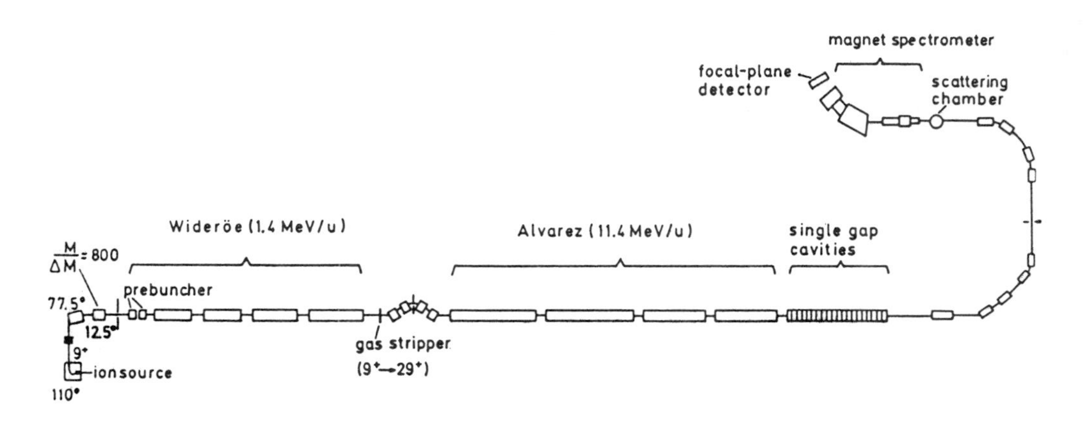

Fig. 2. Schematic layout of the UNILAC accelerator at GSI Darmstadt, with emphasis on those components that are of particular importance in the present accelerator mass spectrometry measurements.

nuclei with good beam characteristics is the UNILAC at the GSI research center in Darmstadt. In the following we briefly describe our AMS measurements using this system.

Figure 2 shows schematically the major components of the UNILAC and the beam transport system used in the present studies. The present system can be viewed as consisting of 3 major components, each of which contributes in a specific way to suppression of unwanted ion species at the final detector: i) the ion-source and injector system with charge-state and mass selection reduce the neighbouring stable isotopes $^{204,206}Pb$, ii) the rf accelerator and beam transport system act as an efficient mass and ion-optical filter in addition to accelerating ions to the necessary kinetic energy, and iii) the heavy-ion detection system allows mass and nuclear charge measurements.

The ion source is a positive-ion penning source pulsed at 50 Hz with a sputter target for the production of the desired ion species. Figure 3 shows a schematic of the source which was originally developed at Dubna.[7] Ar was used as discharge gas during our experiments. The source was mounted in a 110° magnet. The magnetic field was used for both the penning discharge and the separation of the 9^+ charge state. After preacceleration to 2 MeV the beam was analyzed by a 77.5° and 12.5° magnet with a mass resolution of $M/\Delta M \approx$ 800 at mass 200.

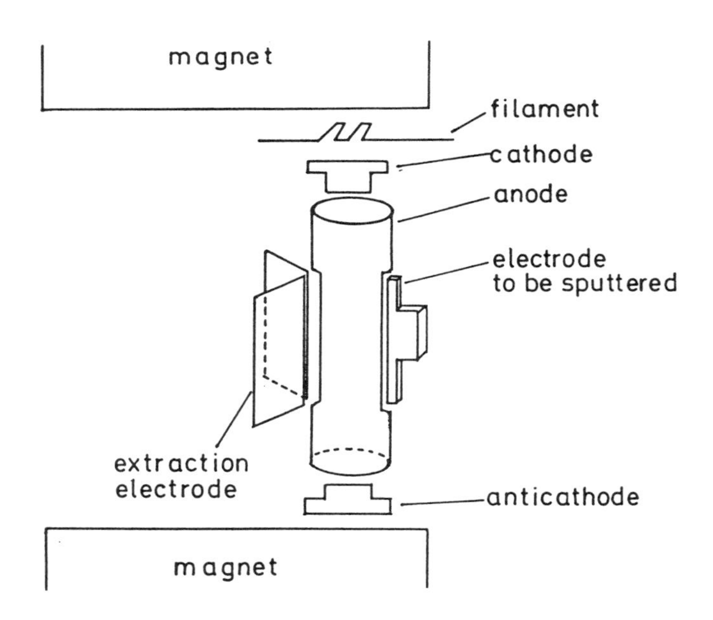

Fig. 3. Positive-ion penning source with sputter target.

After injection into a system of four Wideröe tanks an energy of 287 MeV was reached. In the following section the ions were stripped in a gas stripper, and an achromatic charge analyzing system selected the 29^+ charge-state for further acceleration. A final energy of 2.3 GeV was reached in the Alvarez tanks. One of the 20 single gap resonators was used as a debuncher, resulting in a reduction of the energy spread from 0.3% to 0.1%. A highly dispersive 180° beam-line system ($p/\Delta p \simeq 1000$) transported the beam to the target chamber of the GSI magnetic spectrometer.[8]

For mass and nuclear charge identification the magnetic spectrometer was used in conjunction with a gas absorber positioned in the target chamber in front of the spectrometer and a high resolution detector system as shown in Fig. 4. The flight time of the ions through the spectrometer ($\simeq$ 6.4 m path length) was measured between a channel plate start detector at the spectrometer entrance slits and a scintillator foil stop detector in front of the focal plane ionization chamber. The ion chamber provided energy loss and position signals. The elemental (Z) resolution was mainly achieved by a residual energy measurement of the ions following their traversal of the absorber. The residual energy was determined independently from both time-of-flight and magnetic rigidity measurements, with auxiliary gating on the other detector signals. The gas cell (14 cm long) was filled with a few hundred mbar of CH_4 or isobutane. The expected energy-loss difference between ^{205}Tl and ^{205}Pb was calculated from the Northcliffe-Schilling tables[9] to be about 18 MeV for an energy loss of ~ 1 GeV at an incident energy of 2.3 GeV.

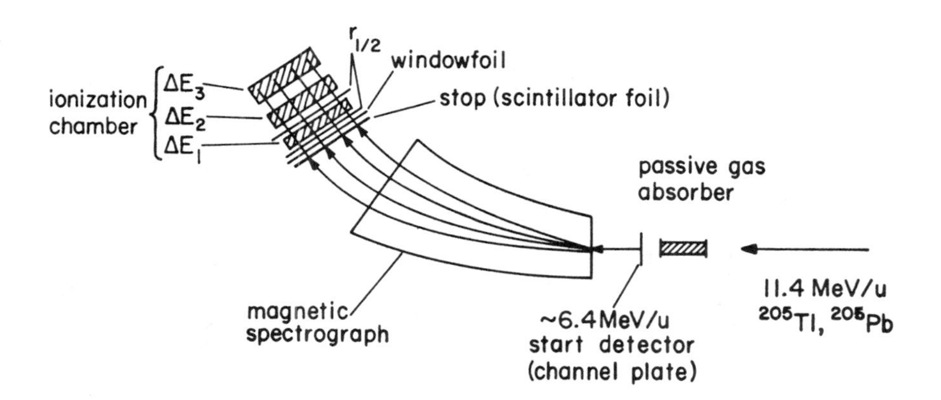

Fig. 4. Schematic of the heavy-ion detection system used in the present measurements. Isobaric ^{205}Pb and ^{205}Tl ions of identical energy experience different energy losses in a gas absorber due to their different nuclear charges. This energy-loss difference is determined by the magnetic spectrograph in a high-resolution measurement of the remaining energy.

RESULTS

In two initial experiments at 2.3 GeV incident energy we investigated the isotopic separation between ^{205}Pb and its stable neighbouring isotopes $^{204,206}Pb$, as well as the elemental separation between ^{205}Pb and ^{205}Tl. Figure 5 shows a mass spectrum obtained from time-of-flight and magnetic rigidity measured in the spectrograph. During this run a mixture of natural thallium and lead was used in the source. First the full accelerator system plus spectrograph were carefully tuned for the ^{205}Tl isotope. Then the extraction magnet at the ion source and the injection system for the UNILAC was tuned to mass 206, leaving the rest of the accelerator and beam line system optimized for mass 205. The source currents during these measurements were about 40 nA $^{206}Pb(9^+)$ and 40 nA $^{205}Tl(9^+)$, respectively. For this latter setting, we measured a count rate for ^{206}Pb ions of 1 Hz and for ^{205}Tl ions of 45 Hz in the spectrograph focal plane detector. Using the previously determined transmission from ion source to detector and comparing these count rates the following suppression between neighbouring isotopes is obtained: injection system $\sim 2*10^{-6}$, accelerator and beam-line system $\sim 4*10^{-8}$. This results in a total suppression between neighbouring masses of $\sim 8*10^{-14}$. Time-of-flight and magnetic-rigidity measurement in the

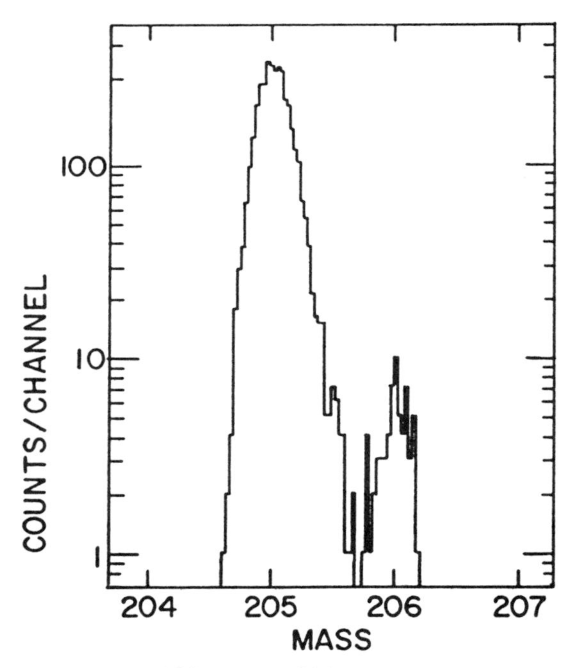

Fig. 5. Mass Spectrum of ^{205}Tl and ^{206}Pb ions as determined from the time-of-flight and magnetic-rigidity signals from the spectrograph focal plane detector.

spectrograph provided for another factor of ~ 10^{-3} resulting in a total suppression between neighbouring masses near A ≃ 200 of 1 in 10^{16}.

In the second experiment a source sample containing ^{205}Pb and ^{205}Tl in a ratio of ~1:10 was used (the actual sample composition was $^{nat}Pb:^{nat}Tl:^{205}Pb = 1:10^{-3}:10^{-4}$). Taking into account that ^{205}Tl and ^{205}Pb have the same magnetic rigidity and therefore identical energy before they pass through the absorber, we can extract from our data (Fig. 6a) a nuclear charge resolution for Z ≃ 82 of $\Delta Z/Z \simeq 10^{-2}$. Making use of the additional information contained in the time-of-flight, energy-loss and total energy signals from the focal plane

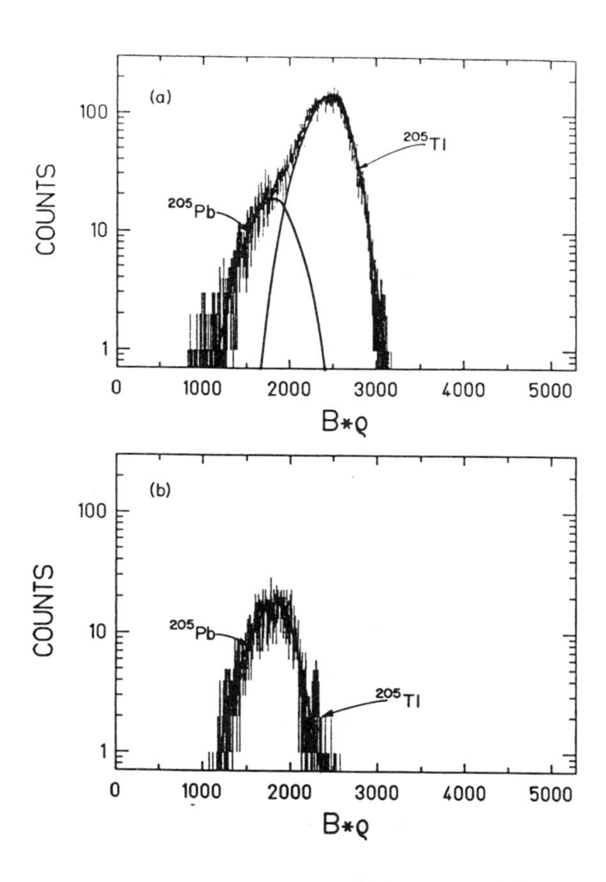

Fig. 6. Magnetic rigidity spectra for ^{205}Pb and ^{205}Tl ions as measured in the spectrograph focal plane detector: a) Total ungated spectrum; b) Spectrum gated with conditions on the various other detector signals for the selection of ^{205}Pb ions. The isobutane gas pressure in the absorber cell was 350 mbar.

detector we are able to achieve a suppression of ^{205}Tl with respect to ^{205}Pb of approximately 1 in 10^3 (Fig. 6b). The fraction of ^{205}Pb ions lost in the gating procedures amounts to less than 8%.

SUMMARY AND CONCLUSION

The suppression of neighbouring isotopes is quite adequate if compared to the requirements defined at the beginning of this paper. On the other hand better charge separation is desirable to further reduce the ^{205}Tl background and to increase the accuracy of the ^{205}Pb identification. This can presumably be achieved by optimizing the absorber thickness with respect to energy straggling versus separation of neighbouring nuclear charges, and by increasing the beam energy. Improving the ΔE signals from the ionization chamber seems also possible by optimizing the gas pressure in the counter.

The major problem identified in our mesurements is the very low efficiency of producing Pb^{9+} ions in the present ion source (10^{-6} to 10^{-5}). The transmission of the accelerator and beam-line system and the detection efficiency contribute another factor of $\sim 10^{-3}$ yielding a total efficiency of 10^{-9} to 10^{-8}. Since only a limited number of ^{205}Pb atoms will be available ($\sim 10^7$), the efficiency of the ion source at present is unacceptable. A solution of this problem may be the preacceleration of a low charge state ion (1^+) in conjunction with a stripping process before entering the main accelerator. The production of this ion beam should be attainable with an efficiency of up to 20%. Preacceleration to an energy of 6 MeV can be achieved for example with a Van de Graff accelerator but other preacceleration systems may be used. A stripping efficiency of $\sim$ 30% for the 8^+ charge state yields a total efficiency of 1.5%, taking into account that only 25% of the beam is accepted by the UNILAC due to its duty cycle.

Another possibility is the use of the recently developed electron cyclotron resonance (ECR) source which provides high charge states with good efficiency.[10] The most attractive feature of these sources seems to be the fact that due to resputtering from the walls of the essentially closed plasma cavity, close to 100% of the original source sample will eventually be emitted as ions.[11] The time needed for this "clean-out" seems to be quite tolerable, typically a few hours to a day. After such a modification of the injector the overall efficiency may allow the detection of the expected 100 events in the spectrometer from about 10^7 ^{205}Pb ions in the ion source sample. Assuming a neutrino capture rate of 430 SNU,[1] this corresponds to measuring neutrino-produced ^{205}Pb atoms from 100 kg of lorandite ($TlAsS_2$) or 20 tons of ore.

From our measurements and the possible improvements, we conclude that accelerator mass spectrometry with a large heavy-ion accelerator seems capable of providing sufficient resolution and sensitivity for identification of neutrino-produced ^{205}Pb. Since, however, both the hitherto unknown neutrino capture cross section and a possible lower solar neutrino flux could result in a considerable lower SNU value, an overall efficiency greater than the estimated 10^{-5} would be desirable. It may not be unreasonable to assume that future developments in ion source, accelerator and detector technology will eventually bring the efficiency up to the 10^{-3} range, already routinely achieved for accelerator mass spectrometry of lighter ions such as ^{14}C.

Operation of a large accelerator like the UNILAC for an experiment in accelerator mass spectrometry is a very difficult task. We would like to express our appreciation to Dr. N. Angert and the operating staff of the UNILAC accelerator for their continuous support and their patient endurance of a multitude of requests from sometimes not so patient experimenters.

This work was supported in part by the U.S. Department of Energy under contract number W-31-103-ENG-38 and in part by the Bundesministerium für Forschung und Technologie, Bonn.

REFERENCES

1. M. S. Freedman, C. M. Stevens, E. P. Horwitz, L. H. Fuchs, J. S. Lerner, L. S. Goodman, W. J. Childs and J. Hessler, Science 193, 1117 (1976).
2. J. G. Pengra, H. Genz and R. W. Fink, Nucl. Phys. A302, 1 (1978).
3. J. N. Bahcall, Rev. Mod. Phys. 50, 881 (1978).
4. M. S. Freedman, Proc. Int. Conf. on Stat. and Future of Sol. Neutr. Res. (G. Friedlander, ed.) Vol. 1, 313, BNL Rep. 50879; M. S. Freedman, Bull. Inst. Chem. Res., Kyoto University, Vol. 57, No. 1, 117 (1979).
5. C. D. Goodman, see contribution to this conference.
6. Proc. First Conf. on Radiocarbon Dating with Accelerators, 1978 (Univ. of Rochester); Proc. of the Int. Conf. on Accelerator Mass Spectrometry, 1981 (Argonne National Laboratory Report); Proceedings of the 3rd Intl. Symposium on Accelerator Mass Spectrometry, 1984 (Zürich), to be published in Nucl. Instr. and Meth. B.
7. A. S. Pasyuk and Y. P. Tretyakov, Proc. of Sec. Intern. Conf. on Ion Sources, (F. Viehböck, H. Winter, M. Bruck, ed.), Vienna, 512 (1972).
8. D. Schüll, F. Pühlhofer, B. Langenbeck, K. Blasche and Th. Walcher, Proc. 6th Conf. Magn. Techn., Bratislava 1977.
9. L. C. Northcliffe and R. F. Schilling, Nucl. Data Table A7, (1970).
10. See for example: R. Geller, B. Jacquot and R. Pauthenet, Rev. Phys Appl. 15, 995 (1980); R. Geller and B. Jacquot, Nucl. Instr. and Meth. 219, 1 (1984).
11. R. Geller, private communication.

^{163}Dy as a Solar Neutrino Detector

C.L. Bennett*
Lawrence Livermore National Laboratory
Livermore, California

ABSTRACT

The possibility of using ^{163}Dy as a low threshold solar neutrino detector is discussed. Solar neutrino absorption cross sections are calculated, and expected capture rates presented.

INTRODUCTION

With respect to nuclear structure, perhaps the most favorable reaction for detecting low energy neutrinos is ^{163}Dy(ν_e,e)^{163}Ho. The reason for considering ^{163}Dy in particular, is that it has the lowest threshold for neutrino reactions of all stable isotopes. Unfortunately, it is probably the most difficult reaction with respect to the chemical separation required. Since some of the relevant nuclear properties for this transition have recently been measured with high precision by our group[1] in connection with a search for neutrino mass effects in the decay of ^{163}Ho, it is worthwhile to consider whether using ^{163}Dy as a solar neutrino detector might be of interest.

The nuclear structure of the ^{163}Dy to ^{163}Ho transition is quite favorable. The ground state of ^{163}Dy is the lowest member of a neutron $5/2^-$[523] rotational band, while the ^{163}Ho ground state is the lowest member of a proton $7/2^-$[523] rotational band. Thus the two ground states differ essentially only by a different coupling of the odd particle's spin to the same underlying intrinsic state, and the fact that the odd particle is a proton in the one case, and a neutron in the other. As a result, the electron capture is a so called allowed unhindered transition, and the nuclear matrix elements are large. Because the spins of the excited states in the ^{163}Ho ground state band are all greater than 7/2 none of them can be populated via an allowed transition. The lowest excited state beyond the ground state which can be reached by an allowed transition is the $5/2^-$ state at 527 keV excitation. This state is beyond reach of the pp neutrinos, and has a negligible cross section for pep neutrinos on a kinematic basis alone. Furthermore, since this state is not a member of the ground state $7/2^-$[523] rotational band, the intrinsic cross section for neutrino capture from the target ^{163}Dy $5/2^-$[523] state will be negligible. The same effects will apply to any of the higher excited states as well, because there is only one allowed unhindered state in ^{163}Ho which can be reached by solar neutrinos, and that is the ground state itself.

MEASUREMENTS

The two recent measurements which are relevant for estimating the solar neutrino cross section are the total half-life[2]

$$T_{1/2}=4570\pm 50 \text{ years (90\% confidence level)} \quad (1)$$

and the ratio of the N to M capture rate[3]

0094-243X/85/1260212-04 $3.00

$$\lambda_N/\lambda_M = 3.82 \pm 0.4 \quad (2)$$

These two measurements can be combined to yield the electron capture Q_{EC} value and the ft value for the transition,

$$Q_{EC} = 2564 \pm 50 \text{ eV} \quad (3)$$

$$ft_{1/2} = 79{,}900\text{sec} \quad (4)$$

The various factors which enter in determining (3) and (4) are given in table I. The notation used is that of ref. 4. It is interesting to note that the corresponding ft value of 86,600sec for ^{161}Ho is very similar to the value of 79,900sec for ^{163}Ho.

One of the uncertainties in deriving the Q_{EC} and ft values from the experimental measurements is the absence of a calculation of the overlap and exchange correction factors for some of the outer shell orbits in dysprosium. The values assumed in deriving (3) and (4) (Table I) are based on interpolation and extrapolation of values from ref. 4. In particular, for the most important unknown overlap and exchange factors, i.e. those for the $3p_{1/2}$ and $4p_{1/2}$ orbitals, it was assumed that both the average of the 3s and $3p_{1/2}$ factors, and the average 4s and $4p_{1/2}$ factors are unity. This assumption leaves much to be desired, but is probably better than using equal factors for s and $p_{1/2}$. In any case, the uncertainty introduced in the total phase space factor for electron capture in ^{163}Ho is only at the level of 1%.

The cross section for neutrino capture in ^{163}Dy is given by[5]

$$\sigma(\nu_e, e) = 2.629\times10^{-41}\text{cm}^2 p_e W_e F_0 L_0 C(W_e)_\nu/(ftC(W_e)_e) \quad (5)$$

where p_e and W_e are the emerging electron's momentum and energy in atomic units, and F_0L_0 is the Fermi function for the ^{163}Ho + ν_e reaction products.

Using our deduced ft value, and correcting for the spin statistics factors, the neutrino capture cross section becomes

$$\sigma(\nu_e, e) = 4.4\times10^{-46}\text{cm}^2 p_e W_e F_0 L_0 \quad (6)$$

With these expressions, approximate average cross sections for some of the more important reactions producing neutrinos in the sun have been calculated, and are listed in table II. For the standard solar model, the capture rate in ^{163}Dy is about 850 SNU's. This is the largest capture rate of any of the isotopes so far considered. The amount of ^{163}Dy needed for one capture per day is about 3.7 Tons. In the form of Dy_2O_3, the amount needed for one capture per day is 17 Tons, or 2.2 m^3.

The equilibrium concentration of ^{163}Ho in natural Dy_2O_3 from the solar neutrino flux alone would be about 1 atom per cm^3. However, since holmium and dysprosium will inevitably be present together geologically, the (p,n) reaction on ^{163}Dy { and the (n,γ) reaction on ^{162}Er} would produce a far larger abundance of ^{163}Ho than the solar neutrinos, and since the naturally occurring actinides are likely to be associated with the lanthanides, there would be no shortage of low energy protons and neutrons. This fact eliminates the possibility of using ^{163}Dy in a geochemical type of experiment.

Finally, with regard to the separation of a few atoms of ^{163}Ho from 17 tons of dysprosium oxide: This is not easy[6]. Although the amount of dysprosium needed is not in excess of the annual production by the rare earth industry, it is a significant fraction. At this stage it is too early to say much about the feasibility of extracting the neutrino generated ^{163}Ho. Suffice it to say that it does not appear totally absurd. It is worthwhile pointing out that in the final detection stage, the resonance ionization schemes being applied to the detection of ^{81}Kr in the ^{81}Br based experimental proposal are far more readily applied to the detection of ^{163}Ho, in fact the first pure rare earth compound laser[7] was based on HoF_3. This could be a great advantage since such a laser would be intrinsically resonant with a holmium atomic transition.

*This work was performed under the auspices of the U.S. Department of Energy by Lawrence Livermore National Laboratory under contract No. W-7405-Eng-48.

REFERENCES

1. C.L. Bennett, A.L. Hallin, R.A. Naumann, P.T. Springer, M.S. Witherell, R.E. Chrien, P.A. Baisden, D.H. Sisson, Phys. Lett. 107B,19 (1981)
2. P.A. Baisden, D.H. Sisson, S. Niemeyer, B. Hudson, C.L. Bennett and R.A. Naumann, Phys. Rev. C28,337 (1983)
3. F. Hartmann, private communication and Ph.D. Thesis (Princeton University, 1985)
4. W. Bambynek, H. Behrens, M.H. Chen, B. Crasemann, M.L. Fitapatrick, K.W.D. Ledingham, H. Genz, M. Mutterer and R.L. Intemann, Rev. Mod. Phys. 49,77 (1977)
5. H. Behrens and W. Buhring, Electron Radial Wave Functions and Nuclear Beta-decay, Oxford University Press (1982)
6. P.A. Baisden, J. Rydberg, private communication
7. D.P. Devor, B.H. Soffer and M. Robinson, Appl. Phys. Lett. 18,122 (1971)
8. P.T. Springer, C.L. Bennett, and P.A. Baisden, to be published.

Table I

Phase Space Factors for ^{163}Ho(7/2$^-$[523])->^{163}Dy(5/2$^-$[523]), Q_{EC}=2564eV

Shell x	Energy[a]	$q_x^2*10^6$	$\beta_x^2*10^3$	B_x	f_x*10^9
$3s_{1/2}$	2029.3	1.10	49.52	1.065	0.909
$3p_{1/2}$	1824.2	2.10	2.605	0.935	0.080
$4s_{1/2}$	406.7	17.83	11.54	1.12	3.620
$4p_{1/2}$	322.6	19.25	0.589	0.88	0.157
$5s_{1/2}$	68.7	23.86	1.71	1.18	0.756
$5p_{1/2}$	33.9	24.53	0.07	0.82	0.022
					Total=5.54

a)These energies include corrections for the shift between characteristic x-rays observed in electron capture as opposed to those observed in photoemission, or electron bombardment[8]

Table II

Solar Neutrino Capture Rates in ^{163}Dy

Source	Cross Section	Flux($cm^{-2}s^{-1}$)	Rate(SNU)
p+p->d+e$^+$+	$1.2*10^{-44}$	$6.1*10^{10}$	740
p+e+p->d+ e	$5.9*10^{-44}$	$1.5*10^8$	9
e+^{7}Be->^{7}Li+ e	$3.0*10^{-46}$	$3.4*10^9$	100
			Total=$850*10^{-36}s^{-1}$

THE INDIUM SOLAR NEUTRINO PROJECT*

N. E. Booth and G. L. Salmon
Nuclear Physics Laboratory, University of Oxford, Oxford, U.K.

D. A. Hukin
Clarendon Laboratory, University of Oxford, Oxford, U.K.

ABSTRACT

The only way to resolve the solar neutrino puzzle is to perform a new experiment. We will show that ^{115}In has unique possibilities as a target for solar neutrino detection. We review our progress in developing a detector based on ^{115}In and outline our future plans.

THE SOLAR NEUTRINO PUZZLE

Basically, the puzzle is how to resolve the discrepancy between theory (the standard solar model) and experiment (the ^{37}Cl experiment of R. Davis et al).
The experimental rate is:

$$2.1 \pm 0.26 \text{ SNU}$$

while the theoretical number is:

$$6.6 \pm 2.7 \text{ SNU}$$

The SNU (solar neutrino unit) is the number of neutrino captures per second per 10^{36} target atoms. Many suggestions have been made to resolve the puzzle - most of them raise more problems than the one they attempt to solve. There remain two front-runners:

(1) the sun is working differently from what we believe,
or
(2) something happens to the neutrinos in the 8 minutes it takes them to reach the earth.

Neutrino production in the standard solar model takes place via the reactions shown in Fig.1. There is also a small contribution from the CNO chain. Perhaps the only inadequacy of the ^{37}Cl experiment is the high threshold (816 keV). The ^{37}Cl capture reaction is sensitive mainly to the neutrinos from the β-decay of ^{8}B, and is insensitive to the large flux of neutrinos from the proton-proton fusion reaction.

Clearly a new experiment must be done with a threshold below the 420 keV maximum energy of the p-p neutrinos.

Any target ^{A}Z for solar neutrino detection must satisfy the following criteria:

1) low threshold (< 420 keV for neutrinos from pp fusion)

---

* Presented by N.E.Booth. Supported in part by the U.K. Science and Engineering Research Council.

2) $\Delta J^P = 0^+, \pm 1^+$
3) means to identify the occurrence of a capture
4) availability of ^AZ.

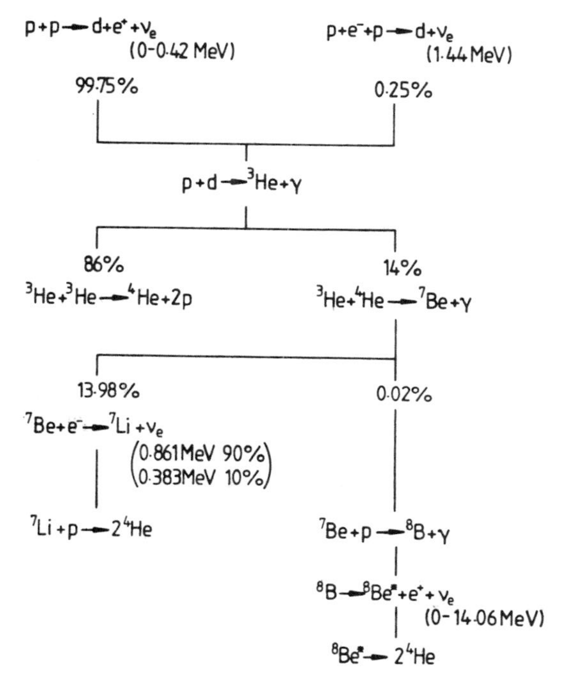

Fig.1. The proton-proton chain of reactions in the standard solar model.

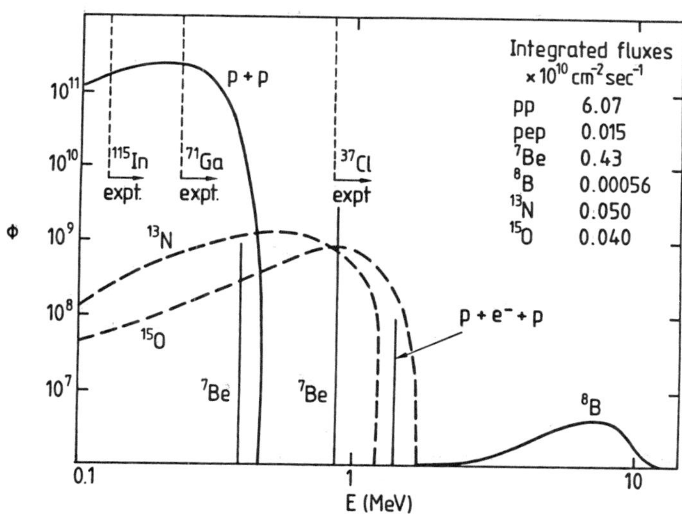

Fig.2. Neutrino fluxes predicted by the standard solar model.

Of the few isotopes which even approximately fulfill these criteria, ^{71}Ga and ^{115}In are being actively pursued. Their thresholds, together with the fluxes predicted by the standard solar model are shown in Fig.2. The model predicts that about 35 tons of natural gallium are needed for 1 capture per day, or 3.5 tons of natural indium. To get an event rate one must of course put in the efficiency.

THE ^{115}In REACTION

Apart from the relatively low target mass required, and the low threshold, ^{115}In has some unique properties. It is the only nucleus with which a real-time electronic experiment can be performed, and it is the only target with which the differential neutrino energy spectrum can be measured. The reaction is

$$\nu_e + {}^{115}In \rightarrow {}^{115}Sn^* + e^-$$

and the relevant energy levels are shown in Fig.3. Neutrino capture with electron emission takes ^{115}In (96% natural abundance) to the second excited state of ^{115}Sn. This state lives for 3 μsec and then emits two γ-rays in coincidence with energies of 115 and 498 keV. In about 1/2 of the captures the 115-keV γ-ray is internally converted,

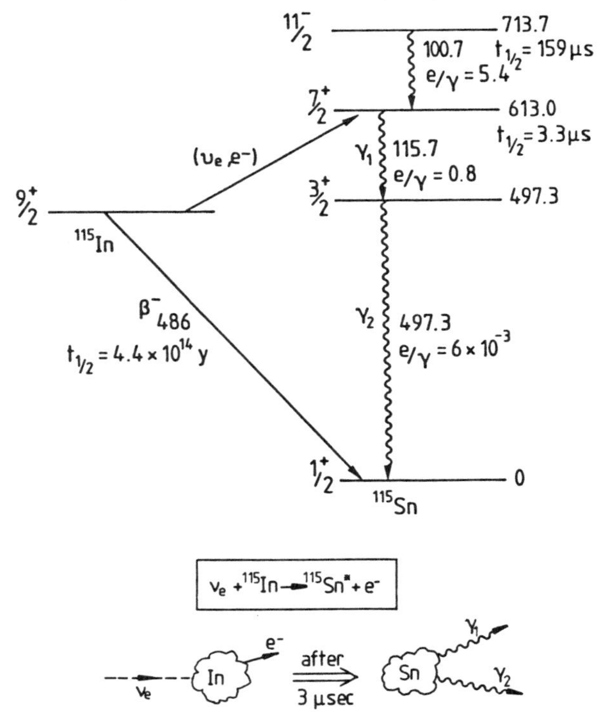

Fig.3. The ^{115}In neutrino capture reaction.

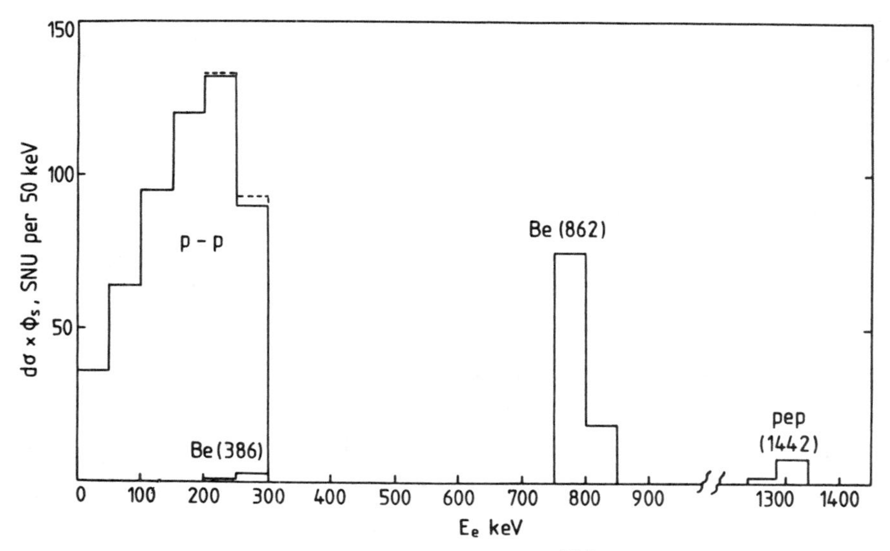

Fig.4. Electron energy distribution from ^{115}In in the standard solar model, plotted in 50 keV bins.

emitting a 90-keV electron plus the characteristic X-rays. Thus, the unique signature for neutrino detection is a pulse from the electron, followed by, on average 3 μsec later, two coincident pulses, one spatially very close to the electron with energy 115 keV, and the second of 498 keV. The energy of the electron is $E_e = E_\nu - 128$ keV, so a measurement of its energy gives the neutrino energy. The expected electron energy spectrum based on the 'standard solar model' is shown in Fig.4. This model predicts a rate of 1 event/day for 3.5 tons of natural indium. Thus for a 1 ton detector running for 1 year we expect 85 events in the pp distribution and 15 events in the ^{7}Be peak.

DESIGN CONSIDERATIONS

Although the ^{115}In reaction has a very characteristic signature, the basic problem in a low rate experiment is to select the signal and reject the various backgrounds. Another problem with ^{115}In is to get the indium in such a form that the detector is sensitive to the low energy ($\langle E_e \rangle \simeq 200$ keV) electrons from the neutrino capture and the 115-keV γ-ray which is internally converted in 45% of the captures. One possibility is to use cells of indium-loaded liquid scintillator, and this has been studied by Raghavan, Pfeiffer and colleagues[1,2], and more extensively by us[3]. We have also constructed prototype detectors consisting of 10 μm thick indium foils, 1 m long, sandwiched between thin sheets of plastic scintillator and viewed by 50-mm diameter phototubes at each end. The problem with these approaches is that to get any kind of energy resolution, say ±25% at 500 keV, one is restricted to about 5% indium, so that one ton of indium requires 20 tons of scintillator (plus about 10,000 phototubes). In addition this approach can solve some of the background problems to be discussed in the next section, but not all of them.

BACKGROUND CONSIDERATIONS

There are several possible backgrounds, the most obvious being accidental coincidences due to the natural radioactivity of ^{115}In. The decay rate is 0.22/sec per gram of indium or 0.22×10^{6}/sec per ton. The rate due to solar neutrinos is only 3.3×10^{-6}/sec/ton.

There are three types of accidental background which we have discussed in detail previously[3,4].

(1) $R_{\beta\beta\beta}$. Here the electron and γ_1 and γ_2 are faked by β-particles. This background can be suppressed below the signal by a combination of
  - (a) dividing the detector into 10^3-10^4 individual cells or elements
  - (b) having good to moderate energy resolution on γ_1 and γ_2
  - (c) having a coincidence resolving time between γ_1 and γ_2 of about 10^{-8}sec.
  - (d) taking advantage of the fact that γ_2 will usually Compton scatter once or twice before being absorbed by the photoelectric effect.

(2) $R_{\beta\beta\gamma}$. Here the electron and γ_1 are faked by β's, and γ_2 is faked by a stray γ-ray from cosmic rays or natural radioactivity. In addition to (a) - (c) of (1) one must minimize the flux of stray γ-rays. The suppression already achieved in several low-level counting laboratories is adequate. However this is not trivial for a 20-ton detector.

(3) $R_{\beta\gamma}$. Here the electron is faked by a β, and the whole delayed event by the plural Compton scattering of a stray γ-ray of approximately 615 keV. This is potentially the most severe background and makes the straightforward scintillator approach unfeasible.

One possibility proposed by Raghavan and Deutsch[5], and looked at again by a group from Saclay[6] during the past year, is to consider only the 45% of captures where γ_1 is internally converted. In this subset of events the second electron comes from the same nucleus as the first. Thus, a detector with very good spatial resolution should be able to select these events above the background. In a practical design this can only be achieved at the almost total expense of energy resolution.

The approach we are taking is to have moderate spatial resolution but shoot for good energy resolution. The best energy resolutions are obtained with semiconductor detectors. What about the semiconductors InP, InAs and InSb? We have done some work with InP without success, but have had some success with InSb in collaboration with Brian England of Birmingham University. At the present time these materials are not of sufficient quality and do not have low enough impurity concentrations to make particle detectors. They are also expensive. As a result we have been forced to develop a new type of particle detector containing indium.

SUPERCONDUCTING INDIUM DETECTORS

Indium is a superconductor with critical temperature $T_c = 3.408°K$ and energy gap $2\Delta(0) = 1.05$ meV at T=0. We will consider temperatures $T \lesssim 0.5\ T_c$. The valence electrons near the Fermi level E_F are bound into Cooper pairs (binding energy 2Δ). In thermal equilibrium pairs are occasionally broken by phonons, and the excitations, called quasiparticles can also recombine with phonon emission as shown in Fig.5(a). The rates of the two processes are equal, but are strongly temperature dependent.

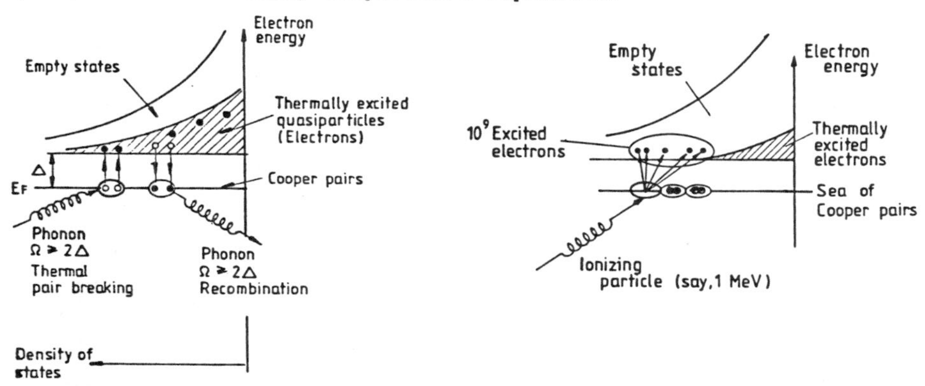

Fig.5(a) Electron-phonon processes in a superconductor in thermal equilibrium.

Fig.5(b) Effect of excitation by an ionizing particle.

When a particle deposits ionization energy the net result is the breaking of Cooper pairs, shown schematically in Fig.5(b), and the production of an excess of quasiparticles and phonons. A much larger fraction of the energy deposited goes into the electronic system in a superconductor as compared to a semiconductor. This is because the phonon spectrum lies below the energy gap in a semiconductor, whereas it is mainly above the gap in a superconductor and is available for breaking Cooper pairs. If no advantage is taken of this situation the quasiparticles will eventually recombine to give a thermal equilibrium situation at a minutely higher temperature.

One method of detecting the excitation is to use a superconducting tunnel junction for example In-I-In where I is a thin insulating layer, the best being In_2O_3. The equilibrium tunneling processes are shown in Fig.6 for a bias voltage $V_B < 2\Delta/e$. Processes a and b have equal rates but are strongly temperature dependent (the main dependence is $\exp(-\Delta/kT)$). Although each of them is an electron tunneling process, the combined result is the transfer of a Cooper pair from S_1 to S_2. The diagrams of Fig.7 show how an excitation increases the tunnel current for an excitation on either side, or alternately for either polarity of the bias voltage.

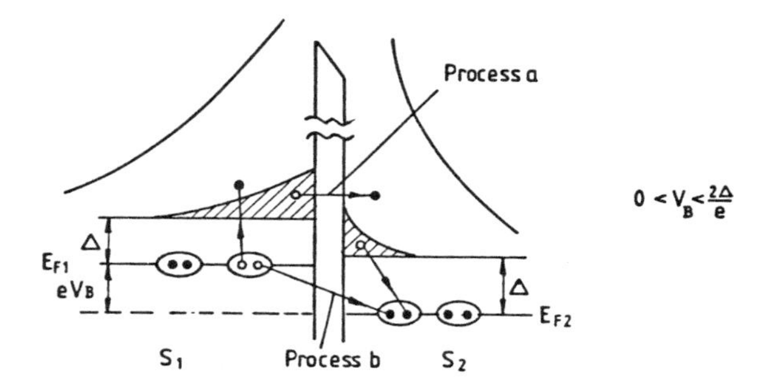

Fig.6. Quasiparticle tunnelling processes between two superconductors.

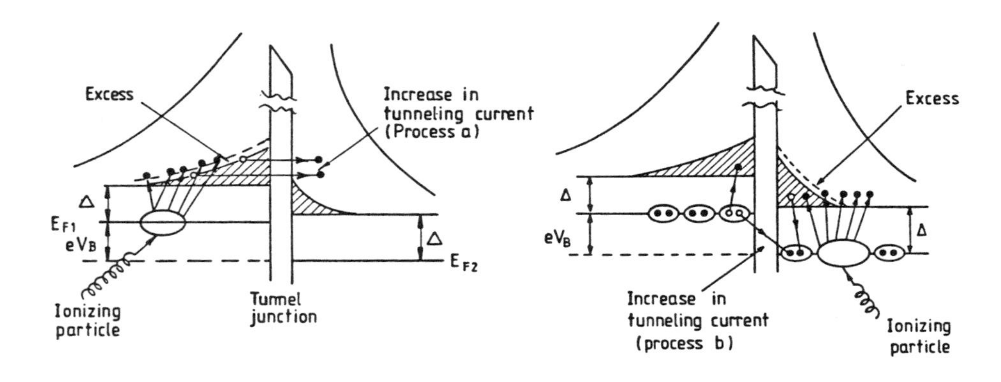

Fig.7. Effect of an excitation on tunnelling current.

The first experiment we did after learning how to make tunnel junctions is shown schematically in Fig.8. The crossed film junction could be irradiated either with α-particles or optical photons from a xenon flash tube. The low impedance transmission lines were used to approximately match the dynamic resistance of the junction at the most horizontal part of the I/V characteristic shown in Fig.9. For good junctions the current scale decreases exponentially with decreasing temperature, whereas the voltage scale remains fixed. At that time we could not go lower in temperature to increase the junction impedance. We indeed saw pulses as shown in Fig.10. The overshoot is due to the transformer coupling. Taking into account the impedance matching we estimate an energy deposition of 10 meV per electronic charge detected. A better result should be obtainable with improved junction geometry and lower temperature. However this number already predicts an energy resolution of 0.15 keV (FWHM) at 500 keV.

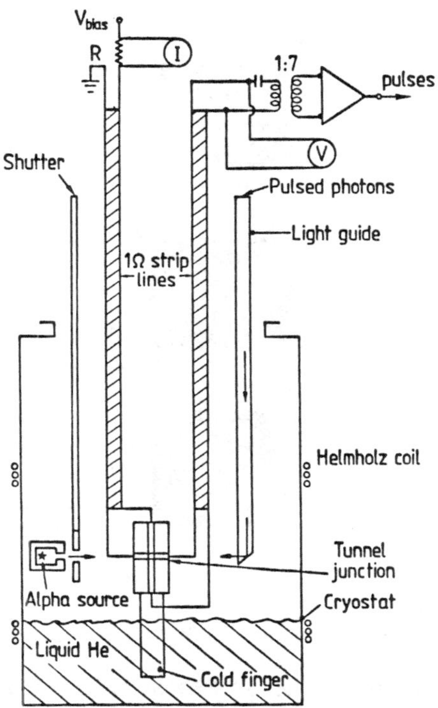

Fig.8. Experimental set-up used in 1983 to detect excitations with a tunnel junction.

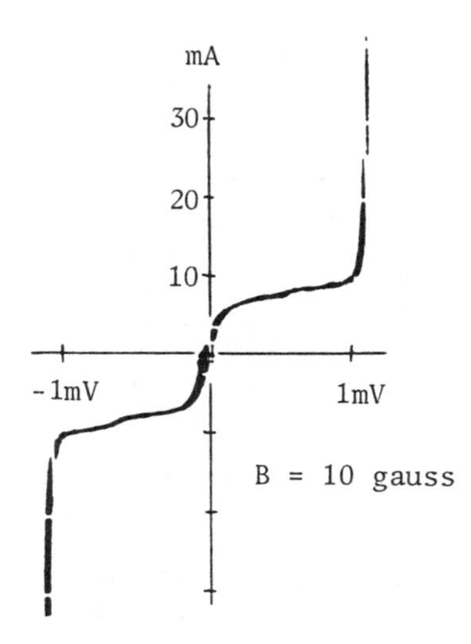

Fig.9. Current-voltage characteristic of a superconducting tunnel junction.

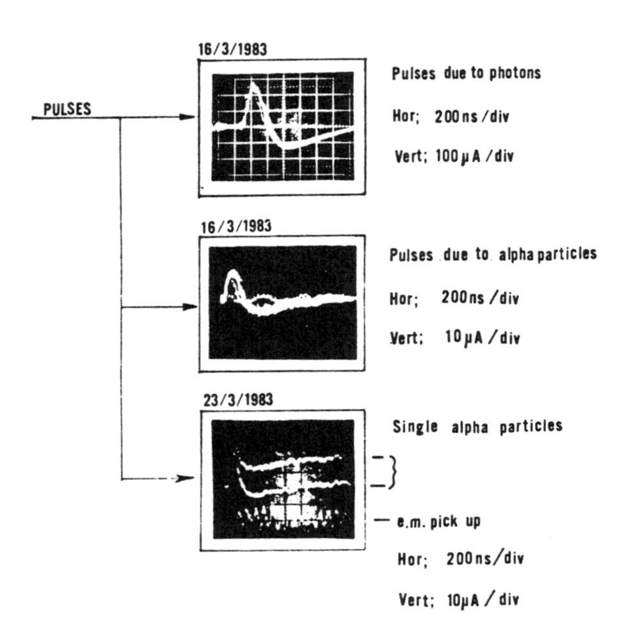

Fig.10. Typical pulses observed with tunnel junction detector.

SCALING UP THE SUPERCONDUCTING DETECTOR

To do the solar neutrino experiment we must have the indium in chunks, at least 1 gram each, and not in the form of thin films. Moreover we need sufficiently good time resolution to resolve the electron from the neutrino capture from the delayed 115-keV γ-ray which will almost always interact in the same chunk. Moreover we need to solve the technical problem of making junctions on the surface of bulk indium, electrical contacts, etc.

Consider a piece of indium of area A, thickness d with a junction of area A_J on one surface. At the present time we are working with A = 100 mm^2 and d = 3 mm. Below about $T = 0.5T_c$ the recombination lifetime τ_R becomes long compared to the pair breaking time by phonons τ_B. Produced quasiparticles propogate by diffusion with a velocity close to the Fermi velocity (1×10^8 cm sec^{-1}) and with a mean free path l_i for impurity scattering. Our indium is in the form of high purity single crystals to give a large l_i. The arrival time and rise-time are proportional to r^2/l_i where r is the source-junction separation. However it will be shown below that the rise time is limited by the junction capacitance. Let us first compute the magnitude of the signal. After a few transit times of the crystal the excess quasiparticles will be distributed uniformly throughout the volume. At this time the signal current is[7]

$$i_s = \frac{n(t)}{2e}\,\frac{1}{d}\,\frac{A_J}{A}\,g_{nn}\,\frac{1}{N(0)}\,f(eV/\Delta)$$

where $n(t) = n_o e^{-t/\tau_s}$, where n_o is the excess number of quasiparticles and τ_s represents the loss due to tunnelling and recombination. Here g_{nn} is the normal state specific conductance of the junction, N(0) is the spin-one density of states at the Fermi surface ($7.68\times10^{21}eV^{-1}cm^{-3}$), and $f(eV/\Delta)$ is a temperature independent factor of order unity at the bias point $V = V_B$. Unfortunately the signal current is inversely proportional to the volume of the crystal[8]. Making A_J large increases the junction capacitance C_J ($\simeq 4\mu F\ cm^{-2}$ for indium oxide) and this increases the signal rise-time. The factor to optimize is g_{nn}. Fortunately very high conductance junctions can be made using indium oxide. Josephson current densities j_c as high as $10^5\ A/cm^2$ have been achieved, although mainly with Pb-In alloy but also with pure In, by workers at IBM[9]. Although the IBM project has been closed down, the technology they have developed is very valuable to us. This gives

$$g_{nn} = \frac{4}{\pi} \frac{1}{(2\Delta/e)} j_c \simeq 10^8 \Omega^{-1} cm^{-2}.$$

Using this number and taking for example a 5-mm^2 junction on a $10\times10\times3\text{-mm}^3$ crystal we estimate a signal current of 1.0 μA. The time constant depends on the impedance level and is 2 μsec for 10 Ω. To get the full energy resolution it is necessary to integrate the signal for about 100 μsec and to work below 0.6°K. We are studying several possibilities for improving this sort of performance[10], for example connecting junctions in series in order to reduce C_J. However, we have not as yet solved all the technical problems.

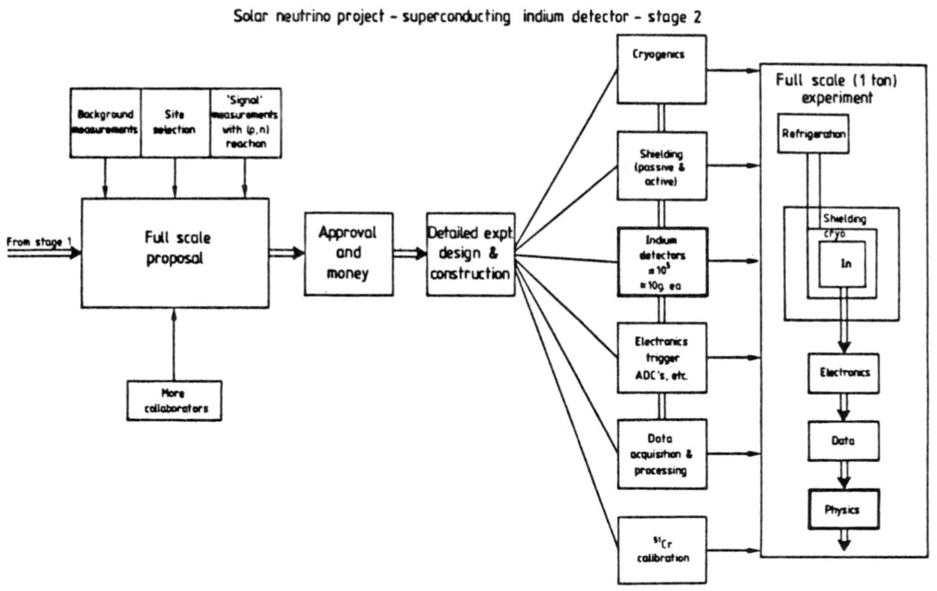

Fig.11. Overall plan for a solar neutrino experiment using superconducting indium detectors.

A SOLAR NEUTRINO EXPERIMENT WITH SUPERCONDUCTING INDIUM

Figure 11 shows how we envisage implementing a full scale experiment once we have solved the problems of the individual detectors. A cubic metre of indium weighs 7.3 tons so the detector is physically small and can be well shielded and have an active veto shield. It should be underground and in an environment with low natural radioactivity. None of the boxes in Fig.11 are easy but most of them can be accomplished with present technology. Happily there has been some recent activity in the "More collaborators" box. In particular a number of French physicists have concluded that the ^{115}In reaction is the best way to detect solar neutrinos. Recently Raghavan reminded us of a statement N. Booth wrote to him in 1980: "this experiment just has to be done". We think we have finally found a way to do it.

REFERENCES

1. R. S. Raghavan, Phys. Rev. Lett. 37, 259 (1976).
2. L. N. Pfeiffer, A. Mills, R. S. Raghavan and E. Chandross, Phys. Rev. Lett. 41, 63 (1978).
3. N. E. Booth and G. L. Salmon, Progress Report, Indium Solar Neutrino Project, Oxford, May 1983.
4. N. E. Booth, "A Programme of Neutrino Physics and Astronomy", Letter of Intent, Oxford, June 1980.
5. R. S. Raghavan, Neutrino '81, eds. R. J. Cence, E. Ma and A. Roberts, p. 27, 1981.
6. M. Cribier, paper presented at Neutrino '84, Dortmund, June 1984.
7. N. E. Booth, "Notes on the Detection of Nuclear Particles with Superconducting Tunnel Junctions", Indium Solar Neutrino Project Note, Oxford, May 1984.
8. G. H. Wood and B. L. White, Can. J. Phys. 51, 2032 (1973).
9. John M. Baker and J. H. Magerlein, J. Appl. Phys. 54, 2556 (1983).
10. N. E. Booth, "Quasiparticle Trapping and Quasiparticle Multiplication", Indium Solar Neutrino Project Note, Oxford, June 1984.

Indium solar neutrino experiment using superconducting grains

A. de Bellefon, P. Espigat
LPC - Collège de France

G. Waysand
GPS - ENS - Université PARIS VII

Presented by A. de Bellefon

Abstract

In this paper we would like to emphasize the revival of interest for Indium experiment in Europe. Properties of metastable superconducting indium grains are presented and our progress towards making an experiment feasible is reviewed.

Introduction

The inverse β decay of ^{115}In to $^{115}Sn^{**}$ has been proposed for a long time by R.S. Raghavan [1)] as a possibility for detecting the solar neutrinos from the proton proton reaction in the sun.

The reaction is

$$\nu_e + {}^{115}In \longrightarrow {}^{115}Sn^{**}_{(7/2^+)} + e_1^- \qquad 1)$$

Followed by

$$Sn^{**} \xrightarrow{\tau_{1/2} = 3.3\ \mu s} Sn^{*}_{(3/2^+)} + \gamma_1 / e_2^-$$

$$Sn^{*} \xrightarrow{\tau = 10^{-10} s} Sn + \gamma_2\ (498\ keV)$$

Reaction (1) has a low threshold of 128 KeV so that the energy of e, is $E_e = E_{\nu_e}$ - 128 KeV and its measurement gives the neutrino energy. The event identification can be accomplished by the following signatures :

- a delayed triple coincidence $e_1 \longleftrightarrow e_2 + \gamma_2$
- a spatial coincidence $e_1 \longleftrightarrow e_2$

If we achieve these identification and if we fight successfully against background mainly due to the ^{115}In spontaneous decay then we have a real time electronic experiment.

Details on the background and accidental coincidence rates have been already discussed in detail elsewhere [2)].

We now want to focus on a new method to detect the event.

Superheated Superconductivity

Type I superconductors such as Sn, In, Al undergo a first order phase transition between the normal and superconductiong states in the presence of a magnetic field.

The general principle of this phase transition is now explained. At a temperature T below the critical temperature T_e for a supercon-ductor material if one sweeps the magnetic field from o to H SUPERCONDUCTIVITY remains above the critical field H_c (T) up to a certain value H_{sh} (T) which is called the superheated critical field.

On the contrary NORMAL RESISTIVE STATE still exist below H_e (T) down to H_{sc} (T) the supercooled critical field below which superconductivity reappears.

It has been shown that if a perfect microsphere of type I superconductor is immersed in a magnetic field at $T < T_c$ it will be in a metastable superconducting condition for

$$H_c(T) < H < H_{sh}(T)$$

where $H_{sh}(T) = \frac{2}{3} \cdot K^{-1/2} \cdot \xi^{-1/4} H_c(T)$ [3] K = Guinzburg - Landau parameter $< 1/\sqrt{2}$

Then a superheated superconducting detector could be a suspension of micro spheres of type I superconductor embedded in paraffin wax or any insulating medium.

We can use as a detection principle the breaking of metastability which can follow two different paths shown on Fig. 1.

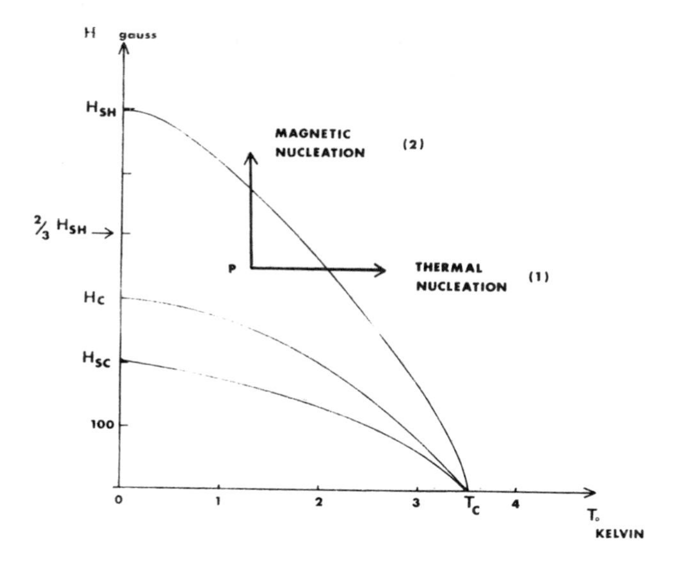

Fig. 1

(1) We can observe that transition by ionizing particle

(2) Observable by sweeping up the field.

Detection

The small amount of energy needed to induce the decay of metastable state can be provided by the energy deposition ΔE of a charged ionizing particles. A track is then formed by grains that flipped back to the normal resistive state.

This type of detector has already been presented by solid state physicists [4)]. A pulse method is now available to detect the signal.

This method is based on the disparition of the Meissner effect. Meissner and Ochsenfeld found that the field distribution around a homogeneous superconducting body always corresponds to a zero internal field. If such a body (micro sphere for example) is brought from normal to superconducting state by cooling down in presence of a magnetic field H_o it is found that the flux penetrating the superconductor is expelled as the normal to superconducting transition takes place. At the same time the external field rises to a value such that internal field is zero.

a) $H(r) = 0$ inside the sphere $r < R$

b) $H(r) = H_0 + D$ (dipole term) outside $r > R$

So the net penetration of the magnetic field resulting from the rupture of Meissner effect produces a flux variation $\Delta\Phi = -\iint \vec{D} \cdot d\vec{S}$ in a pick up coil surrounding the grain. Fig. 2.

This variation takes place within fraction of μs and gives a pulse in the μV range.

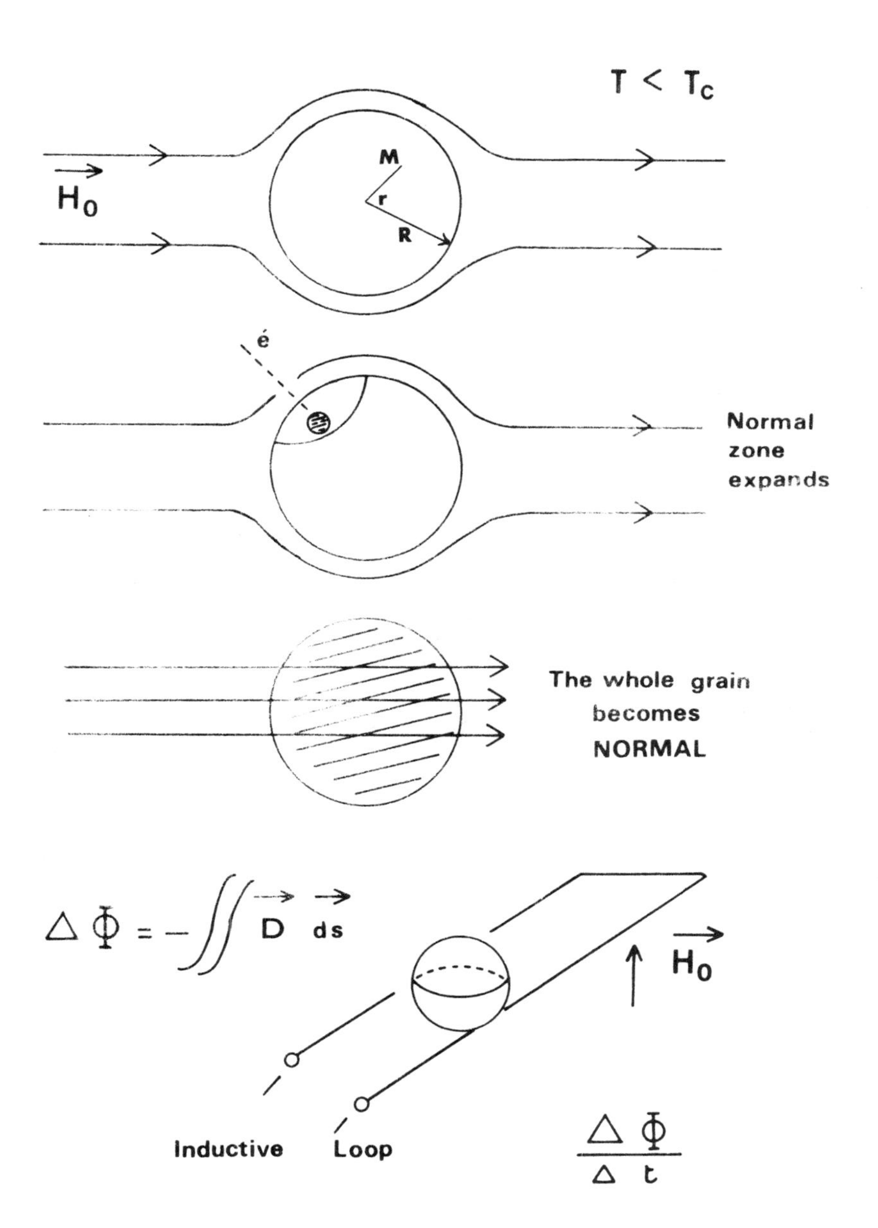

T < Tc
H0
M
r
R
é
Normal
zone
expands
The whole grain
becomes
NORMAL
ΔΦ = −∬ D ds
H0
Inductive
Loop
ΔΦ / Δt

Fig.2

Indium is a type I superconductor and spherical grains of tin or Indium are easily prepared by ultrasonic treatment of a mixture of tensioactive oil with the metal. The superheated state is observed for diameters between 1 and 40 μm.[5)]

In principle this technique should work but the goal we have now is to prove the feasability of such an experiment.

The main questions under study are :

- Energy resolution
- Detection efficiency
- Signal read out
- Destruction of Metastable States by irradiation
- Detection of the 500 KeV photon

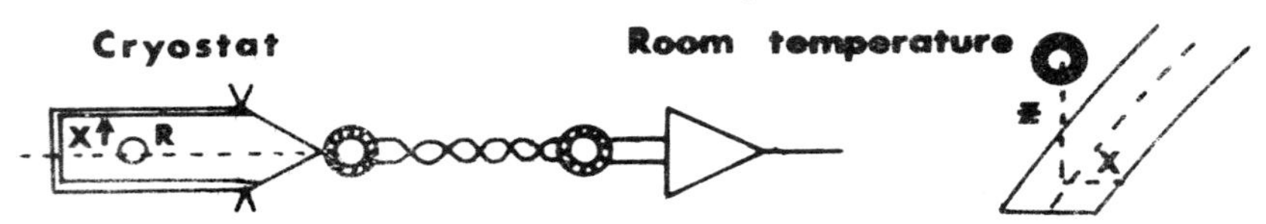

Let R be the radius of grain. L half witdth of the pick up coil and H_0 the applied field then $\Delta\Phi = \Delta\Phi(0,0) \times f(x,z)$

where $\Delta\Phi(0,0) = -2 H_0 \frac{R^3}{L}$ (1) and $f(x,z) \lesssim 1$, $x < .9L$

For a grain R = 2.5 μ L= 1 mm and H_0 = 300 G

$$\Delta\Phi(0,0) \sim \Phi_0 = 2.06\ 10^{-15}\ \mathrm{Wb\ cm}^2$$

Till now we are able to read $\Delta\Phi(0,0) \sim 100\ \Phi_0$

so we have to gain at most two order of magnitude which means to developp new amplifier with a new kind of coupling to the loop. This is going on in Paris at Collège de France [6].

The signal depend also upon the flipping time "dt" which is of the order of a fraction of μs.

In order to detect ν events we have to detect two electrons and a photon. The idea is to pile up stack of grains as on Fig. 4. Suspension of Indium grains is spread upon kapton foils on epoxy cards upon which read out loops are printed.

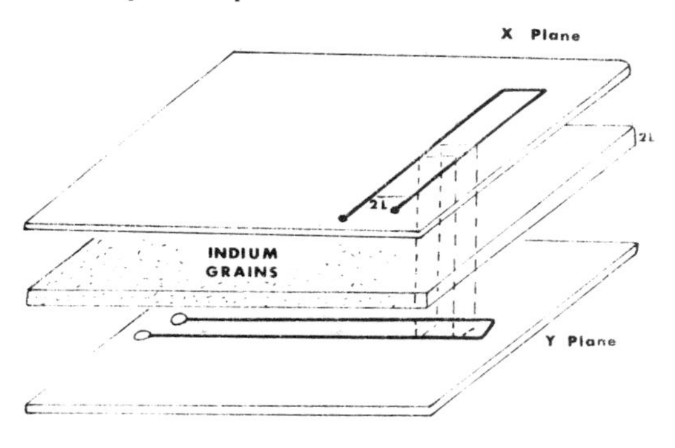

Fig.4

Indium grains embedded in wax with a 30 % filling factor

The number of channels to design for this kind of detector is a function of the width of the loop and of the spatial resolution we want.

The spatial resolution will be given by a cell size which can possibly go down to 25 x 25 μm^2.

The equation (1) implies that we have to decrease L to get a comfortable signal so we have to find a compromise between signal size, resolution requirement and number of channels.

Energy resolution - Detection efficiency

A Monte-Carlo simulation has been developped to answer to these questions and is still in progress.

To perform such a program we needed :

- a model for grains flip [7)]
- a generation of the grains inside paraffin wax
- to track electrons and photons through such a medium
- to generate event as tin desexcitation In β decay compton event

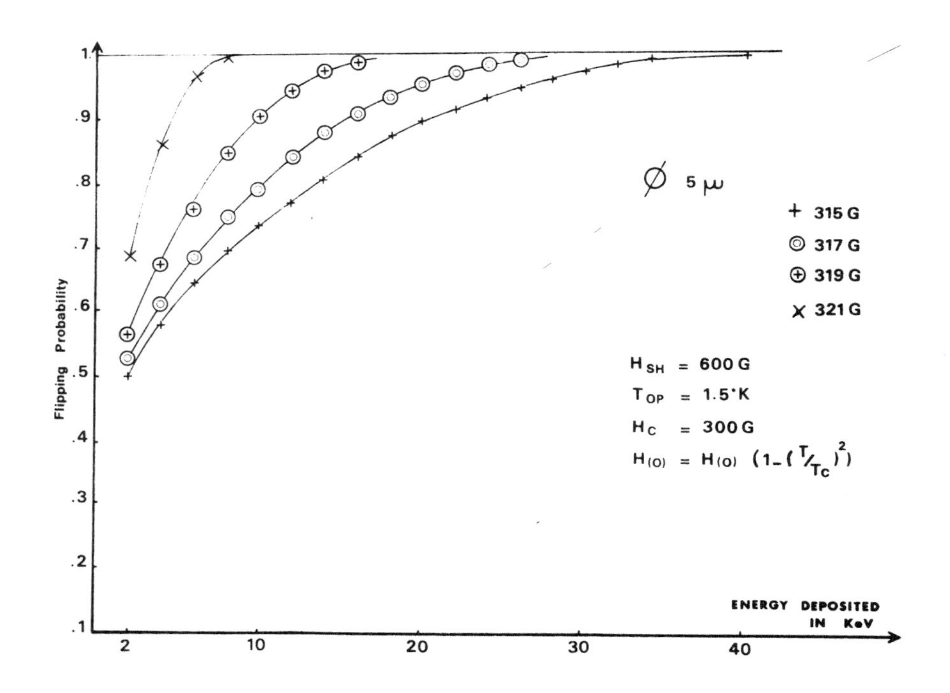

Fig.5

Taking into account energy losses in paraffin and indium, multiple scattering and compton effect we have got as first provisional results :

- number of grains is an increasing function of the energy of the electron and it is quite possible to separate ν_{pp} from ν_{Be} *

- mean value of energy loss per grain of 2.5μm radius is of the order of 6 KeV for the first electron e_1 and 8 Kev for the second electron e_2.

- with a very simple thermodynamical model for grains flip. Flipping probability of a grain with a radius 2.5 μm and for different applied field is shown as a function of energy deposited on Fig.5.

- Around 65 % of energy is lost in Indium.
- A 300 KeV electron covers a distance of the order of 400 μm.

* ν_{pp} comes from $p\,p \rightarrow d\,e^+\,\nu_e$
ν_{Be} comes from $^7Be + e^- \rightarrow {}^7Li + \nu_e$

Work in Progress
Feasibility Study

As I mentionned previously we are developping a Monte Carlo simulation to be able to give an efficiency for this type of detector.

But a superconductiong detector is a new field of detector for high-energy physicist and we have to take great care of the performance for such a detector. For instance I should mention some questions :

- Diamagnetism effect may induce multiple flipping of the grains.

- Pick up of very low signals in the μV range.

- Flipping of grains under irradiation, experimental test is required.

- Metastable state lifetime has to be evaluated.

Conclusion

Indium detector is unique among the proposed solar neutrino detectors since it is a direct counting experiment and it covers the pp reaction.

In the spring of 84 after a meeting with Raghavan a group of Europeans physicists decided to start a new study of the feasability of such an experiment with Indium [8].

In the present talk a superconductiong In grain detector has been described but I should mention also work done on superconducting bulk crystals of Indium [9] and Organo metallic In compound detector.

References

1. R.S. Raghavan. P.R.L. 37 (1976). 259.
2. R.S. Raghavan. Brookhaven Solar Neutrino Conference BNL 50879 VII, VC 34b (1978).

   R.S. Raghavan. International Conference on Neutrino Physics and Astrophysics. Hawaï (1981).

3. J. Matricon, D. Saint-James. Phys. Lett. 24 A (1967) 241.
4. Hueber, Valette, G. Waysand. Cryogenics. July 1981. 387.
5. Astrophysique et Interactions fondamentales. Cargèse. July 1983.
   G. Waysand, Proceedings of the Rencontre de Moriond on Massive Neutrinos - La Plagne. 15-21 janvier 84.

6. Internal note. LPC Collège de France. R. Bruère-Dawson, D. Broskiewicz.

7. Private communication. J.L. Basdevant.
8. CERN Courrier V24 Octobre 1984.
9. N. Booth. Conference on Solar Neutrinos and Neutrino Astronomy. 23-25 August 1984. Lead, South Dakota.

THE SYDNEY UNIVERSITY SOLAR NEUTRINO PROGRAM

A.M. Bakich and L.S. Peak
Falkiner Nuclear Department, University of Sydney,
N.S.W. 2006, Australia

ABSTRACT

The Sydney University Falkiner Nuclear Deparment is planning to install a solar neutrino detector at Broken Hill at a depth of 1230 m. Preliminary measurements have been made and the general features of the final detector have been decided upon.

The detector will utilise some 100 Tonnes of pure water in the final configuration, and will measure the neutrinos directly by means of their scattering off electrons in the water. The cherenkov radiation of the scattered electrons will be picked up by banks of photomultipliers.

INTRODUCTION

For some time now our group has been assessing the feasibility of the direct detection of solar neutrinos by their elastic scattering in pure water. We propose to detect the cherenkov photons emitted by the recoil electrons, thus utilising the water as both target and detector.

Though we have been progressing rather slowly (the experiment is by no means trivial, and may be only marginally possible), we have made some advances and come to various conclusions which we summarise below. The fact that there is some directional information present in the process (and thus the hope of roughly tracking the sun) provides an additional incentive to embark on this difficult project.

We now have a chamber set aside at a depth of 1230 m (3300 m.w.e) in an active silver lead and zinc mine at Broken Hill. Broken Hill is a city of some 30,000 people about 1100 km west of Sydney, which is well serviced by plane, train and coach from both Sydney and Adelaide.

Figure 1 shows the present shape and size of the chamber. The current dimensions are 12 m x 5 m (> 3 m height). It is furnished with power, rail tracks, ventilation and a concrete floor as shown in the diagrams. In a short while, we hope to excavate the chamber to its final size of 20 m x 2 m x 5 m (height). When this is achieved, we will have a long tunnel-like underground laboratory with the capability of housing various underground research programs in addition to the solar neutrino effort.

MEASUREMENT OF BACKGROUND

We have taken detailed measurements of the gamma ray background in our chamber. For this purpose we have used a cubical sodium iodide crystal of side length 10 cm together with lead and boronated paraffin shielding. This has enabled us to confirm that our main high energy background is indeed gamma radiation emanating from the rock walls of the chamber. The detailed results of these measurements are soon to appear in Nuclear Instruments and Methods. Figure 2 shows the high energy differential energy spectrum compared with that of Sobel et al. taken with a much larger liquid scintillator in a gold mine at a much greater depth. The good agreement probably indicates the general nature of the origin of these high energy gammas for quite varying circumstances. It is most likely that these gammas originate from the uranium and thorium in the rock; either directly from spontaneous fission, or from (n,γ) captures of the fission neutrons in substances such as silicon, or even by alphas from uranium and thorium decay causing neutrons (eg. in the $^{29}Si(\alpha,n)^{32}S$ reaction) which are themselves subsequently captured in silicon etc.

SHIELDING

Water is undoubtedly the purest and cheapest shielding material. Unfortunately, it is also the least efficient, leading to a requirement of some 5 - 7 metres shielding in all directions; and a chamber of this size is not within the resources of the Sydney group.

On the other hand, lead seems a hopeful prospect. It is better than concrete as neutron capture in silicon gives gammas with energy ranging up to 10.5 MeV. It is also better than steel, as similar neutron capture leads to gammas ranging up to 10.16 MeV. With lead, however, the maximum gamma energy from neutron capture is only 7.38 MeV.

Figure 3 shows the total annual background for 100 tonnes of water in a spherical configuration, and shielding by various thicknesses of lead (with uranium content assumed to be 10^{-7} gm/gm). We define a background event as occurring whenever an incoming gamma knocks on or produces an electron (pair) of energy greater than 6 MeV. It is clear that a constant background is reached of around 800 events per year per 100 tonnes for a lead thickness greater than about 30 cm - and this is caused by the uranium fission in the shield itself.

As a consequence of the above, we plan to line our chamber with between 25 - 30 cm of lead, and are currently seeking to borrow or lease the necessary amount (around 1500 tonnes).

MULTIPLE SCATTERING AND DIRECTIONAL RESOLUTION

As the angular resolution of the detector becomes more precise, one can obviously tolerate greater background. Figure 4 shows the annual background that can be tolerated as a function of annual signal, for various angular accuracies; $\Delta\theta$ is the error around the current solar direction with which we can assign the incoming neutrino direction for any given event. The curves have been calculated for a required significance of two standard derivations within one year. For our expected annual signal of 30 events in 100 tonnes we see that our background of 800 events would be acceptable for an angular resolution of around 50°.

At first sights one might expect to improve the situation by designing a detector with high angular resolution. This advantage turns out to be illusory, however when multiple scattering is considered. We have performed extensive monte carlo simulations on the photon distributions arising from low energy scattered electrons and find:

(a) that there are no well defined ring patterns - making ring imaging impracticable;

(b) that about 70% of the photons are spread over an angular range of about 50° with respect to the initial electron direction;

(c) that the remaining 30% of these photons range over large and backward angles, and are virtually unusable.

Because of the above (and to some extent the kinematics of the original scatter), it is pointless aiming for high angular accuracy in any planned detector. We plan to simply divide the sky into six directions using a cubical shape detector with angular accuracy given by $2\pi(1-\cos\Delta\theta) = 4\pi/6$ leading to $\Delta\theta = 50°$. With this angular accuracy and an annual signal of 30 events, we can predict the allowable annual background for various levels of significance and years of operation. This is summarised in the table below.

Standard Deviations	Years of Operation				
	1	2	3	4	5
1	3960	8063	12163	16260	20363
2	888	1913	2938	3963	4988
3	319	774	1230	1686	2141
4	120	376	632	888	1144

Thus, for our expected background of 800 events we can achieve two standard deviations significance in one year (as mentioned before); but if we run for four years, this can be improved to four standard deviations.

DETECTOR DESIGN

Although our final design is not completely determined, there are certain features of this design which have already emerged.

We can list these as follows:-

(a) It will be modular and of cubical shape. Our first module will contain something like 10 tonnes of water viewed by a moderate number of phototubes (about 30-40). This first module will also help us to assess our capability to control the background, to reconstruct angles and to estimate energies.
An additional advantage of the modular design is evident when one considers the need to veto after the passage of muons through the tank. Muon initiated interactions can give rise to long-lived isotopes of nitrogen (such as ^{16}N with a half-life of 7.4 s and maximum electron energy of 10.4 MeV). It is therefore a distinct advantage to only turn off one module (say 10% of the experiment) after the passage of each muon.

(b) We need to keep the dimensions down to less than a few metres, and to trap as many of the ultraviolet photons as possible. Figure 5 shows the number of cherenkov photons from a 6 MeV electron that can be detected after having travelled through various thicknesses of water. The plots are for a wavelength window with upper bound fixed at 700 nanometers and variable lower bound. From the graph it can be seen that we can expect only about 600 photons in the visible region (400 - 700 nanometres). If this lower bound is extended down to 200 nanometres, this number swells to 2000 - but this advantage is quickly lost after passage through only a few metres of water because of the high attenuation of ultraviolet photons in water.

(c) The use of photomultipliers to directly detect these cherenkov photons is inefficient on two accounts. Firstly there is a wavelength mis-match between the spectral distribution of the photons and the spectral response of the photomultiplier cathods. Secondly, the photons are so widely spread out (after multiple scattering) that most would be missed by the tubes. We plan to improve on this situation by utilising wavelength shifting panels to define the faces of the cube; and to view each panel with sufficient tubes to reduce the number of accidental coincidences to an acceptable level. This will probably require about 5 tubes per panel. Figure 6 shows the situation in schematic detail.

The ultraviolet photons are absorbed by the panel and re-emitted in the visible. These visible photons are then channelled by total internal reflection to the edges of the panel, and then transferred to the tubes (which could be well away from the detector) by means of light guides or bundles of optical fibres.

Unfortunately, when one considers the total efficiency of the process (involving absorption and re-emission, coupling to the tubes, solid angle of the forward cone, attenuation, coupling losses etc.), the number comes out to be approximately 1%. If we have around 1500 photons impinging upon the panel, we will therefore only have 15 photons to distribute to 5 tubes; or 3 photons/tube for a 6 MeV electron. Obviously this is too small a number around which to plan an experiment, and our first priority is to see by how much we can improve the response of the panels. This could be achieved by viewing more than one edge, and perhaps adding diffuse reflector to the broad face - as shown in figure 6. If the response can be improved by a factor of at least three (giving 10 photons/tube), the method becomes a promising proposition.

CONCLUSION

It is clear that the direct detection of neutrinos from their scattering in water will be, at best, a long and difficult procedure. It is nevertheless true to say that the scientific merit of the measurement to be made warrants the investment of considerable time and effort. The experimental program that we have proposed above would be complementary to any search using the inverse beta reaction, and would give us valuable understanding about the operating conditions in the centre of the sun.

ACKNOWLEDGEMENTS

We gratefully acknowledge the excellent research facilities provided within the School of Physics by Professor H. Messel and the Science Foundation for Physics. This continuing work is being made possible by grants from the Australian Research Grants Scheme.

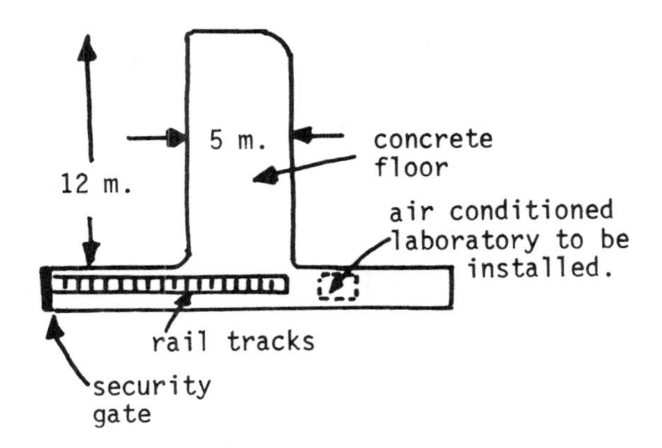

PLAN VIEW

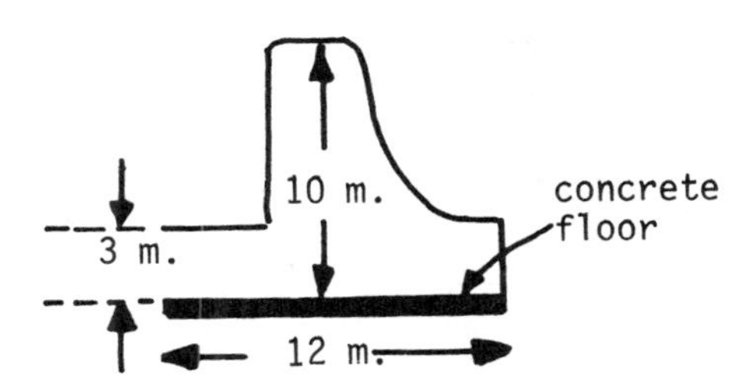

ELEVATION VIEW

Fig. 1 Plan and elevation view of our existing chamber at Broken Hill. The chamber is at a depth of 1230 metres or 3300 m.w.e.

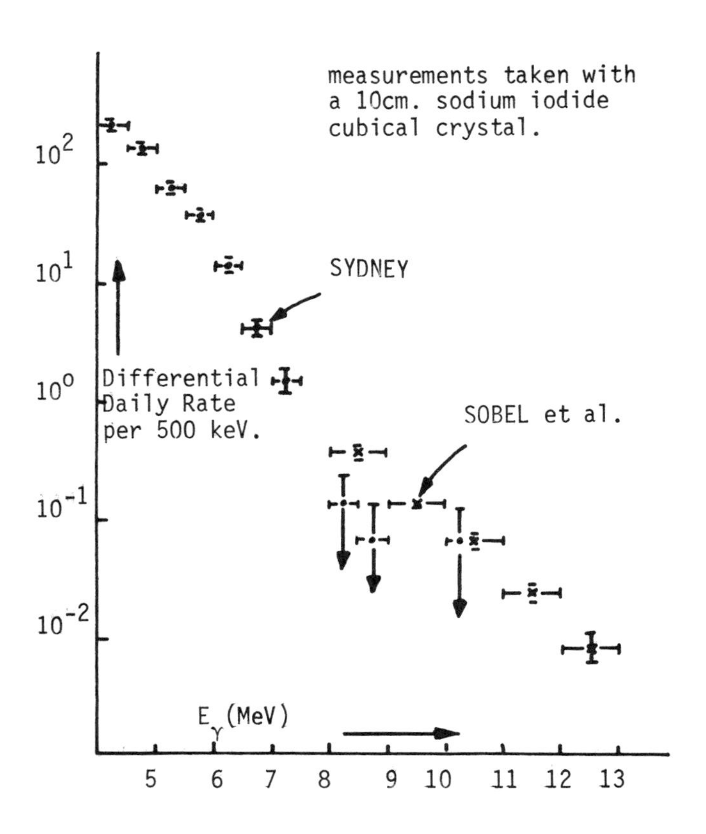

Fig. 2 Differential energy spectrum for gamma rays measured in our chamber with a 10 cm sodium iodide crystal. The results are compared with those of Sobel et al. Phys. Rev. C7 , 1564-1579 (1973).

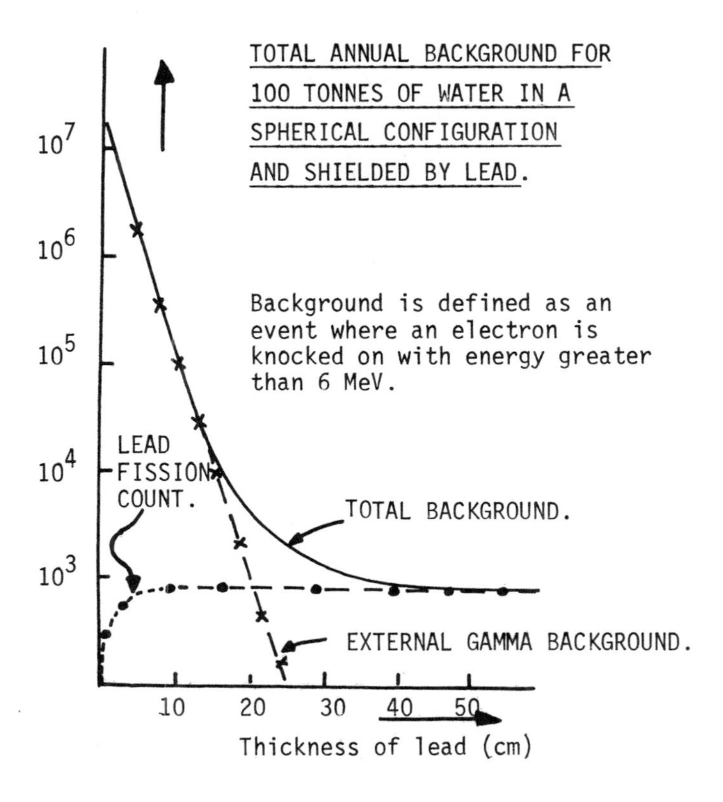

Fig. 3 Total annual background for 100 tonnes of water in a spherical configuration and shielded by various amounts of lead. Background is defined as an event where an electron is knocked on with energy greater than 6 MeV.

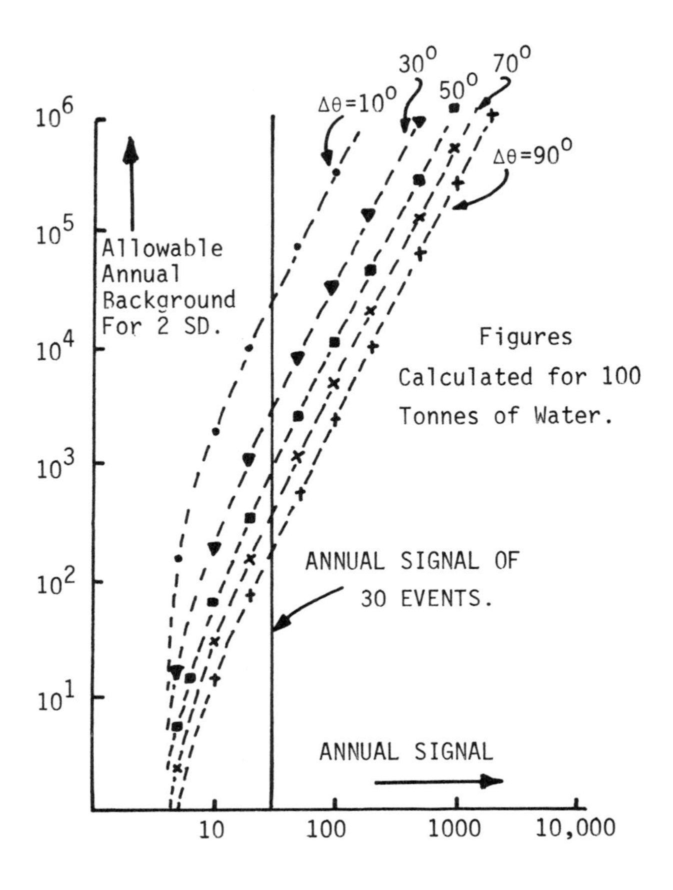

Fig. 4 Annual background that can be tolerated as a function of annual signal and angular sensitivity for 100 tonnes of water. The curves have been calculated assuming the requirement of two standard deviation significance within one year.

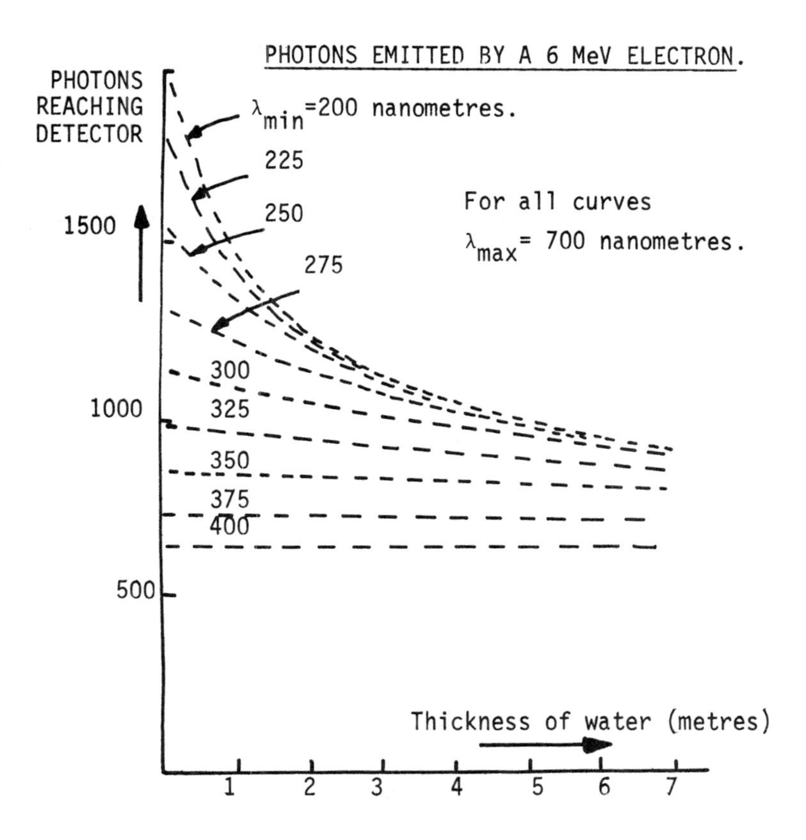

Fig. 5 Measurable number of photons emitted by a 6 MeV electron after the cherenkov radiation photons have passed through various thicknesses of water. The curves have been calculated for various wavelength windows.

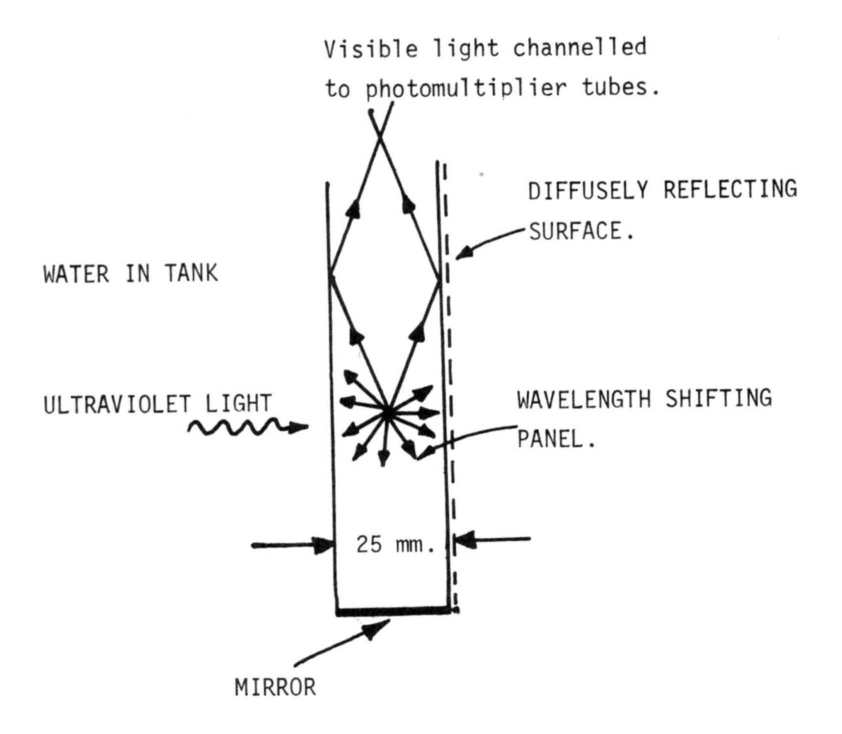

WITHOUT AIR JACKET FORWARD CONE = 5.7% OF TOTAL.
WITH AIR JACKET FORWARD CONE =16.7% OF TOTAL.

Fig. 6 Schematic diagram of a wavelength shifting panel. Various techniques may be necessary to enhance the efficiency of the panel response.

ELECTRONIC DETECTORS FOR THE STUDY OF ^{8}B SOLAR NEUTRINOS *+

(with sensitivity to energy and direction)

Herbert H. Chen

Dept. of Physics, University of California, Irvine, CA 92717

ABSTRACT

The statistical requirement for any directional ^{8}B solar neutrino experiment with an initially large background to signal ratio (about 1000) indicates the need for extremely large detectors. However, this detector scale can be decreased substantially if the detector mass can be made radioactively clean so that only external backgrounds remain. Then the detector can be operated in a self-shielding mode. Among a number of detector options, the water Cherenkov approach being used in the search for proton decay satisfies these requirements. With the neutrino deuteron reaction via the use of heavy water rather than light water, the background to signal ratio can be further improved and one can demonstrate observation of the ^{8}B solar neutrino much before one can demonstrate its directionality.

I. INTRODUCTION

The solar neutrino "problem", i.e. that there are fewer neutrinos from the sun as observed in the chlorine/argon radio-chemical experiment of Davis et al. [1] than that predicted by the "standard" solar model [2], has prompted a variety of solutions ranging from neutrino oscillations [3], neutrino decay [4], to a very large variety of non-standard solar models [5]. These have been discussed widely over the past decade, and the discussions continue here at this conference. The new radio-chemical experiments: ^{71}Ga [6], sensitive to neutrinos from the pp reaction in the sun; ^{81}Br [7], sensitive to the ^{7}Be neutrino; and the geo-chemical experiment: ^{98}Mo [8], sensitive to the ^{8}B flux averaged over the last several million years; will add greatly to our knowledge when they are carried out.

It is clear, however, that radio-chemical and geo-chemical experiments need to be complemented by direct-counting experiments, particularly those sensitive to neutrino direction as well as to energy. With sensitivity to direction and to energy, direct-counting experiments can demonstrate unambiguously

* Research supported in part by the National Science Foundation.
+ Research supported in part by the U.S. Department of Energy.

that solar neutrinos from a particuliar reaction chain have been detected.

Thus, the pioneering effort by Davis et al. should be complemented by a variety of efforts, i.e. radio-chemical, geo-chemical, and direct-counting, to understand and to resolve this longstanding problem. The absence so far of solar neutrino experiments other than the chlorine/argon experiment of Davis et al. is a partial testiment of the difficulty of the task to be accomplished.

II. STATISTICAL REQUIREMENTS

We begin by considering the statistical requirements of any directional and observational experiment. The statistical requirements for direction and observation may differ depending on the reaction used, i.e. ν_e, e^- elastic scattering or inverse-beta reactions.

For neutrino energies much larger than the electron mass, the directional characteristics of ν_e, e^- elastic scattering arises purely from kinematics. In this case, the direction and observation requirements are identical, and are given by:

$$N_e = n_\sigma^2 \, [\, 1 + (B/S)_e \, (\, 1 + 1/M_e \,)\,]$$

where: N_e is the required number of ν_e, e^- events,

n_σ is the desired number of standard deviations,

$(B/S)_e$ is the background to signal ratio for ν_e, e^- scattering,

M_e is the ratio of solid angle outside of the forward cone normalized to the forward cone.

The directional characteristics of "allowed" inverse-beta reactions result from the Fermi (vector) and Gamow-Teller (axial-vector) matrix elements of the transition and the two component (left-handed) neutrino. The differential cross section for these inverse-beta reactions in this energy range is:

$$\frac{d\sigma}{d\Omega} = \overline{\frac{d\sigma}{d\Omega}} \, (1 + \alpha \, \cos\theta)$$

where α is determined by the Fermi and Gamow-Teller matrix elements:

$$\alpha = (\langle 1\rangle^2 - 1/3 \, g_A^2 \, \langle\sigma\rangle^2)/(\langle 1\rangle^2 + g_A^2 \, \langle\sigma\rangle^2)$$

and g_A is the ratio of axial-vector to vector coupling strengths, $g_A = 1.25$. So α has a value between $-1/3$ and 1. θ is the angle between the incident neutrino and the outgoing electron. Thus, there is a backward/forward asymmetry depending on α. This asymmetry is not as distinctive as the forward peaking of the elastic scattering reaction, but the higher event rate of these inverse-beta reactions partially compensates so that the directional capabilities of such inverse-beta processes with large values of α can be competitive with elastic scattering [9]. A particularly favorable case is the neutrino deuteron inverse-beta reaction where the cross section is large (the neutron is almost free) and α equals -1/3 (Gamow-Teller transition). To be explicit, we will consider here only the neutrino deuteron inverse-beta reaction, with deuterons contained in heavy water.

The backward/forward asymetry is maximized by comparing events in the backward/forward thirds of the total solid angle. The statistical requirement in this case is (direction):

$$N_d = n_\sigma^2 \; [\, 1 + (B/S)_d \,] \; [\, 27/(8\,\alpha^2\,) \,],$$

where: N_d is the number of inverse-beta events required for direction determination,

$(B/S)_d$ is the background to signal ratio for the inverse-beta reaction,

α is the asymmetry parameter defined above.

For observation of the inverse-beta reaction, the statistical requirement is decreased, but one has to include the necessity of a target out measurement (observation):

$$N_o = n_\sigma^2 \; [\, 1 + (B/S)_d \,(\, 1 + 1/M_o \,)\,] \; [\, 1 + M_o \,]$$

where: N_o is the number of inverse-beta events required for observation,

M_o is the run time with target out normalized to that with target in. This is optimized for a given $(B/S)_d$ ratio, and it is given by:

$$M_o = [\, (B/S)_d \,/(\, 1 + (B/S)_d \,)\,]^{\,0.5}$$

The number of events needed for a 3 standard deviation effect for the two directional cases is shown in Table 1. The event rate for the two reactions is calculated below so that the live-time needed to observe the effect for a given detector mass, or the detector mass needed to observe the effect in 6 months live-time (one year run-time) is also given in Table 1.

Assuming a detector mass of a thousand tons of heavy water, we give rates for the elastic scattering reaction,

$$\nu_e + e^- \rightarrow \nu_e + e^-$$

Table 1

Statistical requirements for a directional experiment with a given B/S ratio. Rates are calculated assuming a ^{8}B flux of $2x10^6$ cm^{-2} sec^{-1}, and a detection threshold of 7 MeV.

$(B/S)_i$	Number of Events (Required for 3 std dev effect)		Time (Days using a 10^3 ton detector)		Mass (Tons for an effect in 6 mo)	
	ν_e,e	ν_e,d	ν_e,e	ν_e,d	ν_e,e	ν_e,d
0.1	10.	300.	15.	30.	80.	160.
0.3	12.	360.	18.	35.	100.	195.
1.0	19.	550.	29.	55.	160.	300.
3.0	40.	1090.	60.	110.	330.	600.
10.	110.	3010.	170.	300.	920.	1650.
30.	320.	8480.	480.	850.	2610.	4640.
100.	1040.	27600.	1560.	2760.	8530.	15100.

and the inverse-beta reaction,

$$\nu_e + d \rightarrow p + e^- + p$$

The flux of ^8B solar neutrinos is taken to be,

$$F_{\nu_e}(^8B) = 2 \times 10^6 \text{ cm}^{-2} - \text{sec}^{-1},$$

the maximum number consistent with the observation of Davis et al. [1], and about a factor of three lower than that predicted by the "standard" solar model. The number of electron and deuteron targets in a thousand tons of heavy water is:

$$N_e = (10/20)x(6x10^{23})x(10^9) = 3.0x10^{32},$$

$$N_d = (\ 2/20)x(6x10^{23})x(10^9) = 0.6x10^{32}.$$

The event rates are:

$$R_e = F_{\nu_e} \sigma_e N_e \varepsilon_e$$

$$= (2x10^6 \text{ cm}^{-2}\text{-sec}^{-1})x(0.7x10^{-43} \text{ cm}^2)$$

$$x(3.0x10^{32})x(0.2)x(8.64x10^4 \text{ sec-day}^{-1})$$

$$= 0.7 \text{ day}^{-1}$$

$$R_d = F_{\nu_e} \sigma_d N_d \varepsilon_d$$

$$= (2x10^6 \text{ cm}^{-2}\text{-sec}^{-1})x(1.2x10^{-42} \text{ cm}^2)$$

$$x(0.6x10^{32})x(0.7)x(8.64x10^4 \text{ sec-day}^{-1})$$

$$= 9 \text{ day}^{-1}.$$

The cross section for the deuteron reaction [10] is about 20 times larger than that for the elastic scattering reaction. The former reaction has been measured at LAMPF using a 6 ton heavy water Cherenkov detector [11] and the latter reaction is currently being studied at LAMPF [12]. For a detection threshold, 7 MeV is chosen here for both reactions. Note that the detection efficiency for the deuteron reaction is about 70%, while that for the elastic scattering reaction is about 20%. For the inverse-beta reaction, the high efficiency occurs because the outgoing electron carries essentially all the incident neutrino energy (minus the Q value) since the outgoing protons balance momentum without absorbing much energy; while for elastic scattering, the low efficiency occurs because the scattered electron shares energy with the outgoing neutrino, thus it has a soft recoil spectrum.

The high detection efficiency at relatively high thresholds (7 MeV) for the inverse-beta reaction as compared to the elastic scattering reaction is helpful not only in providing a higher event rate in an energy region with lower backgrounds but also in contributing to the possibility of a more accurate neutrino flux measurement, since this rate is quite insensitive to uncertainties in the detector energy calibration (thresholds).

Fig. 1, 2, and 3 shows the required detector mass as a function of background to signal ratio (B/S) for a three sigma effect in the six month live-time (one year run-time) specified in Table 1, i.e. for observation/direction using the elastic scattering reaction in light water, and for observation and for direction using the deuteron reaction in heavy water. An immediate (obvious) conclusion is that the observation experiment, using the deuteron reaction in heavy water, can be accomplished much faster, or alternatively it requires a much smaller detector mass.

To determine which reaction might be better for a directional experiment, one has to compare background to signal ratios. For a given background environment, an estimate is that the background to signal ratio for the deuteron reaction in heavy water is about a factor of two smaller than that for the elastic scattering reaction in light water. The higher rate for the former (factor of about 16) more than compensates the smaller solid angle for the latter (factor of about 8). The net result of such a comparision is that the two directional approaches are about equal in sensitivity [9].

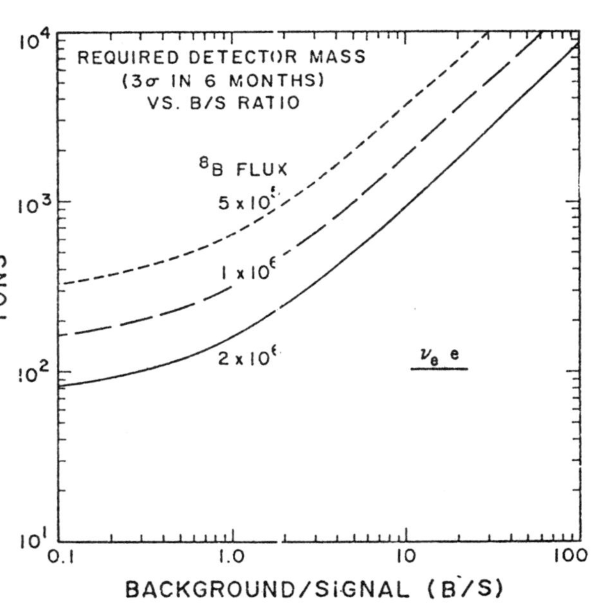

Fig. 1

The required detector mass for the observation and the directional determination of ^{8}B solar neutrinos, using ν_e,e^- scattering, is given as a function of the background to signal ratio (B/S).

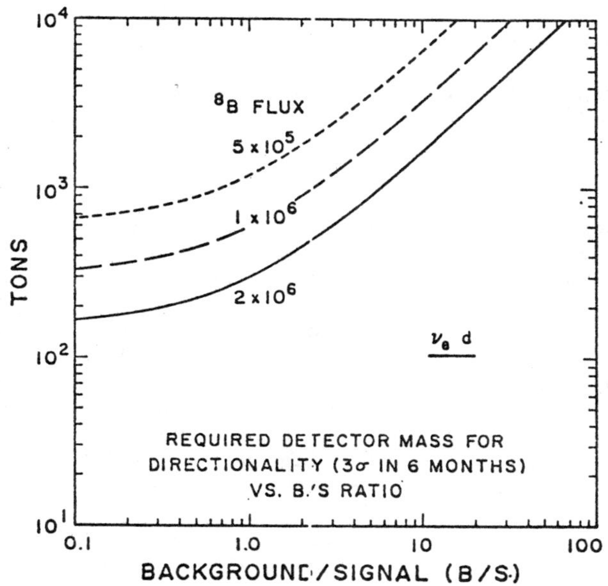

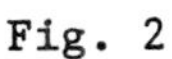

Fig. 2

The required detector mass for the directional determination of ^{8}B solar neutrinos, using the ν_e,d inverse beta reaction, is given as a function of the background to signal ratio (B/S).

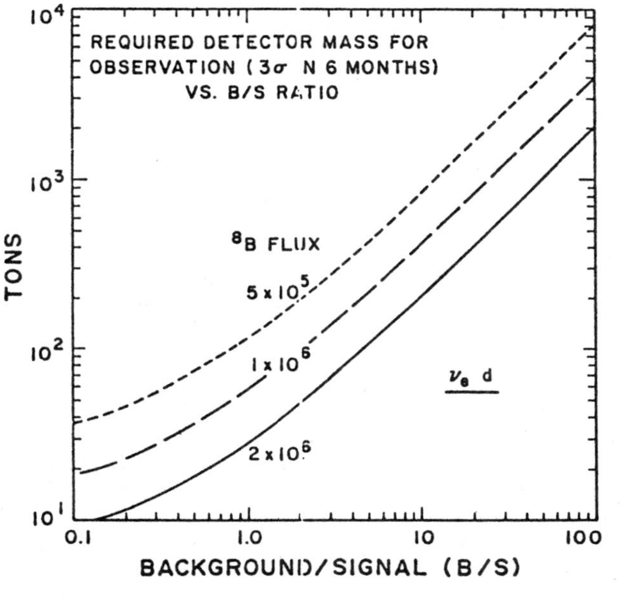

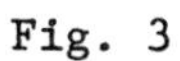

Fig. 3

The required detector mass for the observation of ^{8}B solar neutrinos, using the ν_e,d inverse beta reaction, is given as a function of the background to signal ratio (B/S).

III. THE SELF-SHIELDING DETECTOR

From Figures 1 to 3, it is clear that the required detector mass grows linearly when the background to signal ratio (B/S) is much larger than one, and that the mass scale for a directional experiment exceeds 100,000 tons at the ^{8}B flux level of $2x10^{-6}$ cm^{-2} sec^{-1} for B/S at about 1,000, the value which has been achieved in direct-counting experiments so far [13]. Figures 4 to 6 shows the ^{8}B flux sensitivity for a number of detector masses as a function of B/S. If this background is uniformly distributed throughout the detector volume, then the large detector mass required will likely preclude the possibility of any directional ^{8}B solar neutrino experiment unless the B/S ratio is reduced.

Ultimately, if the target/detector medium can be made sufficiently clean (of radioactivity) so that all residual backgrounds originate from outside the target/detector, then the target/detector medium can serve as a clean shield. For a given initial detector mass, it is useful to consider how the background to signal ratio would improve as the fiducial mass decreases. The B/S reduction factor contains three terms:

1. The signal dependence on fiducial mass,
2. The solid angle factor for backgrounds entering the fiducial volume, and
3. The attenuation of backgrounds entering the fiducial volume.

These three terms combine to give:

$$(B/S) = (B/S)_i \; [\; (M_i/M)\; (D/D_i)^2 \; \exp-(D_i-D)/2\lambda]$$

where: the subscript i indicates initial values,
M is the fiducial mass,
D is the fiducial volume diameter, and
λ is the attenuation length for backgrounds.

The factor within the square brackets gives the B/S reduction factor. Table 2 gives an example of possible B/S reduction factors for an initial detector mass of 3,200 tons of water (16 m diameter) assuming a factor of 5,10 and 10 attenuation per meter. For example the attenuation length of 10 MeV photons is about a factor of 10 per meter of water. Fig. 7 and 8 show the detector mass as a function of this B/S reduction fctor for an initial detector mass of 3,200 tons and 1,360 tons (12 m diameter) of water. An overlay of either of these two Figures over Figures 1, 2, or 3 will show the ^{8}B flux sensitivity for any initial B/S condition. The results of such an exercise are also shown in Fig. 4 to 6. It is clear that the clean self-shielding detector can achieve much greater flux sensitivity, and the gain in sensitivity depends critically on the attenuation length for backgrounds (photons and/or neutrons).

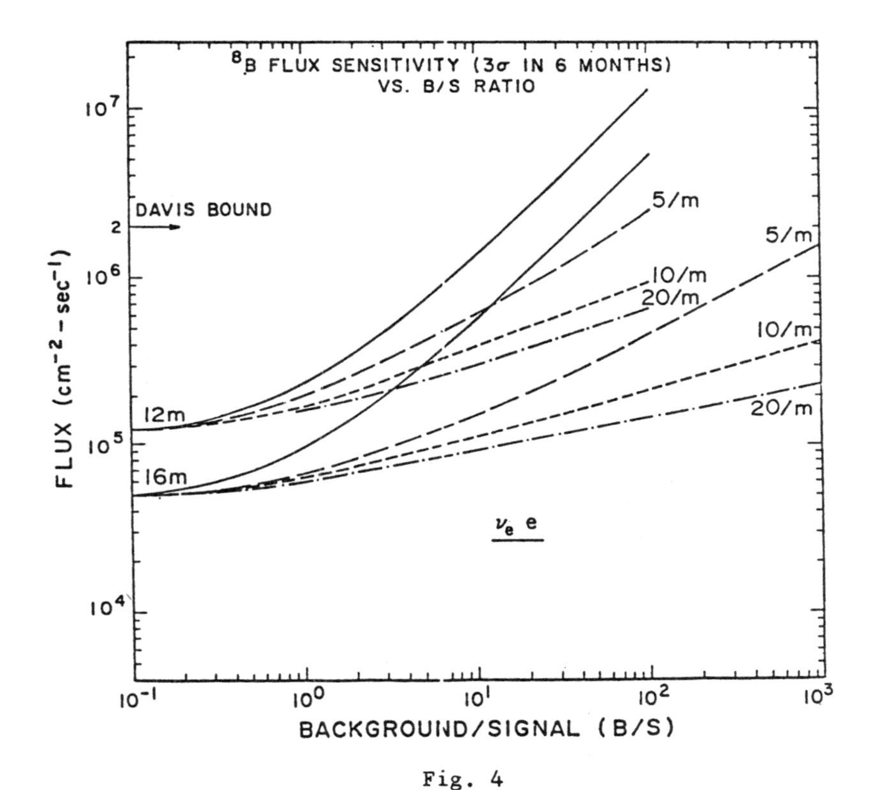

Fig. 4

The ^{8}B flux sensitivity, for right circular cylinder detectors of 12 m and 16 m, using ν_e, e^- scattering, is given as a function of the B/S ratio. The solid line gives the sensitivity for a volume distributed background; the three types of dashed lines are given for the cases where the background is external to the detector and where water is assumed to attenuate the background by factors of 5/m, 10/m, and 20/m.

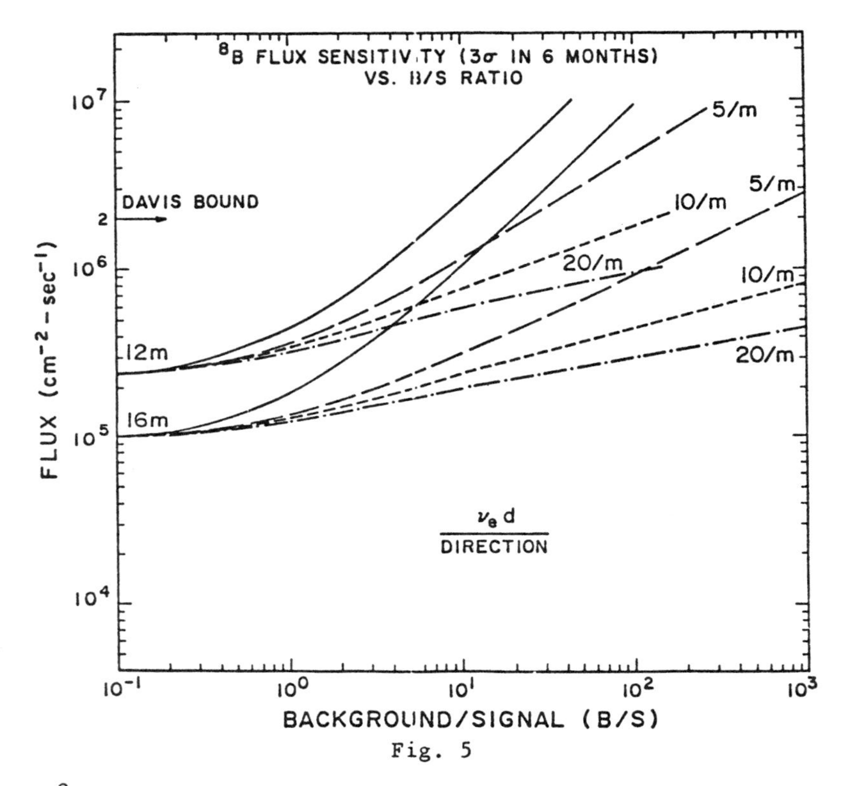

Fig. 5

The ^8B flux sensitivity for direction, with right circular cylinder detectors of 12 m and 16 m, using the ν_e,d inverse beta reaction, is given as a function of the B/S ratio. The solid line gives the sensitivity for a volume distributed background; the three types of dashed lines are given for the cases where the background is external to the detector and where water is assumed to attenuate the background by factors of 5/m, 10/m, and 20/m.

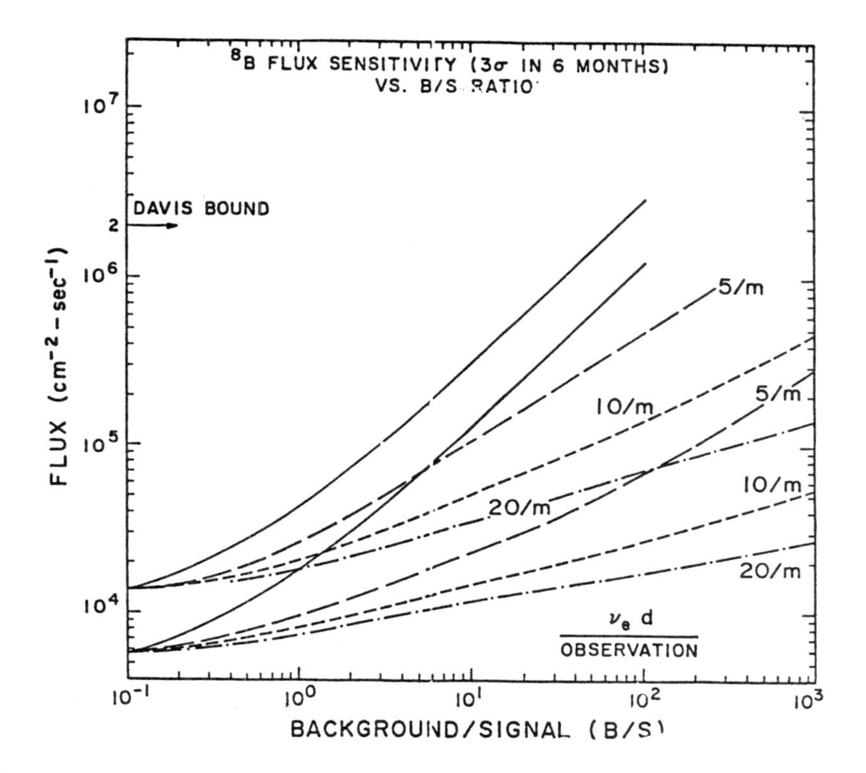

Fig. 6

The ^{8}B flux sensitivity for observation, with right circular cylinder detectors of 12 m and 16 m, using the ν_e,d inverse beta reaction, is given as a function of the B/S ratio. The solid line gives the sensitivity for a volume distributed background; the three types of dashed lines are given for the cases where the background is external to the detector and where water is assumed to attenuate the background by factors of 5/m, 10/m, and 20/m.

Table 2

Detector mass versus B/S reduction factor for a 3200 ton water Cherenkov detector of 16 meter diameter.

Mass (tons)	Diameter (meters)	B/S reduction factor (attenuation factor per meter) (5/m	10/m	20/m)
M_o= 3200	D_o= 16	1.	1.	1.
2160	14	$2.4x10^{-1}$	$1.1x10^{-1}$	$5.7x10^{-2}$
1360	12	$5.3x10^{-2}$	$1.3x10^{-2}$	$3.3x10^{-3}$
780	10	$1.3x10^{-2}$	$1.6x10^{-3}$	$2.0x10^{-4}$
400	8	$3.2x10^{-3}$	$2.0x10^{-4}$	$1.3x10^{-5}$
170	6	$8.5x10^{-4}$	$2.7x10^{-5}$	$8.3x10^{-7}$
50	4	$2.6x10^{-4}$	$4.0x10^{-6}$	$6.2x10^{-8}$
6.3	2	$1.0x10^{-4}$	$8.0x10^{-7}$	$6.2x10^{-9}$

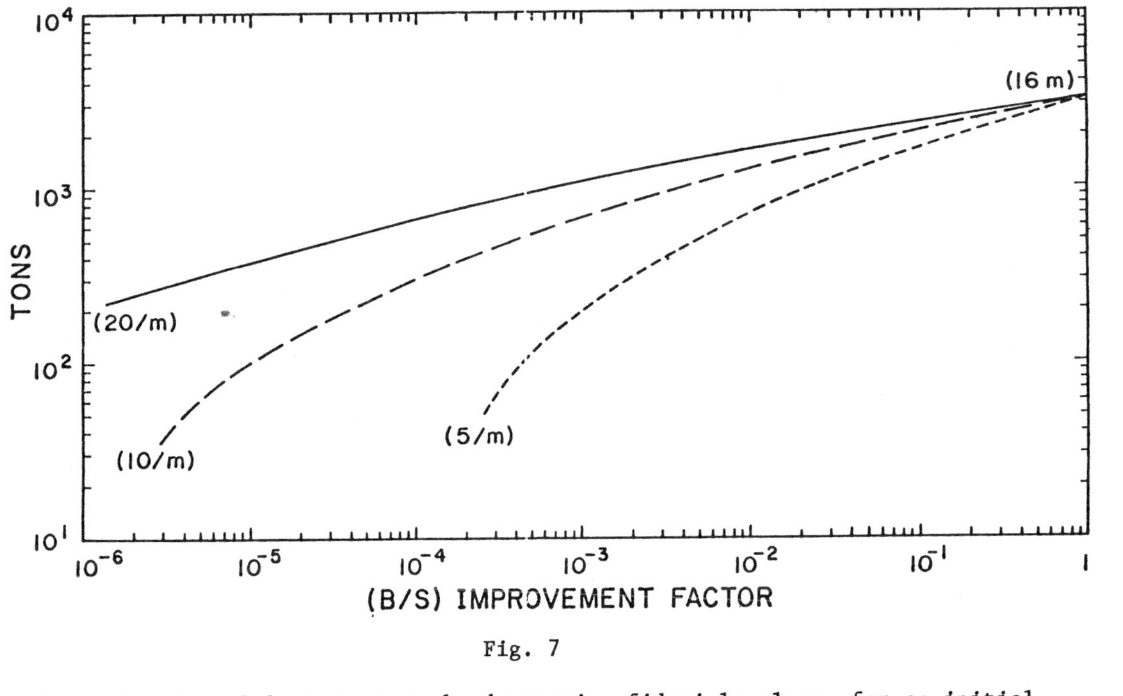

Fig. 7

The reduction of detector mass by decreasing fiducial volume, for an initial 12 m right circular cylinder detector, is given as a function of the background to signal improvement factor, where water is assumed to attenuate the external background by factors of 5/m, 10/m, and 20/m.

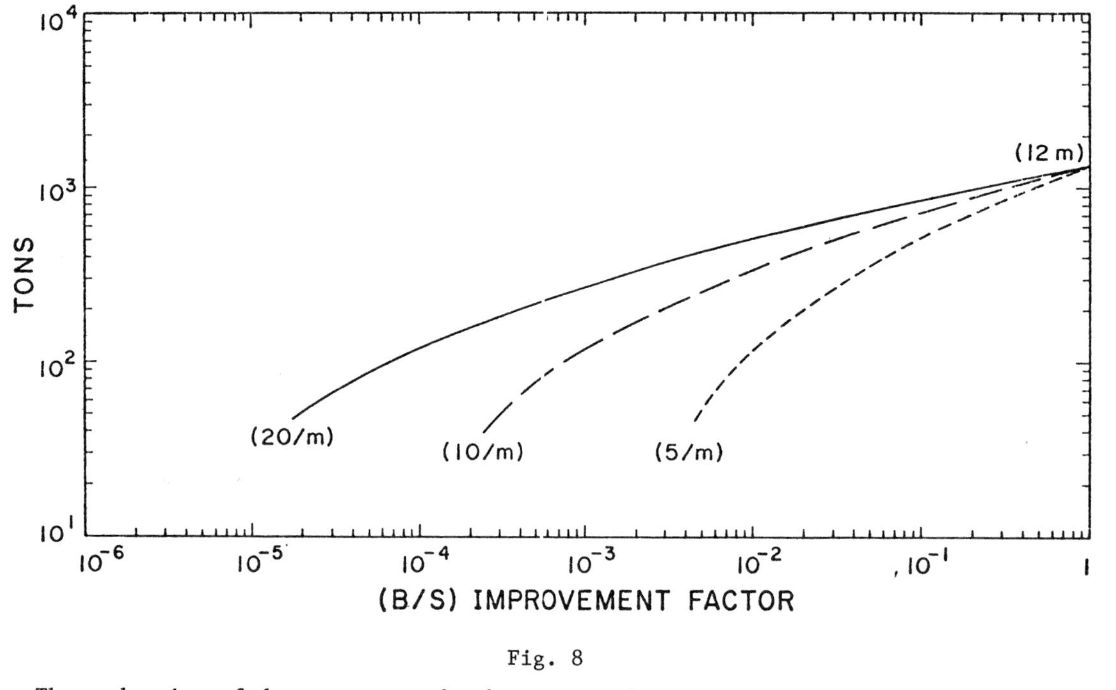

Fig. 8

The reduction of detector mass by decreasing fiducial volume, for an initial 16 m right circular cylinder detector, is given as a function of the background to signal improvement factor, where water is assumed to attenuate the external background by factors of 5/m, 10/m, and 20/m.

IV. POSSIBLE DETECTORS

Three categories of electronic detectors have been considered in the search of ^{8}B solar neutrinos with directional and energy sensitivity. They are: the sandwich variety; the high density time projection chamber (TPC); and the water Cherenkov.

Sandwich detectors, used for many neutrino experiments at accelerators, have the possibility of tracking, energy measurement, and particle identification. However, the fine granularity and the large mass required for a directional experiment without having a reasonable confidence in achieving the required low internal radioactivity environment so that there is the possibility of selfshielding forces one to make progress in understanding backgrounds incrementally. This suggests a very long term effort for the sandwich approach.

The high density TPC, especially liquid argon which has been under development for some time [14], promises all of the capabilities of the sandwich approach, and it has the potential of a much finer granularity than sandwich detectors. Fig. 9 and 10 are two examples of 10 MeV electrons generated using EGS3 [15]. The segmentation for the TPC readout in these Figures is 2 mm. Track definition, even with multiple scattering, appears sufficient at this level of segmentation. For this case, the available charge in each segment is about 1 fC (6000 electrons from the deposition of about 200 keV/mm with recombination and the sharing of charge in the two orthogonal readout sets included) and electronic noise becomes the serious problem when large anode capacitance is considered. The relatively few electrodes in a TPC as compared with a sandwich detector, and the high purity of the liquid required for a TPC makes such detectors approach the ideal self-shielding detector. If tracking is not required, or not possible, then operating the detector as an ionization chamber, with coarse segmentation for background rejection as well as for lower anode capacitance, may provide the means to detect the pp neutrino via ν_e, e^- elastic scattering.

The water Cherenkov detector has a number of essential features that encourages consideration of its use. Water is a medium that can be made very pure so that one can hope to decrease its concentration of radioactive contaminants, e.g. uranium, thorium, to a level well below 1 ppb. Reverse osmosis [16], filtration [17], and distillation are techniques which all offer hope that this is possible. Spontaneous fission of uranium and thorium produces high energy gammas [18] which would look like electrons in the water Cherenkov detector, and which, therefore, must be removed. The clarity of pure water also allows the use of PMT's at the surface of the detector volume so that there would be no additional radioactive contaminants in the fiducial volume. The Cherenkov technique automatically gives directional information,

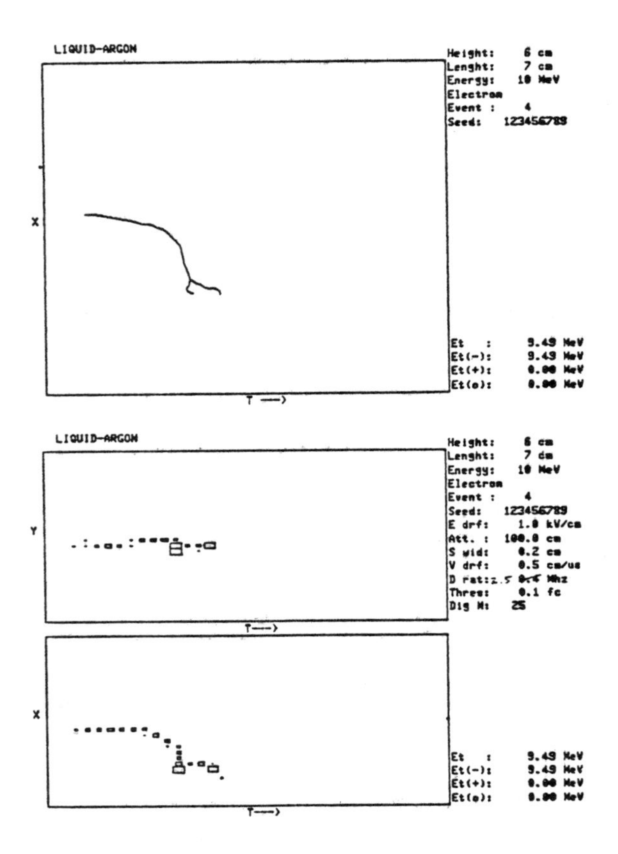

Fig. 9

A Monte Carlo simulation of a 10 MeV electron in liquid argon using EGS3 is shown in the top figure. A TPC readout of the event with 2 mm resolution is shown for the two projections in the bottom two figures.

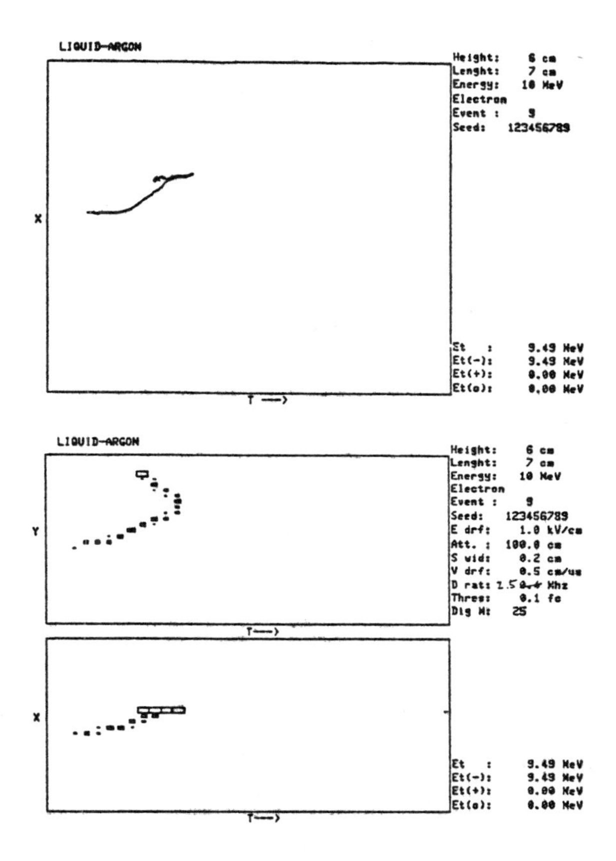

Fig. 10

A Monte Carlo simulation of a 10 MeV electron in liquid argon using EGS3 is shown in the top figure. A TPC readout of the event with 2 mm resolution is shown for the two projections in the bottom two figures.

but the low light level has been the serious stumbling block. However, recent developments of this technique for proton decay experiments forces a detailed re-examination of this approach.

V. WATER CHERENKOV DETECTORS

The existence of large water Cherenkov detectors to search for proton decay and the development of large moderately fast photomultiplier tubes [19] for the Kamioka collaboration, in particular, has led to serious consideration of adapting these large water Cherenkov detectors to search for ^{8}B solar neutrinos via the ν_e, e^- elastic scattering reaction [20,21]. The adaptation of an existing detector is clearly a most cost effective approach to a very difficult problem, and should be attempted. However, it is not at all clear that the low energy backgrounds (~ 7 MeV) in such detectors can be adquately suppressed, particularly at the relatively shallow depths and in an environment not designed with low background as an apriori requirement.

In order to be sensitive to the low energies associated with ^{8}B solar neutrinos, a minimum of 20% to 40% of the surface of an appropriately large volume of water would have to be covered by photo-cathodes, most likely by using the Kamioka 20" PMT's (Hamamatsu also has 16" PMT's which are being considered by the DUMAND collaboration [22]). These PMT's would have to be instrumented to digitize both signal amplitude and time, a la IMB, to facilitate event reconstruction. Fig. 11 and 12 show the response to an 11.4 MeV electron by a water Cherenkov with 20% and 40% coverage, respectively, for a 3 kiloton version of the IMB setup. Fig. 13 shows histograms of the number of phototube hits with 20% and 40% photocathode coverage for elastic scattering and for deuteron events. With 40% coverage (using 2,000 20" PMT's in a detector the size of Kamioka), a 7 MeV electron will result in about 55 photo-electrons [23] in about 40 PMT's [24]. A 40 tube pattern should be adequate for event recognition [25].

A detector about the size of Kamioka, somewhat smaller than IMB, has much to recommend it, e.g. single track reconstruction in such detectors have position errors of order one meter [18]. Figures 14-16 show the difference between the initial position and the fitted position, the fitted position along the initial track direction, and the fitted position perpendicular to the initial track direction, respectively, for 40% photocathode coverage. Fig. 17 shows the angular difference (in degrees) between the fitted track and the initial track. Multiple scattering is primarily responsible for this large difference. The choice of a fiducial volume two meters from the PMT plane allows the attenuation of external backgrounds while allowing event reconstruction. Thus, the self-shielding directional water Cherenkov detector leads naturally to a detector volume with a linear dimension of more than ten meters, i.e. approaching the

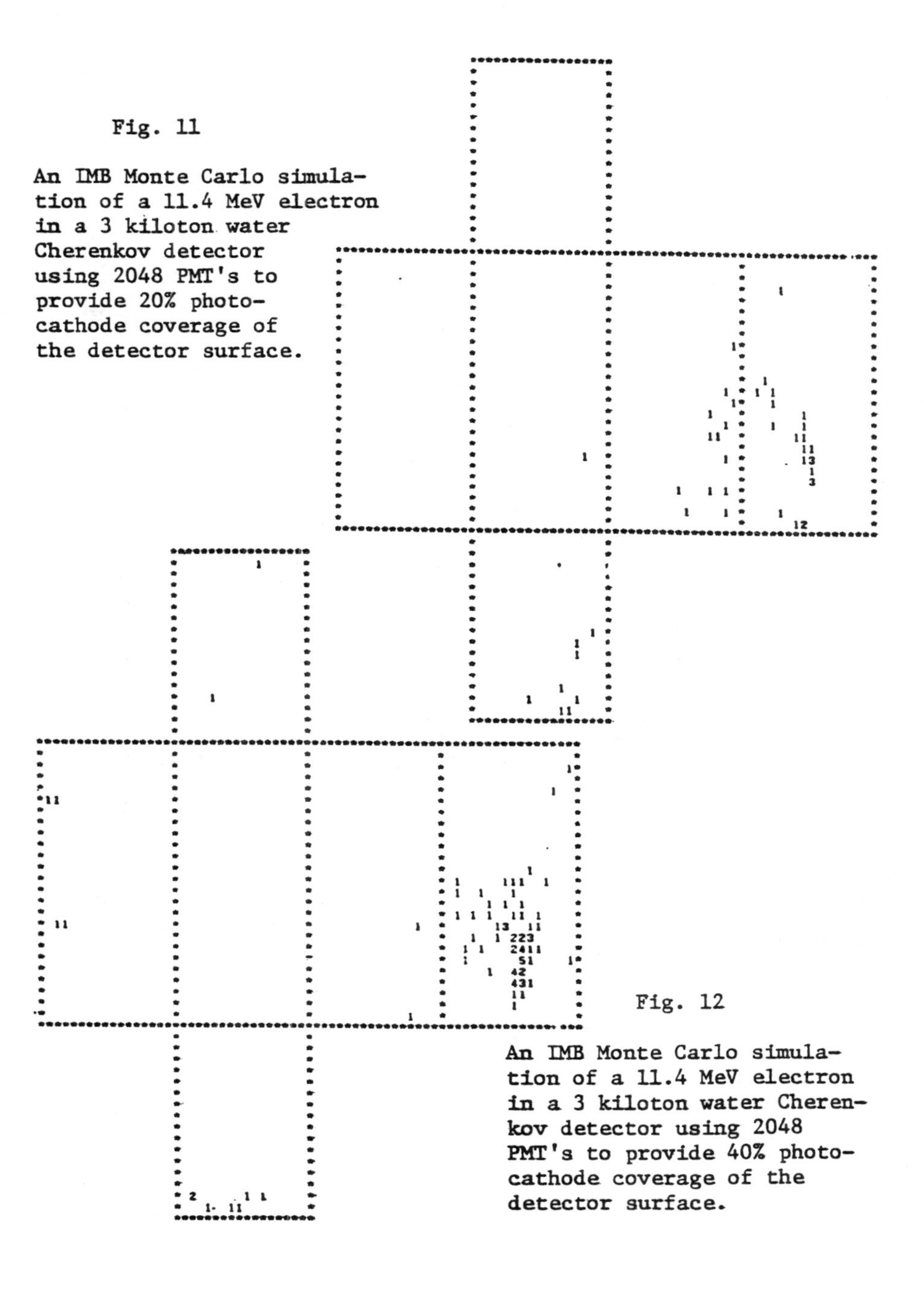

Fig. 11

An IMB Monte Carlo simulation of a 11.4 MeV electron in a 3 kiloton water Cherenkov detector using 2048 PMT's to provide 20% photocathode coverage of the detector surface.

Fig. 12

An IMB Monte Carlo simulation of a 11.4 MeV electron in a 3 kiloton water Cherenkov detector using 2048 PMT's to provide 40% photocathode coverage of the detector surface.

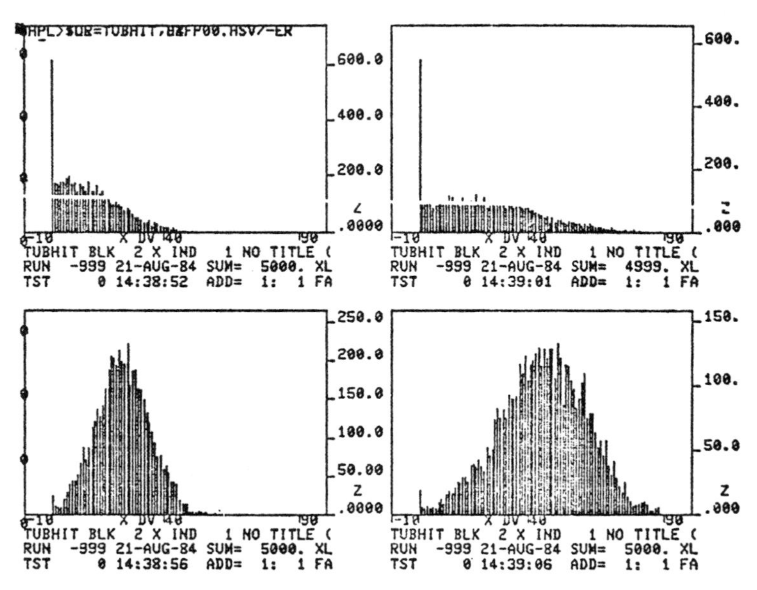

Fig. 13

An IMB Monte Carlo simulation of the number of PMT's hit for ν_e, e^- scattering and for the ν_e,d reaction with detectors using 2048 PMT's giving 20% and 40% photocathode surface coverage.

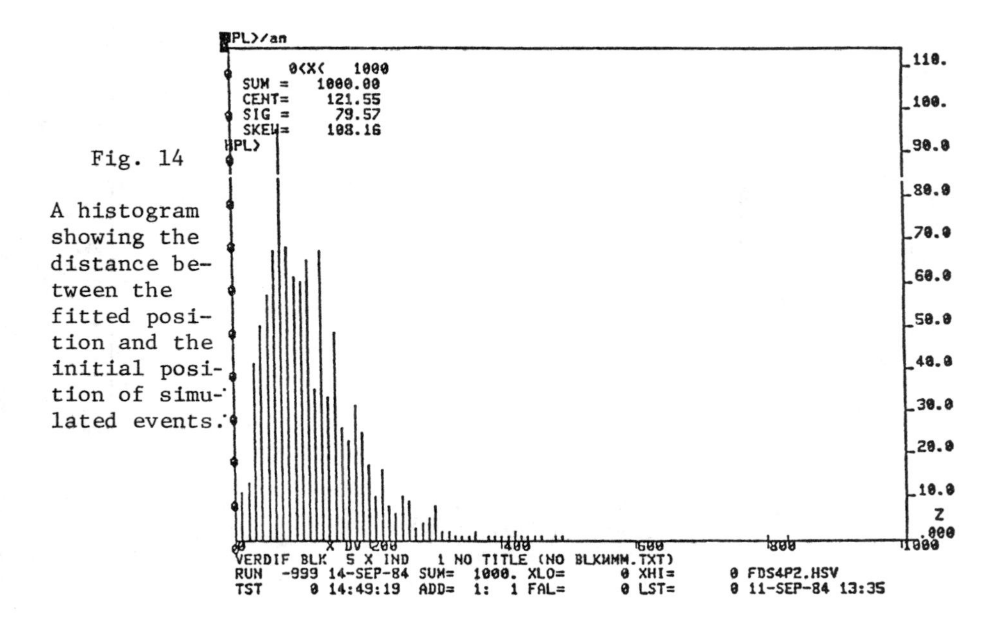

Fig. 14

A histogram showing the distance between the fitted position and the initial position of simulated events.

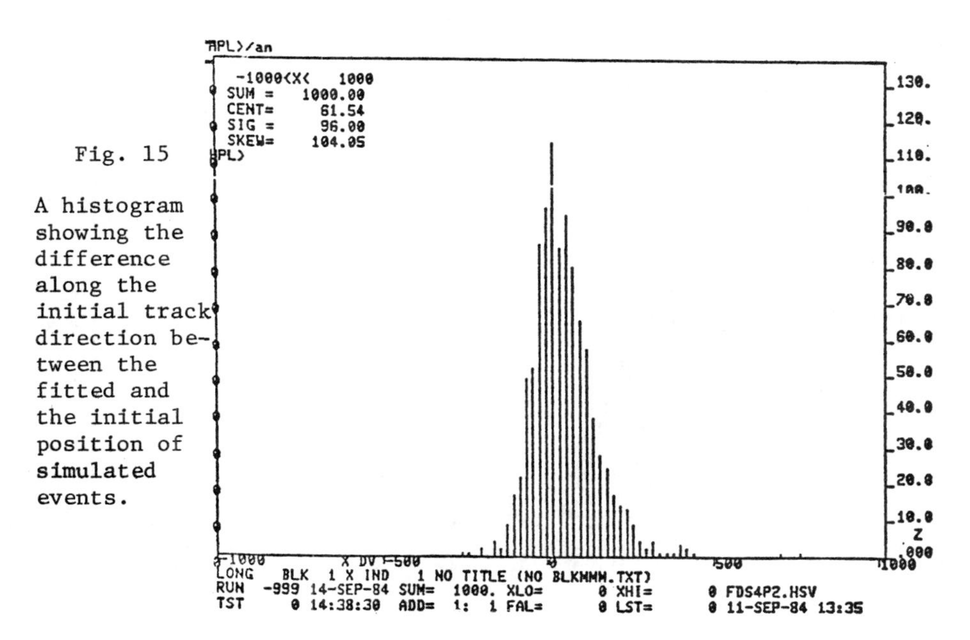

Fig. 15

A histogram showing the difference along the initial track direction between the fitted and the initial position of simulated events.

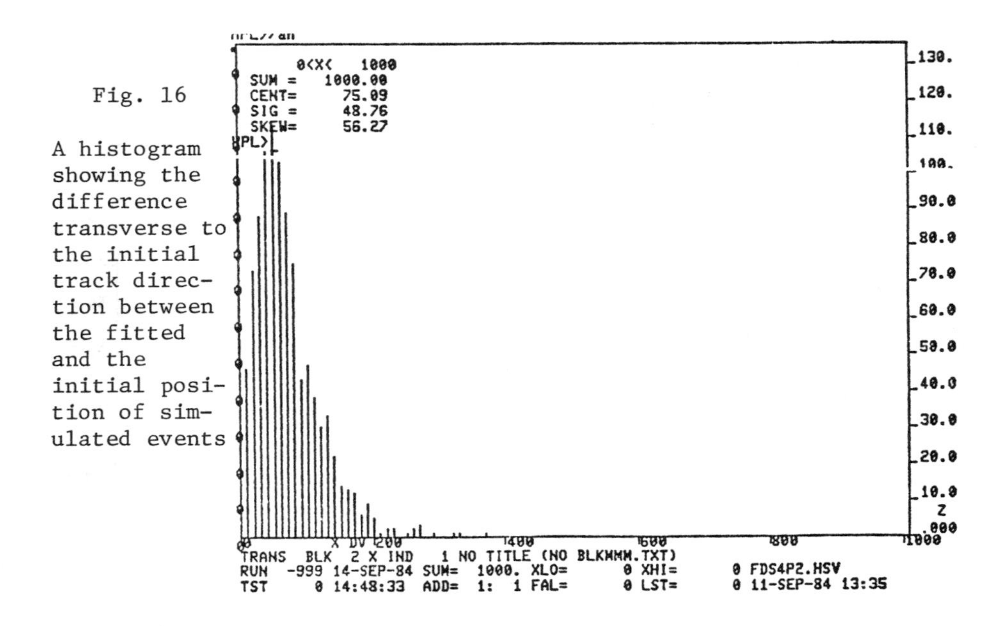

Fig. 16

A histogram showing the difference transverse to the initial track direction between the fitted and the initial position of simulated events

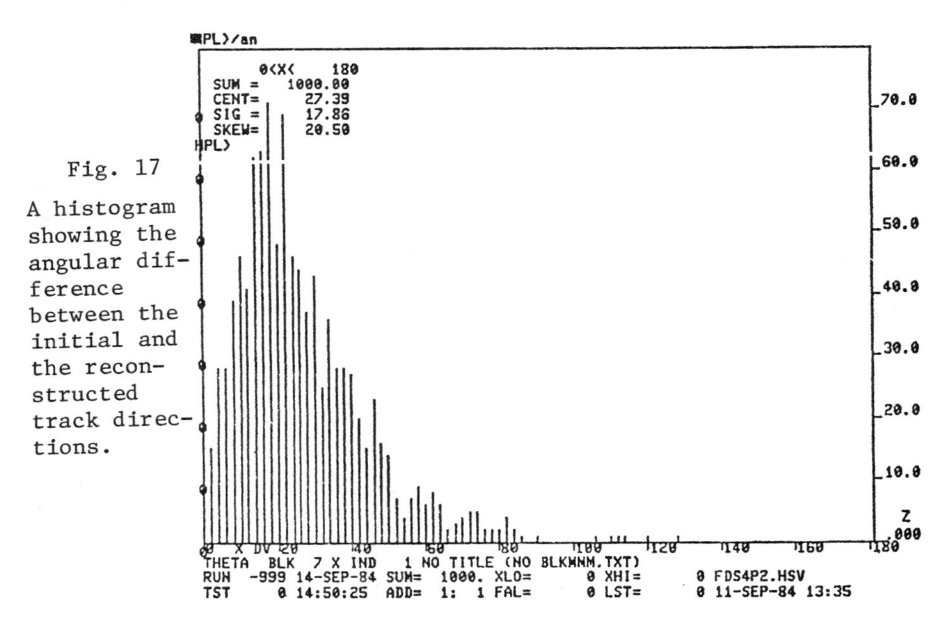

Fig. 17

A histogram showing the angular difference between the initial and the reconstructed track directions.

scale of present proton decay detectors, but with somewhat better PMT coverage in order to be fully sensitive to ^{8}B solar neutrino events.

The possibility of a water Cherenkov detector filled with heavy water (D_2O) in order to use the large cross section of the deuteron reaction for the detection of ^{8}B solar neutrinos was proposed many years ago prior to the present generation of large water Cherenkov detectors for proton decay [17,26].

Obviously, the availability of several thousand tons of heavy water for such an experiment is a critical issue. The existence of many reactors using heavy water as moderator, e.g. CANDU power reactors in Canada, encourages the belief that such volumes of heavy water not only exists, but also can be borrowed [27]. To proceed with such an experiment, one would have to be well educated about the handling of large volumes of heavy water.

In any case, a heavy water experiment will likely operate initially with a light water fill. The light water fill allows not only a thorough test of the water containment system, but also represents a "target out" mode of operation. The capability to operate in this "target out" mode is very important. In this mode one would not neglect to use the elastic scattering reaction to carry out a directional search for ^{8}B solar neutrinos. However, the neutrino flux measurement would benefit from the use of heavy water, as mentioned earlier.

A further possible test that solar neutrinos have been observed would be to demonstrate that the flux of neutrinos is correlated with changes of the sun/earth distance. The anticipated event rate of 9 day^{-1} per thousand tons of heavy water at the ^{8}B flux level of 2×10^6 cm^{-2} -sec^{-1} is somewhat marginal. But this rate is useful for a consistency check and for monitoring the existence of non-standard phenomena, e.g. neutrino oscillations, neutrino production by solar flares, or of other transient phenomena, e.g. super-novas.

VI. NOISE AND BACKGROUNDS

We consider noise and backgrounds for the water Cherenkov approach only, and assume a Kamioka sized detector instrumented with 2,000 20" diameter PMT's so as to cover 40% of the surface area, i.e. twice the photo-cathode coverage of the Kamioka detector. The detector tank would be a vertical cylinder 16 m diameter by 16 m high, for a total volume of 3,200 m^3. Defining the fiducial region to be 2.5 m from the tank wall results in a fiducial volume of 1,050 m^3. (A central transparent leak-tight vessel to contain only heavy water for the fiducial volume would minimize the mass of heavy water required. The design and construction of such a container may not be easy). This tank would be located in a cavity which would be flooded using light water and this water would be made sensitive by the use of more PMT's to provide an additional active shield several meters thick to further reduce external backgrounds, both cosmic rays and radioactivity. The cavity would then be about 20 m in diameter and 20 m high or larger. An implicit assumption is that such a cavity can be located at an appropriate depth to minimize cosmic ray backgrounds, and its walls might be lined with low background concrete to minimize radioactivity nearby. A cavity with such dimensions or larger is planned [28] for the Gran Sasso laboratory in Italy [29], and it also appears to be possible [30] for the Creighton mine in Sudbury, Canada [31].

VI.A. NOISE

Since the 2,000 PMT's in this detector would be operated at a threshold below one photo-electron, the noise in each phototube is substantial (about 20 kHz at 1/4 p.e. [14]). A coincidence window of 80 ns, the flight time for light across the detector, is necessary. Then, in any trigger, 3.2 tubes are accidentally lit. If one demands an n tube coincidence, the accidental rate is [32]:

No. Tubes	18	21	24	27
Rate(day^{-1})	1.5x10^5	734	2.3	.005

The trigger rate at the 21+ PMT level is tolerable since the number of PMT's lit for interesting events would be greater than about 40. Noise triggers can be further reduced in analysis by placing tighter constraints on time and position. Thus the noise rate can be tolerated, although quieter tubes would simplify the analysis somewhat.

VI.B. COSMIC RAY

Backgrounds from cosmic rays decrease as a function of depth. The IMB detector, at a depth of 1,700 mwe, has a cosmic ray muon rate of about 3 sec^{-1} (260,000 day^{-1}). The Kamioka detector, at a depth of 2,700 mwe, has a muon rate an order of magnitude lower for a detector which is about a factor of three smaller. The muon rate through a Kamioka sized detector at Gran Sasso (4,000 mwe) and at Sudbury (5,600' level or 5,000 mwe) are 2,000 day^{-1} and 500 day^{-1}, respectively. Greater depths are perhaps possible at Sudbury since the mine goes to the 7,000' level, but it is not clear whether this is necessary. The muon flux and associated stopped muon decays are useful for calibration. At the 7,000' level, the cosmic ray muon rate is reduced to 100 day^{-1}.

With the cosmic ray muon rate only a factor of 3 to 70 times larger than the neutrino deuteron rate (50 to 1,000 times larger than the elastic scattering rate), backgrounds generated by cosmic rays should not be too serious. The existence of the active water shield further contributes to minimizing the magnitude of this problem. Thus the production of long-lived fragments from oxygen which give energetic betas, or the production of neutrons that capture to give energetic gammas are reduced to low levels.

VI.C. RADIOACTIVITY

The background problem which remains is that from natural radioactivity, both from the surrounding rock and from the detector components. Much care would have to be exercised in the choice of location for the experiment, and the use of low radioactivity materials. Then, liberal use of light water for the additional active shield would reduce residual backgrounds, presumably to the required level. This apriori condition on a low background environment plus vigilence in its implementation, and the self-shielding approach gives the only reasonable assurance for success.

To have an initial impression of how severe the radioactive background problem might be, one searches for relevant data:

1. An earlier attempt to detect ^8B solar neutrinos by H. Sobel et al. in a S. African gold mine [32] saw a gamma flux of 3.3 hr^{-1} above 8 MeV in a liquid scintillation detector of 19 m^2, shielded by 1 m wax. A Kamioka sized detector with an area of 1,200 m^2 and corresponding shield will have a rate of 210 hr^{-1} (5,000 day^{-1}). Three additional meters of water shielding would reduce this to about the required level.

2. An experiment in the Mont Blanc tunnel to search for neutrino bursts from stellar collapse using 100 tons of liquid scintillator is now just beginning to turn on. Several 1.5 ton modules have been operated and a rate of 1.1×10^{-4} sec^{-1} above 7 MeV per module is quoted [34]. This rate includes cosmic ray muons since no veto was used. Shielding cleanliness and other details were not specified. Each module has an area of about 8 m^2. Thus, the 1,200 m^2 area of a Kamioka sized detector will have a rate of 0.0165 sec^{-1} (1,400 day^{-1}). Subtraction of the cosmic ray associated component will be possible in the near future; but this rate is already more than an order of magnitude lower, at a lower threshold, than the rate from Sobel, requiring only perhaps two meters of water for shielding.

VII. CONCLUSION

With a laboratory located deep underground and an adequate volume of heavy water, a directional ^8B solar neutrino experiment using the self-shielding water Cherenkov approach appears to have the potential to succeed, particularly if enough care is taken to control radioactivity in the detector and its environment. Much detailed work still needs to be done to fully demonstrate that this potential can be realized so that such a project can proceed with confidence.

VIII. ACKNOWLEDGEMENTS

All the useful information on large water Cherenkov detectors have come from the pioneering IMB and Kamioka experiments. The Irvine members of the IMB collaboration, as well as other members of the UCI Neutrino Group, have been most helpful in many discussions. In particular, R.C. Allen, K. Ganezer, T. Haines, A. Lash, and K.C. Wang have provided useful Monte Carlo simulations. T.W. Jones and J. van der Velde were the first to consider in detail the problems of detecting ^{8}B solar neutrinos via elastic scattering in a water Cherenkov [35]. They considered an upgrade of the IMB proton decay detector (installing 256 20" PMT's on one wall). More recently, this problem has been re-considered by M. Koshiba, B.G. Cortez, A.K. Mann, et al. for the Kamioka detector and they will attempt to detect the ^{8}B solar neutrino after completion of an upgrade.

REFERENCES AND FOOTNOTES

1. R. Davis Jr., D.S. Harmer, and K.C. Hoffman, Phys. Rev. Lett. 20, 1205 (1968); J.N. Bahcall and R. Davis Jr., Science 191, 264 (1976); R. Davis Jr., BNL Solar Neutrino Conf. BNL 50879. 1, 1 (1978); also, see J.K. Rowley's report at this conf.

2. J.N. Bahcall, W.F. Huebner, S.H. Lubow, P.D. Parker, and R.K. Ulrich, Rev. Mod. Phys. 54, 767 (1982).

3. B. Pontecorvo, Sov. Phys.-JETP 26, 984 (1968).

4. J.N. Bahcall, N. Cabibbo, and A. Yahil, Phys. Rev. Lett. 28, 316 (1972); also see F. Reines, H.W. Sobel, and H.S. Gurr, Phys. Rev. Lett. 32, 180 (1974).

5. R.T. Rood, BNL Solar Neutrino Conf. BNL 50879. 1, 175 (1978).

6. J.N. Bahcall, B.T. Cleveland, R. Davis Jr., I. Dostrovsky, J.C. Evans Jr., W. Frati, G. Friedlander, K. Lande, J.K. Rowley, W. Stoenner, and W. Weneser, Phys. Rev. Lett. 40, 1351 (1978).

7. G.S. Hurst, C.H. Chen, S.D. Kramer, M.G. Payne, and R.D. Willis, Science Underground, AIP Conf. Proc. 96, 96 (1983).

8. G.A. Cowan and W.C. Haxton, Science Underground, AIP Conf. Proc. 96, 105 (1983).

9. H.H. Chen, internal report, UCI-Neutrino No. 15, July (1975).

10. F.J. Kelly and H. Uberall, Phys. Rev. Letters 16, 145 (1966); also, see S.D. Ellis and J.N. Bahcall, Nucl. Phys. A114, 636 (1968); For ν_e, e^- elastic scattering, Kelly and Uberall assumed the V-A cross section. The Weinberg/Salam model gives a cross section about 60% that of V-A.

11. S.E. Willis, V.W. Hughes, P. Nemethy, R.L. Burman, D.R.F. Cochran, J.S. Frank, R.P. Redwine, J. Duclos, H. Kaspar, C.K. Hargrove, and U. Moser, Phys. Rev. Lett. 44, 522 (1980); Errata. Ibid., 45, 1370 (1980).

12. LAMPF experiment E225, the Irvine/Los Alamos/Maryland collaboration, H.H. Chen, spokesman.

13. W.R. Kropp, invited talk at the APS meeting, January (1979); see UCI-10P19-134, February (1979).

14. H.H. Chen, P.J. Doe, and H.J. Mahler, Science Underground, AIP Conf. Proc. 96, 182 (1983).

15. R.L. Ford and W.R. Nelson, SLAC-210, UC-32 (1978).

16. H.W. Sobel, private communication, July (1984).

17. A.M. Fainberg, BNL solar Neutrino Conf. BNL 50879. 2, 93 (1978).

18. J. Brooks, Ph.D. thesis, U.C. Irvine (1972).

19. H. Kume, S. Sawaki, M. Ito, K. Arisaka, T. Kajita, A. Nishimura, and A. Suzuki, Nucl. Instr. & Meth. 205, 443 (1983).

20. T.W. Jones and J. van der Velde, Research Note, July (1981).

21. M. Koshiba, ICOBAN, January (1984); Also, A.K. Mann, Conf. Intersections Part. & Nucl. Phys., Steamboat Springs, Colo., May (1984).

22. J.G. Learned, private communication, July (1984).

23. M. Koshiba, ICOBAN, January (1984).

24. B.G. Cortez, UCI Seminar, May (1984).

25. The IMB collaboration uses a minimum 40 PMT cut in their analysis.

26. T. Jenkins and F. Reines; see F. Reines, Proc. Roy. Soc. A301, 159 (1967).

27. C.K. Hargrove, private communication, June (1984). In a phone conversation with G. Hanna, it appears that a thousand tons of heavy water exists at Chalk River Laboratory, and may be available on loan if approved by the Canadian government.

28. E. Bellotti, private communication, June (1984).

29. A. Zichichi, Science Underground, AIP Conf. Proc. 96, 52 (1983).

30. C.K. Hargrove, private communication, July (1984).

31. W.F. Davidson, ICOBAN, January (1984).

32. The "Learned" formula for accidental rates with n phototubes is used:

$$R_n = k^n e^{-k}/[(n-1)!\tau],$$

where $k = Nr\tau$

N is the total number of phototubes (2,000),
r is the phototube noise rate (20 kHz),
τ is the time resolution (80 ns).

33. H.W. Sobel thesis (1968).

34. O. Saavedra, NEUTRINO '84, June (1984).

35. The suggestion for a "Neutrino Eye" by K. Lande, BNL Solar Neutrino Conf. BNL 50879. 2, 79 (1978), has all of the essential ideas, but was insufficient in so far as photocathode area coverage and/or the number of PMT's is concerned. He was using 5" PMT's, and proposed using 1,000 PMT's. The large 20" PMT's were still to be developed.

ATMOSPHERIC NEUTRINO FLUXES AT LOW ENERGY*

T.K. Gaisser and Todor Stanev+
Bartol Research Foundation of The Franklin Institute
University of Delaware, Newark, DE 19716

INTRODUCTION

Since the IMB nucleon decay detector [1] has become operational two years ago the detection of neutrinos of atmospheric origin became an experimental routine. IMB and at least four other smaller detectors are increasing the worldwide neutrino statistics by more than 1.5 events/day. The neutrino interactions with products contained within the fiducial volume of the detectors, ($E_\nu < 2$ GeV) represent the major background for proton decay and are the subject of careful studies.

A calculation of the neutrino flux at 0.2 - 2.5 GeV was made[2,3] to estimate this background and to use the atmospheric neutrinos for calibration of the different detectors. The atmospheric neutrino flux varies:

a. with time as the solar activity modulates the primary cosmic ray flux.

b. with the location of the detector because of the geomagnetic cut-off. In the vicinity of the geomagnetic poles the cut-offs are close to the production threshold and do not seriously affect the downward flux. Close to the geomagnetic equator the cut-offs reach values of tens of GV and significantly decrease the downward flux. Upward fluxes come from a large portion of the Earths surface and thus appear about the same in each detector.

The geomagnetic effects, together with the modification of the decay paths in the atmosphere as function of the zenith angle, determine the angular distribution of the neutrino flux.

A comparison of our predictions with the rate and angular distribution in the IMB detector, as well as with the total rate in Kamiokande[4] shows good agreement within the limits of the accumulated statistics[3]. Although the neutrino flux at Kamioka is geomagnetically surpressed, the total counting rate in that detector is higher than at IMB by a factor of about 1.5 (180 neutrino interactions per kT.yr as compared with 120 at IMB) because the detection threshold is lower at Kamioka ($E_e > 30$ MeV as compared to 225 MeV in IMB). The high sensitivity of this detector, which may be increased to $E_e > 10$ MeV, motivated us to expand the calculation for $E_\nu > 10$ MeV.

*Talk given by Todor Stanev at the Conference on Solar Neutrino and Neutrino Astronomy, Lead, South Dakota, August 84.

+On leave of absence from the Institute for Nuclear Research and Nuclear Energy, Sofia 1184, Bulgaria.

ATMOSPHERIC NEUTRINO FLUX ABOVE 10 MeV

The existing programs for simulation of atmospheric cascades were able to accomodate a neutrino energy threshold of 10 MeV. The inelastic hadron - Air nuclei interactions close to the production thresholds were modeled as a mixture of resonances production and the parametrization, described in Ref. 3. Special attention was paid to the energy loss processes and decay kinematics down to decay at rest. As the neutrino direction cannot be experimentally determined, because of the big lepton--neutrino angle at these energies, only geomagnetic cut-offs[5] averaged over all angles were used and the corresponding average flux was calculated.

The crudest approximation in the calculation is the use of one-dimensional cascade calculation, i.e. the assumption that all neutrinos follow the direction of the primary particle. This leads to an overestimate of the neutrino flux from stopping particles by a factor of 2. In fact such decays are isotropic and half the neutrinos go backwards. For this reason our fluxes should be considered as upper limits at $E_\nu \cong 25$ MeV, which approach the true flux as the neutrino energy increases.

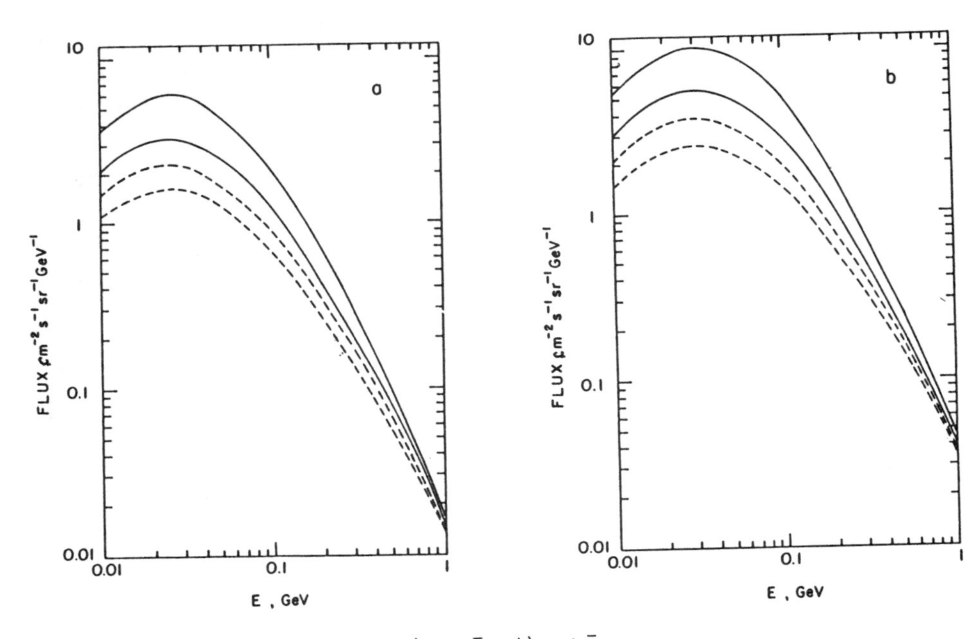

Figure 1 - Atmospheric neutrino fluxes: a) $\nu_e + \bar{\nu}_e$ b) $\nu_\mu + \bar{\nu}_\mu$.
Solid lines are for the location of IMB, dashed - for Kamioka. The upper curve in each set is for minimum, lower one - for solar maximum.

Fig. 1 shows the fluxes of electron and muon neutrinos (we have calculated the sum of neutrino and antineutrino flux) at the locations of IMB (Morton Salt Mine near Cleveland, Ohio) and Kamioka. The first location is representative for all North American experiments.

ATMOSPHERIC BACKGROUND FOR THE HOMESTAKE SOLAR NEUTRINO EXPERIMENT

To estimate the atmospheric background for the Homestake solar neutrino experiment we folded the neutrino faction of the fluxes (0.55 for electron and 0.5 for muon neutrinos) with the $^{37}Cl \rightarrow ^{37}Ar$ cross-sections of Domogatskii[6]. The neutrino energy response is shown on Fig. 2. The main contribution comes from neutrinos with energy above several tens of MeV, where the cross-sections used seem to be in good agreement with the precise values of J.N. Bahcall[7] and those from an independent calculation of N. Itoh and Y. Kohyama[8].

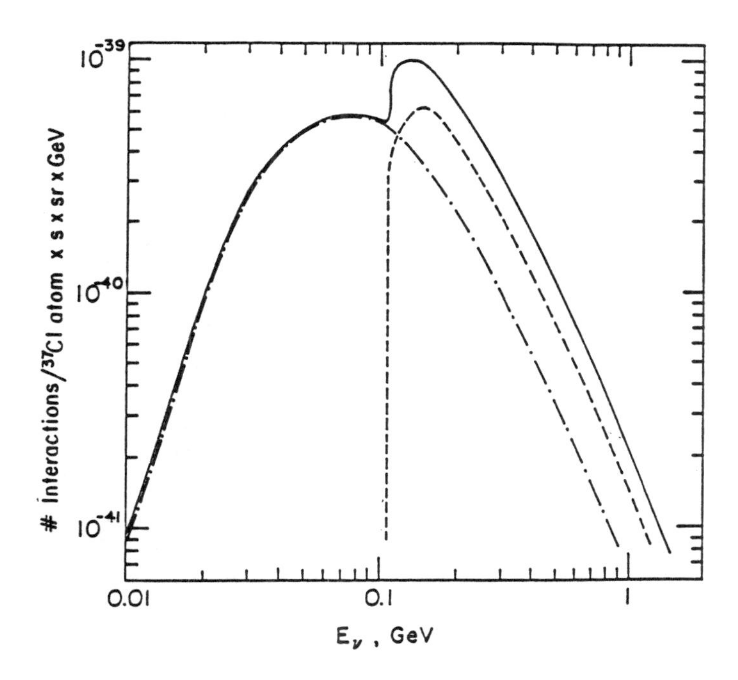

Figure 2 - Neutrino energy response to the $^{37}Cl \rightarrow ^{37}Ar$ reaction.

Dot-dash line is for ν_e, dash for ν_μ, solid line - total.

Electron neutrinos contribute 1.2×10^{-40} and muon neutrinos 1.3×10^{-40} interactions per ^{37}Cl atom.s.ster for a total rate in the detector of 3×10^{-4} ^{37}Ar per day as compared to an observed rate of $\cong 0.4$/day. Thus the background due to atmospheric neutrinos is negligible in agreement with previous estimates.

THE KAMIOKA EXCESS

Although the total rate of neutrino interactions at Kamioka is in good agreement with our predictions an attempt to explain the detailed structure of the detected events was not entirely successful. The account for the experiment reported at ICOBAN '84[4] showed separately "S-type" single ring events, which might be interpreted as elastic, charged current interactions of $\nu_e(\nu_e)$. We estimated the expected number of such events as function of the electron energy and found a significant experimental excess in the lowest energy bins. Table I shows the comparison.

Table I. Comparison of the Kamioka "S-type" single ring events with the calculated elastic charged current $\nu_e(\nu_e)$ interactions.

P_e (MeV/c)	30 - 50	50 - 100	100 - 200
"S-type"	3 (±1.7)	5 (±2.2)	4 (±2)
calculated	0.06	0.54	2.2

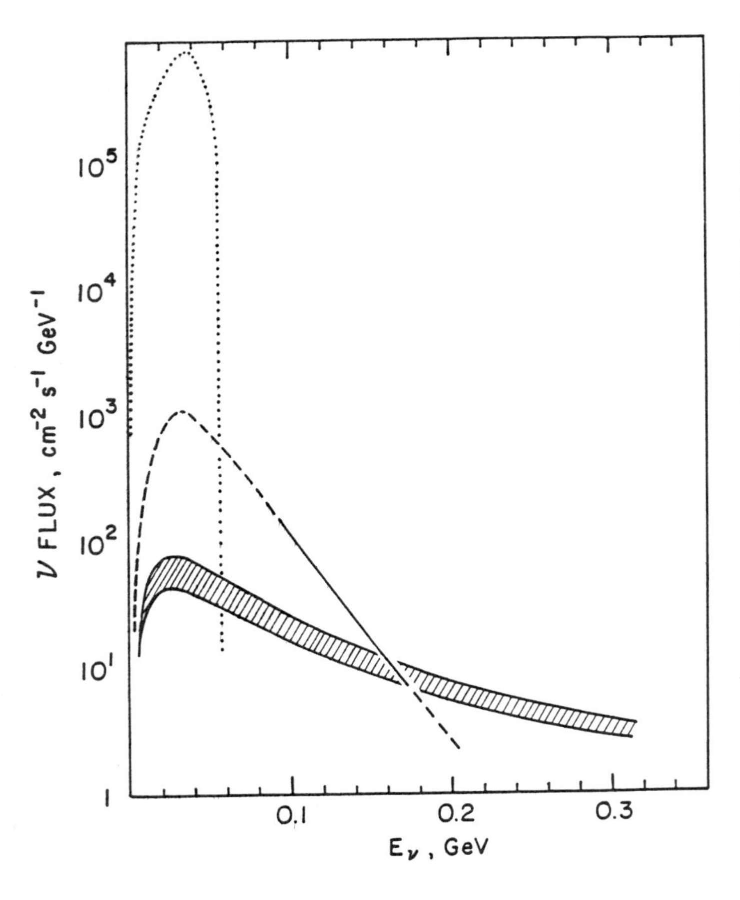

Figure 3 - Backgrounds for solar neutrino experiments: shaded stripe is the atmospheric ν_e flux limited by solar minimum and maximum, solid/dash line-Kamioka ν_e excess in case of normal neutrino/antineutrino ratio & the dotted line represents the flux of 2×10^4 $cm^{-2}s^{-1}$, which corresponds to the hypothesis that the whole excess at 30-50 MeV is due to ν_e from stopped μ.

Since the elastic charged current cross-sections on free nucleons significantly overestimate the cross-sections on bound nucleons, we have used the cross-sections for ^{16}O calculated by Bugaev et. al.[9]. This calculation gives two effects: the cross-section for fixed E_ν is lower, and a given electron energy requires a neutrino energy higher on the average by about 60 MeV. If we attempt to work back from the "S-type" rate to express the excess in terms of neutrino flux we shall obtain a $\nu_e + \nu_e$ flux higher by a factor of 20-80 at E_ν 90-110 MeV and by 5-12 at 110-160 MeV. Fig. 3 shows the neutrino fraction of such "excess flux". It will produce a background of 2×10^{-3} events per day in the Homestake tank.

There is, however, another way to interpret the excess. If one attributes all three events with electron energy 30-50 MeV to ν_e from stopped μ^+ decay (as it would happen in case of monopole catalysis of proton decay within the sun[10] and uses the cross-section for free neutrons, then they will require a ν_e flux of 2×10^4 cm^{-2} s^{-1} [11]. The background rate from such flux will be 0.37 ^{37}Ar/day. This hypothesis does not provide any explanation for the observed excess om the 50-100 MeV bin.

CONCLUSIONS

With the exception of the high rate for "S-type" single ring events in Kamioka, which might have an easier experimental explanation, the observed rates of neutrino interactions is well understood and agrees with predictions based on the current knowledge of the cosmic ray flux, particle interactions and geomagnetic field.

ACKNOWLEDGEMENTS

This research was supported in part by the U.S. Department of Energy (T.K.G.) under contract DE-AC02-78ER05007 and by the U.S. National Science Foundation (T.S.).

REFERENCES

1. R.M. Bionta et. al., Phys. Rev. Lett. 51, 27 (1983).
2. T.K. Gaisser, Todor Stanev, S.A. Bludman and H. Lee, Phys. Rev.Lett. 51, 223 (1983) and Proc. Fourth Workshop on Grand Unification, University of Pennsylvania (Birkhauser, Boston), eds. A. Weldon, P. Langacker and P.J. Steinhardt, p. 87 (1983).
3. T.K. Gaisser and Todor Stanev, to appear in Proc. ICOBAN '84, Park City, Utah, January 1984.
4. M. Koshiba, talk presented at ICOBAN '84, Park City, Utah, January 1984.
5. D.J. Cooke, Phys. Rev. Lett. 51, 320 (1983).
6. G.V. Domogatskii and R.A. Eramzhyan, Bull. Acad. Sci., USSR (phys.ser.) 41, 9, p. 169 (1978).
7. J.N. Bahcall, Rev. Mod. Phys. 50, 881 (1978).
8. N. Itoh and Y. Kohyama, Nucl. Phys. A306, 527 (1978).
9. E.V. Bugaev et. al., Nucl. Phys. A324, 320 (1979).
10. J. Arafune, M. Fukigita and S. Yanagita, RIFP-531, Uni. of Kyoto preprint.
11. See Ref.4, also H.H. Chen, UCI Internal Report Neutrino-120, 1984.

Neutrinos from Stellar Collapse

Adam Burrows
Department of Physics, State University of New York,
Stony Brook, N. Y. 11794

INTRODUCTION

When a star more massive than about 10 $M_{\odot}$ has exhausted its thermonuclear fuel, its white dwarf-like core collapses dynamically. Collapse is halted only after nuclear densities are achieved, at which time the core matter stiffens, bounces, and drives a shock into the outer stellar envelope. In the standard collapse scenario[1], this bounce-shock ejects the massive envelope with the accompanying optical pyrotechnics we associate with a Type II supernova. This optical outburst lasts for months and can rival the luminosity of its parent galaxy, but involves only $\sim 10^{49}$ ergs. Most of the energy of the supernova, approximately 10^{51} ergs, resides in the kinetic energy of the ejectum and is not "seen" at all. However, the hot, electron-rich remnant of collapse should evolve into a cold neutron star, possibly a pulsar, with a binding energy ($2 - 4 \times 10^{53}$ ergs) that exceeds by more than two orders of magnitude the supernova energy quoted above. This binding energy is radiated in a burst of neutrinos of all species ($\nu_e, \bar{\nu}_e, \nu_\mu, \bar{\nu}_\mu, \nu_\tau$, and $\bar{\nu}_\tau$) that lasts between one and ten seconds and dwarfs in magnitude even the spectacular supernova itself. The 10^{58} neutrinos emitted with average energies between 10 and 20 MeV are so central to the energetics of collapse and neutron star formation because they are both copiously produced and, due to their small interaction cross-sections, readily freed. Though the opacity, τ_ν, of the core to neutrinos may approach $\sim 10^{6}$, that for photons is at least fourteen orders of magnitude larger.

With little effort, this neutrino signature can be made to seem quite exotic. The prompt burst[2] of electron-type neutrinos that signals the bounce-shock's breakout through the "neutrinosphere" and lasts ~1 millisecond approaches a luminosity of 1 $M_{\odot}c^2$ per second and, for that instant, is within an order of magnitude of the total optical luminosity of the observable universe. A stellar collapse every 50 years implies that, when averaged over the eons, the average neutrino luminosity of our galaxy is comparable to its photon luminosity. What is more, it can be shown that, with reasonable assumptions as to the magnitude of the $\bar{\nu}_e$ component, this neutrino burst is lethal to humans within ~ 20 A.U..

The neutrino signature is the only direct diagnostic of the internal dynamics of massive star death. Information on the mix of neutrino types, their spectrum, and the time behavior of the emission will enable us to track the progress of the collapse, bounce, and relaxation phases and keep the theorists honest. Unfortunately, detectors now operating or currently envisioned will not be sensitive enough to tell us more than the integrated emission, the average neutrino energy, the rough mix of neutrino species, and, maybe, the direction of the event. However, considering that there is as yet no non-solar data in this astronomy, any nugget would be a gold mine.

In this contribution, we will first describe the context and dynamics of stellar collapse. Then the role of neutrinos will be explained and their signature, in the standard model, sketched. Finally, we will discuss detectors and the detectability of the integrated neutrino signal. Despite temptation, no attempt will be made to deviate from the standard model of spherically symmetric evolution involving classical neutrinos and physics.

THE HYDRODYNAMICS OF COLLAPSE

In Figure 1, we depict the various scales of the collapse problem. The maximum radius of the photosphere of the subsequent supernova is seven orders of magnitude larger than that of the ~1.4 $M_\odot$ core that implodes to initiate it. The dynamical time of this destablized core is so short (~1 second) that the outer mantles of heavy elements, helium, and hydrogen are effectively frozen during much of the relevant action. They participate only after the consequent shock encounters and ejects them.

Figure 1 : A telescoping depiction of the scales of the supernova photosphere, the progenitor structure, and the collapsing core. Note that most of the interesting evolution transpires in the deep interior.

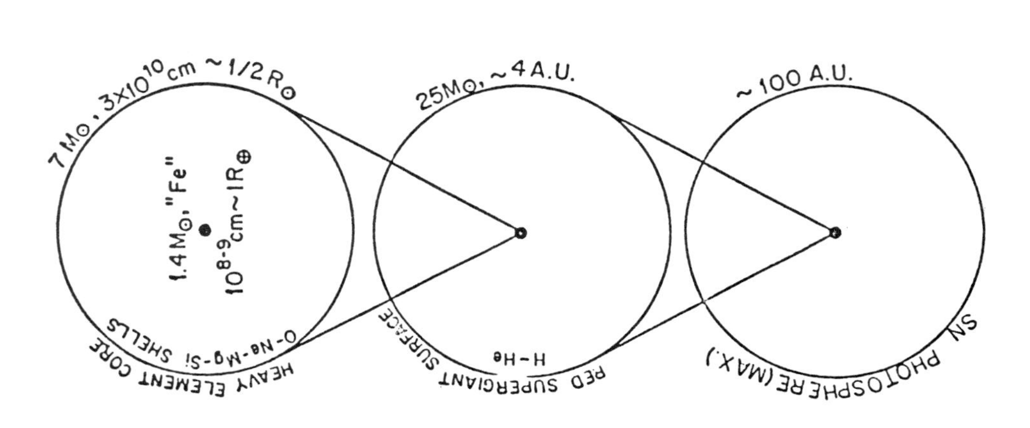

Figure 2 is a snapshot of the radius versus both the velocity of infall and the speed of sound at an intermediate stage of collapse before bounce. As we can see, the flow separates into an inner subsonic core and an outer supersonic reverse wind. This inner core is collapsing homologously ($v \alpha r$) as a unit. The maximum velocity of infall during the entire evolution approaches one fourth the speed of light.

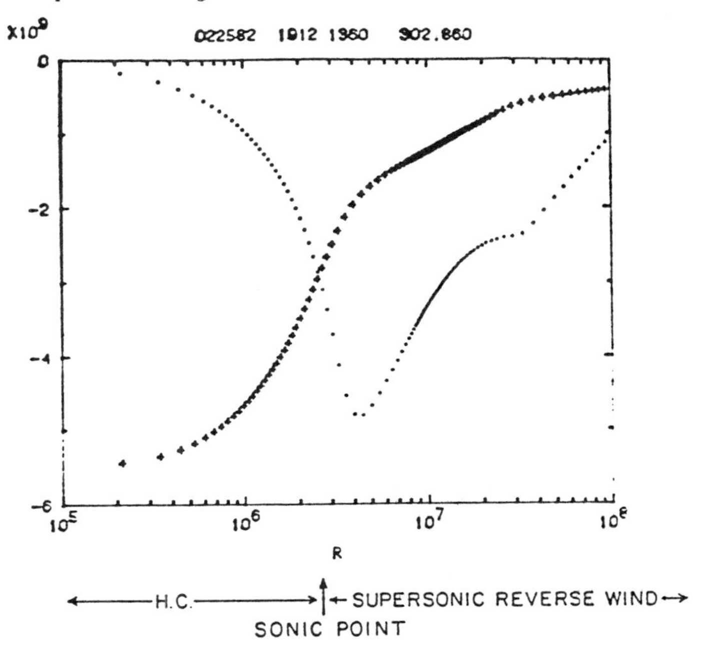

Figure 2 : Velocity (...) and sound speed (+++) versus radius at an intermediate stage of collapse. Note that, for convenience of comparison, the speed of sound is negative. The sonic point is at the intersection of the two curves. The units of velocity are 10^9 cm/s and of radius are cm. H.C. stands for homologous core.

When the central density reaches nuclear density (~2.7×10^{14} gm/cm^3), the matter stiffens and strong pressure waves are radiated, which steepen into a shock at the sonic point. The homologous core rebounds as a unit, does PdV work on the shock, and, within ~1 millisecond, settles into hydrostatic equilibrium. In the standard scenario, the energized shock slams into the infalling supersonic outer core, reverses the outer core's velocity at some mass cut, and ejects much of the star. The residue is a nascent neutron star. In Figure 3, we graph the matter temperature (in MeV) versus the radius at shock formation. Temperatures as high as 20 MeV are achieved, not in the center, but near the 20 kilometer point (~0.6 $M_\odot$).

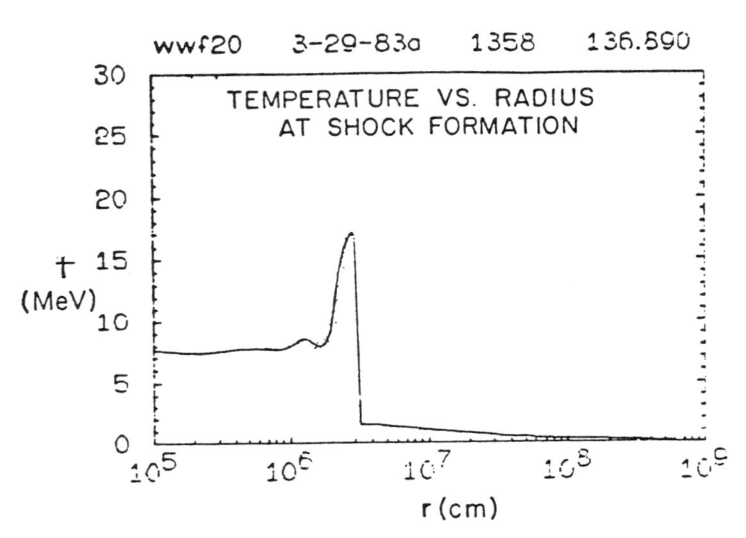

Figure 3 : Temperature versus radius at shock formation.

Figures 4 and 5 show the entropy profile before and ~1 millisecond after shock formation. The huge entropy spike in Figure 5 clearly delineates the shock-heated region. The entropy steps in Figure 4 mark the fossil iron, silicon, and oxygen burning shells of the progenitor.

Figure 4 : Initial entropy per baryon per boltzmann's constant versus enclosed mass in grams. This model is taken from Weaver, Woosley, and Fuller 1983.[3]

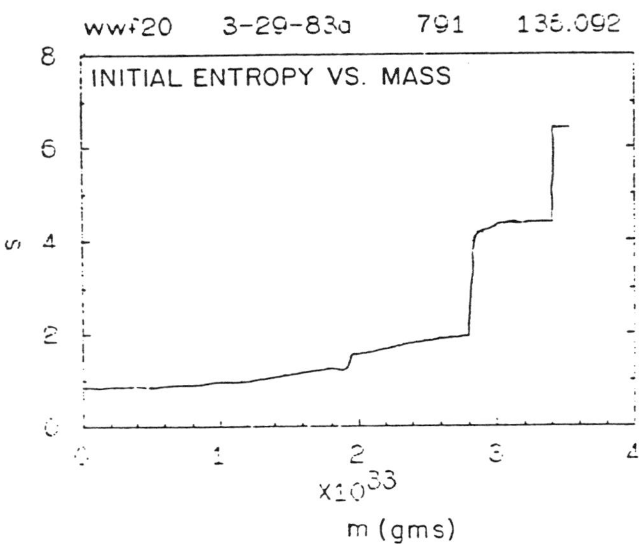

Figure 5 : Same as Figure 4, but 1 millisecond after shock formation. Note the change of entropy scale.

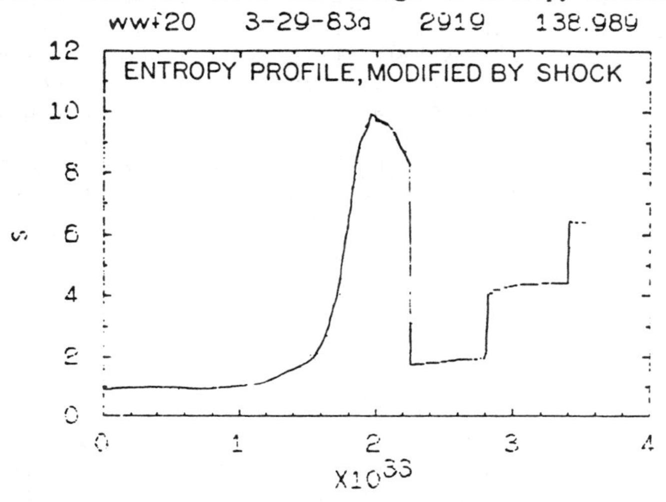

A representative glimpse of the post-shock density structure of the core is given in Figure 6. Note both the large density range of the core and the fact that the shock, despite its strength, is merely a perturbation on this structure. The inner core is largely unaffected by the bounce-shock. An important difference between this explosion and others (eg. TNT) is the

Figure 6 : Density (rho) in units of gm/cm^3 versus radius (r) in units of centimeters just after shock formation. The position of the shock is clearly marked.

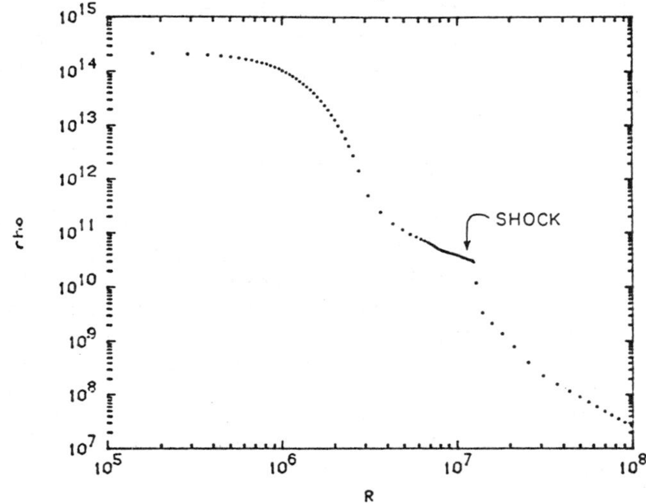

presence and centrality of gravity. The post-shock material is never far from hydrostatic equilibrium, to which much of it rapidly settles. An overview of the progress of the shock is provided by the velocity versus radius plots of Figure 7. In this figure are superposed snapshots of the velocity profile at consecutive times during the evolution. The position and motion of the shock is evident, as is the strong post-shock sound wave generated by the second bounce of the core. If the simple bounce-shock mechanism of Type II supernovae works, the shock proceeds outward, ejects the mantle, and leaves behind a shocked, lepton-rich hydrostatic protoneutron star.

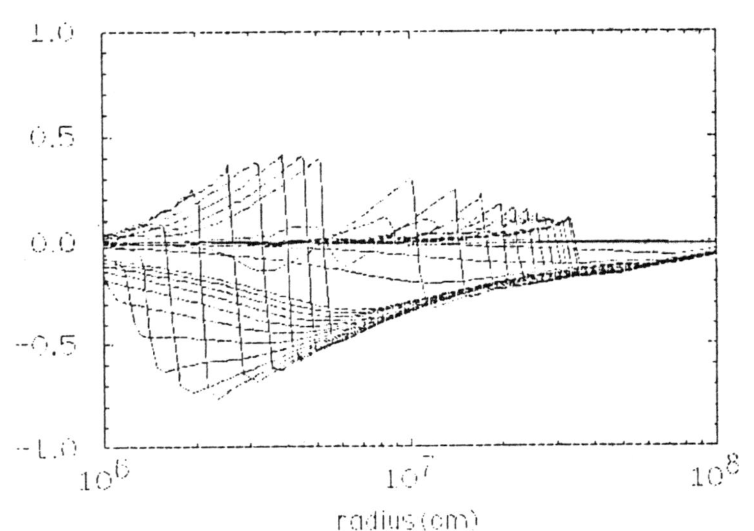

Figure 7 : An overlay of velocity versus radius plots at various times during the collapse and bounce phases. Snapshots are taken every decade in central density from 10^{10} to 10^{14} gm/cm^3, then every tenth of a millisecond (14 times), and finally every millisecond until 15,000 timesteps have been executed.

THE ROLE OF NEUTRINOS

When the central density of the collapsing star exceeds $\sim 10^{11}$ gm/cm^3, the opacity of the core to electron-type neutrinos produced by electron capture rises abruptly. These neutrinos no longer stream out, but are trapped in the matter[4]. After trapping, the inverse reaction of neutrino absorption on neutrons, in or out of nuclei, quickly leads to chemical (beta) equilibrium. Relevant thermodynamic variables, conserved during the subsequent dynamical phases, then become the lepton fraction, Y_L, which is the sum of the electron and electron-type neutrino fractions, and the entropy per baryon. Figure 8 depicts schematically the time-scales of trapping. The essential point is that the curves cross significantly below nuclear densities. Just when the characteristic capture and dynamical rates are becoming rapid, neutrino loss is choked off. The specific sources of opacity are electron scattering, charged-current absorption on neutrons, neutral current scattering off free nucleons, and, importantly, the coherent neutral current scattering off nuclei[5].

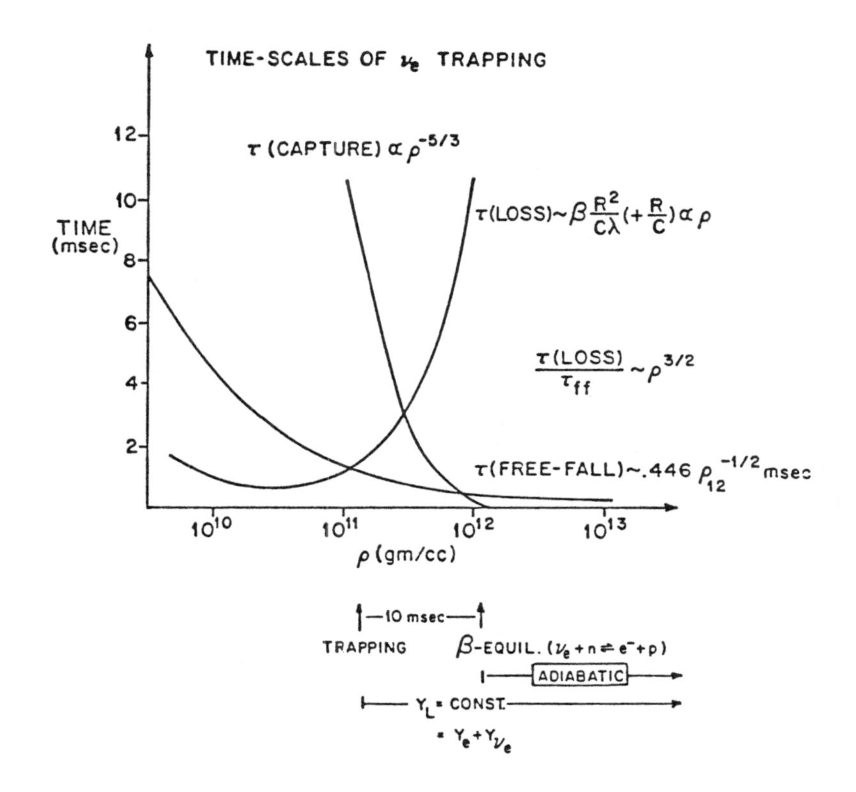

Figure 8 : The time-scales of ν_e trapping. τ_{ff} is the dynamical time ($\rho/[d\rho/dt]$), τ(loss) is a measure of the ν_e escape time, and τ(capture) is a measure of the electron capture time. After $\sim 10^{12}$ gm/cm^3, β-equilibrium and adiabaticity obtain.

The lepton fraction, instead of plunging precipitously to neutron star values (0.03 - 0.05) on dynamical times, is frozen at high values (0.35 - 0.4). Figure 9 communicates the essentials of the evolution of the fractions, with and without trapping.

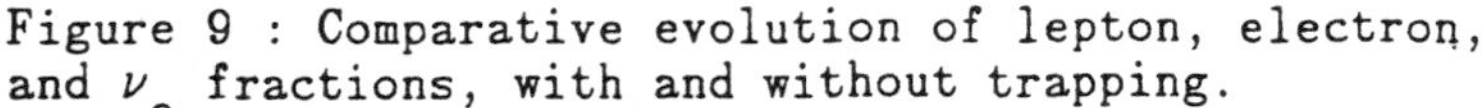

Figure 9 : Comparative evolution of lepton, electron, and ν_e fractions, with and without trapping.

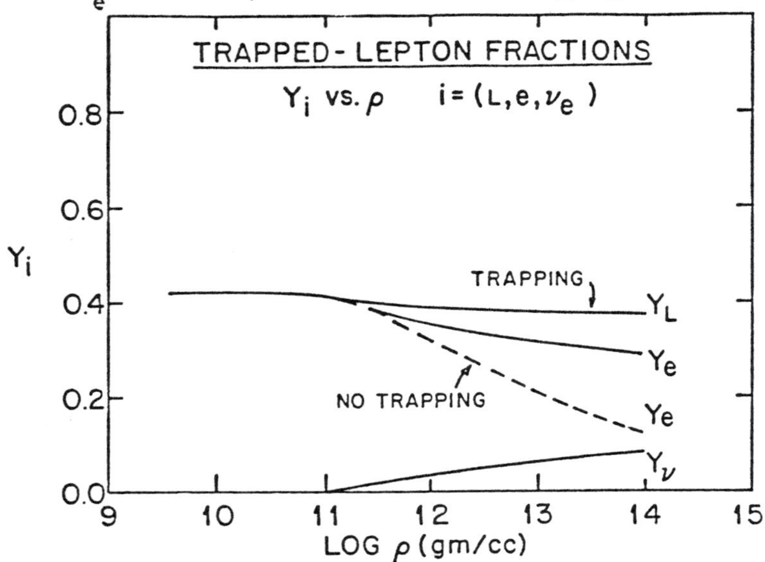

The consequences of trapping are manifold. Among these are that the matter is <u>not</u> neutronized during the collapse and bounce phase, that the leptons (with $\Gamma \sim 4/3$), not the baryons, dominate the pressure, and that the gravitational energy of collapse is pumped, not into heat or the baryons, but predominantly into the degenerate lepton (e^- and ν_e) Fermi seas. As the leptons are trapped, the gravitational energy of collapse is trapped. However, after the core has bounced and settled into hydrostatic equilibrium, the neutrinos will start to leak out. This occurs not on a timescale of milliseconds, but hundreds of milliseconds to seconds[6].

The collapse energy during this relaxation, or Kelvin-Helmholtz stage, suffers various conversions. After the initial conversion via compressional work of the gravitational energy into lepton degeneracy energy, the ν_e's start to diffuse out. As they do, electron capture proceeds in an attempt to compensate for the loss. The net result is a diffusive loss of lepton number, i.e. neutronization. Since the average ν_e energy in the interior can approach 150 MeV, and the neutrinos escape with only 10 - 20 MeV, the process of deleptonization is accompanied by the degradation of degeneracy energy into heat. Electron neutrino energy downscattering during diffusion is analogous to resistive heating in a current carrying conductor. During and after the deleptonization of the remnant, the core cools by the diffusive loss of pairs of neutrinos of all species generated by the thermalization. Most of the binding energy of the resulting neutron star is radiated during this cooling phase, which may last as long as 10 seconds. A schematic of the evolution of the neutrino luminosity of the various species is given in Figure 10 (taken from Burrows 1984[7]).

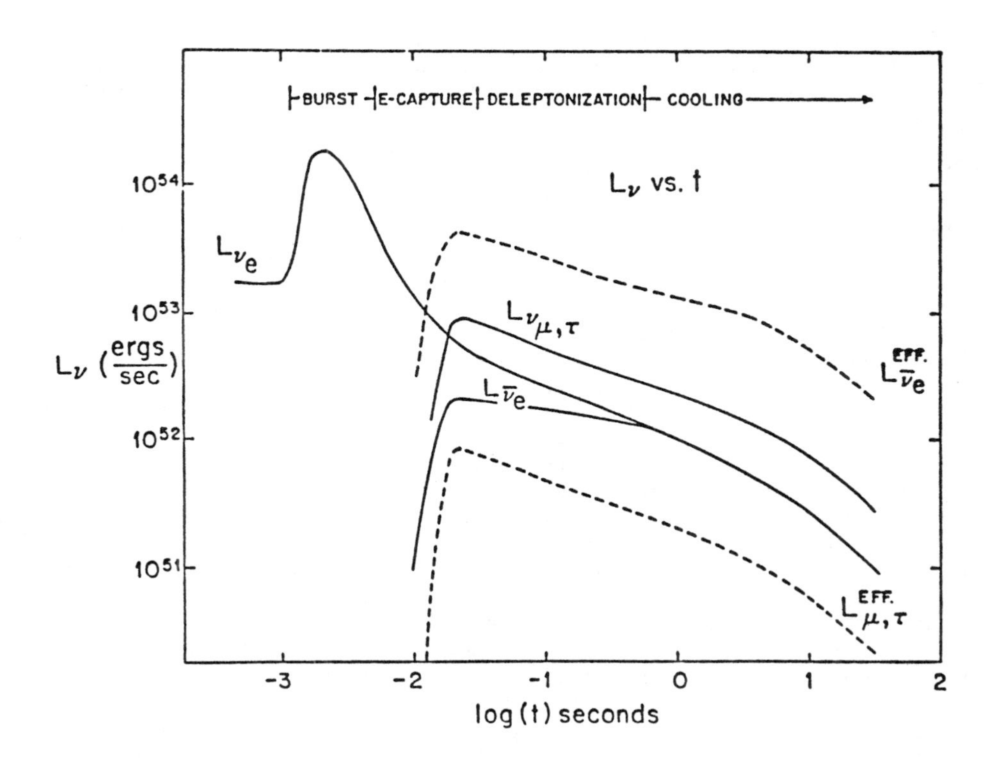

Figure 10: A schematic of the luminosity versus time for the various neutrino species. L_i^{eff} is the electron neutrino luminosity that could mimic the signal in Cherenkov detectors of the actual neutrino luminosity of species i. Note the logarithmic time axis. The initial spike in the electron neutrino luminosity occurs when the shock breaks out of the neutrinosphere. (t = 0 at bounce)

DETECTION

If the total binding energy of the resulting neutron star is ~2×10^{53} ergs and we assume that the average neutrino energy is ~ 10 MeV, then, with reasonable assumptions, we can derive the fluence 1 kiloparsec from a collapse in the various neutrino types. The results are 2.4×10^{13} ν_e's/cm^2, 1.8×10^{13} $\bar{\nu}_e$'s/cm^2, and

$7.2\times10^{13}\nu_\mu$'s/cm^2, where here ν_μ stands for both mu and tau neutrinos and anti-neutrinos. The $\bar{\nu}_e$ number is the most ambiguous and should be used with caution. As both scintillation and Cherenkov detectors are most sensitive to this $\bar{\nu}_e$ component via the charged-current absorption on protons, future theoretical research should focus on improving our understanding of the $\bar{\nu}_e$ signature.

For the purposes of this discussion, we will compare, with two different burst model assumptions (Burrows 1984), the integrated detector responses to collapse neutrinos of both radiochemical and Cherenkov detectors. The radiochemical detectors (the Brookhaven ^{37}Cl (Davis et al 1978[8]) and hypothetical gallium and bromine solar neutrino experiments) are sensitive to only the ν_e component, whereas the Cherenkov detectors (the Irvine-Michigan-Brookhaven (IMB) proton decay H_2O pool and the Homestake burst device (now defunct) (Cherry et al 1982[9])) respond to all neutrino types, but are most sensitive to anti-electron neutrinos. Burst model A assumes the above numbers for the average neutrino energy and total burst energy and burst model B assumes 1.5 times these numbers. Using signal-to-noise considerations, we derive the detector ranges, the fraction of the galaxy sampled, and the mean time between detections. The results are given in Table 1 and graphically illustrated for the chlorine experiment in Figure 11. Similar calculations can be performed for other potential burst detectors such as the Park City (Cherenkov) proton decay (D. Cline, private communication) and the Mt. Blanc (scintillator) (P. Galeotti, private communication) experiments. Indications are that the latter experiments may be the best extant burst detectors. Note that, as Table I indicates, the detector ranges are sensitive to the burst spectrum and that neither the gallium nor the bromine solar neutrino experiments will be useful for burst detection.

Neutrino astronomy now has only about 500 neutrinos to its credit after more than 20 years of effort. Those few neutrinos have stimulated an incredible amount of discussion and physics over the years. A galactic supernova could double that neutrino inventory overnight.

Acknowledgement

The collapse calculations were done in collaboration with Amos Yahil and Jim Lattimer. Thanks are extended to T-6 division at Los Alamos and the National Magnetic Fusion Energy Computing Center (NMFECC) for Cray time. This work was supported in part by the Department of Energy under grant no. DEAC02-80ER10719.

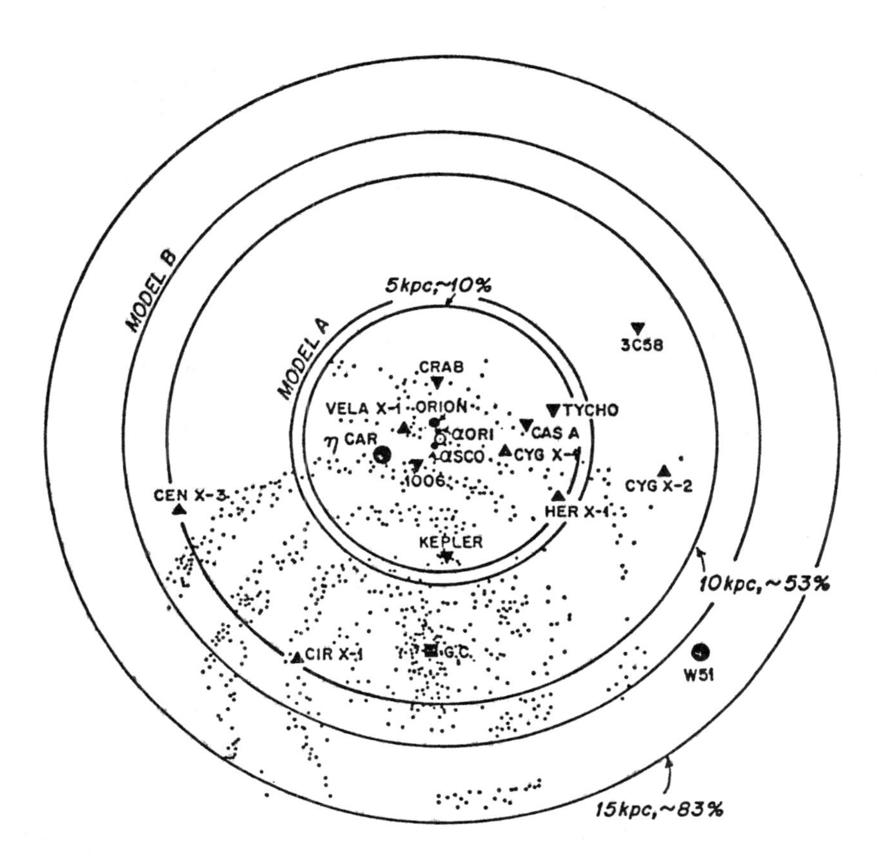

Figure 11: A face-on view of the galaxy, centered on the sun, depicting schematically the range of the ^{37}Cl detector. G.C. is the galactic center, whose distance to the sun was assumed to be 8.5 kpc (Tammann 1982[10]). The small dots mark, in an approximate way, the positions of O-B associations, HII regions, open associations, and concentrations of HI. With these tracers, the spiral structure of the galaxy stands out clearly. The percentages on the 5, 10, and 15 kpc circles refer to the fraction of the galaxy enclosed therein, as calculated by Bahcall and Piran (1983). Historical supernovae, some of which may have been Type II's, are marked by inverted filled triangles. Some X-ray sources are included for context. The giant molecular cloud complex, W51, η Carina, α Orionis, and α Scorpii are included because they are familiar objects often considered potential sites of stellar collapse.

Table I: Detector Specifics

	(S/N) at 1 kpc		Range (kpc)		Fraction of Galaxy Sampled		Mean Time Between Detections (YRS)	
Model:	A	B	A	B	A	B	A	B
Brookhaven ^{37}Cl	32	140	5.6	12	.12	.7	80	15
I.M.B.	70	158	3.7	5.6	.05	.15	200	67
Homestake Čerenkov	70	158	8.4	13	.38	.75	26	13
50 ton ^{71}Ga			∿ .5		∿ .001		10^4	
600 ton ^{81}Br			∿ 1		∿ .003		$3x10^3$	

In model A, we assume that the average neutrino energy is 10 Mev and the total neutrino flux is $2x10^{53}$ ergs. This is the standard model. Model B stretches these numbers to 15 Mev and $3x10^{53}$ ergs, respectively. The numbers for the fraction of the galaxy sampled and the rate of collapse (1/10 years) are taken from Bahcall & Piran (1983)[11]. Their number for the collapse rate may be too optimistic by as much as a factor of 4.

REFERENCES

1. G. E. Brown, H. A. Bethe, G. Baym, Nucl. Phys. **A375**, 481 (1982).
2. A. Burrows, and T. J. Mazurek, Nature **301**, 315 (1983).
3. T. A. Weaver, S. E. Woosley, and G. Fuller, B.A.A.S. **14**, 957, (1982).
4. T. J. Mazurek, Nature **252**, 287 (1974).
5. D. Z. Freedman, Phys. Rev. **D9**, 1389 (1974).
6. A. Burrows, T. J. Mazurek, and J. M. Lattimer, Ap. J. **251**, 325 (1981).
7. A. Burrows, Ap. J. **283**, 848 (1984).
8. R. Davis, J. C. Evans, and B. T. Cleveland, in Proceedings of the International Neutrino Conference at Purdue (1978), p. 53.
9. M. L. Cherry, M. Deakyne, T. Daily, K. Lande, C. K. Lee, R. Steinberg, E. J. Fenynes, J. Phys. G: Nucl. Phys. **8**, 879 (1982).
10. G. Tammann, in "Supernovae: A Survey of Current Research," edited by M. J. Rees and R. J. Stoneham (Dordrecht:Reidel, 1982), p. 371.
11. J. N. Bahcall, J. N. and T. Piran, Ap. J.(Letters) **267**, L77 (1983).

SEARCH FOR STELLAR COLLAPSE BY MULTIPLE NEUTRINO INTERACTIONS DURING JANUARY 11 - FEBRUARY 11, 1984 USING THE HPW DETECTOR (For the HPW Collaboration)

David B. Cline
Physics Department, University of Wisconsin, Madison, WI 53706

ABSTRACT

The HPW proton decay detector has been used to search for stellar collapse events during a short time interval. The search is sensitive to the collapse of rotating stars of smaller energy release that gives greater than 3 neutrino interactions of 10 MeV or more in a time interval of 20 seconds. The detector is very sensitive to a range of stellar collapse phenomena.

The observation of the $\bar{\nu}_e$ burst from a collapsing star in our galaxy is one of the most important tests of the theory of element formation and stellar collapse. The expected $\bar{\nu}_e$ flux is by now well known.[1-3] If the star is rotating before the collapse the $\bar{\nu}_e$ flux is expected to be reduced by about an order of magnitude.[4] It is possible that the gravitational radiation will be strongly suppressed in this case.[5] It has been pointed out that the neutrino emission from the collapsing star may significantly reduce the gravitational radition,[5] thus the detection of the $\bar{\nu}_e$ burst may be the only way to observe these processes in regions of the galaxy that are optically thick.[5]

In Table 1 we compare the expected number of events in the HPW detector for the cases:

1. $M_s > 2M_\Theta$ stars - non-rotating
2. $M_s > 2M_\Theta$ stars - rotating

In Fig. 1 is shown the expected position spectrum from the reaction[1-3]

$$\bar{\nu}_e + p \rightarrow e^+ + n \qquad (1)$$

from stellar collapse ($M_s = 2M_\Theta$).

In Fig. 2 is shown the neutrino and anti neutrino flux from a stellar collapse. To complete our summary we show some of the expected cross sections for low energy neutrinos and anti neutrinos in Fig. 3.

In this report we compare the number of $\bar{\nu}_e$ events expected in the HPW detector at Park City, Utah and the 100(130) ton detectors

0094-243X/85/1260295-12 $3.00

operating in the USSR[6,7] (Fig. 4). We use the $\bar{\nu}_e$ flux from J. R. Wilson.[8] The calculations are typical for any ~100 ton detector of scintillation material and detector of ~1000 tons (water or perhaps cryogenic detectors in the future) operating at the low threshold of the HPW detector. We may remark that a previous search for stellar collapse was carried out by Lande, et al.[9]

We assume a cutoff of 8 MeV for the detectors. Fig. 4 shows the cleanliness of the electron spectrum from

$$\mu^+ \rightarrow e^+ + \nu_e + \bar{n}_\mu$$

in the HPW detector in the energy range covered by the dominant position spectrum of reaction 1. (From A. Mores´ Thesis, University of Wisconsin-Madison, 1984.) We conclude that the HPW detector gives a clear response to individual electrons in the appropriate energy range.

The detection of stellar collapse events above a background of noise in these detectors requires the observation of K position counts in a time interval of 10-20 sec. For the two scintillation detectors K ranges from 7 (Baksan) to 15 (Artymomousk). For the HPW detector we estimate $K \sim 4$ although it will be as small as 3. Comparing the K value with the expected rates in the various detectors (Table 1) we reach the following conclusions.

1. The HPW detector is sensitive over the full 30 kilo par sec of the galaxy for rotating and non-rotating stars.
2. The scintillation detectors are sensitive to 20 kilo par sec for non-rotating stars and to less than 10 kilo par sec for rotating stars. Thus a large part of the galaxy is missing from their observations.

We conclude that the HPW detector "sees" the full galaxy for the most general type of collapse of stellar objection. If smaller objects collapse giving $\geq 1/10$ the flux of a non-rotating Stellar collapse the detector is sensitive to such objects throughout the entire galaxy.

We now turn to a preliminary search for the stellar collapse signature in the HPW detector during the period of January 11, 1984 and February 11, 1984. Fig. 6 shows the uncut spectrum of events within a time window of 20 seconds during this period. We now apply cuts to enhance $\bar{\nu}_e$ interactions on all events with the $N_{umw} > 3$-- the result is shown in Fig. 7. Most events are due to μ's clipping the corner of the detector with a combination of noise. The cuts reduce the possible signal by an enormous amount. We have studied all events (after the cuts) with $N_{umv} > 3$. No events remain in the sample. Fig. 8 compares the level of sensitivity of this search to that of previous searches

(Homestake) and the significance of the search is reported--Table 2.

Using the full detector allows us to search for stellar collapse over the entire galaxy, and most of the galaxy even if a reduction in flux factor is included due to the initial rotation of the star. We are now continuing the search using data from May 1983-August 1984.

In the future the trigger will be operated to reduce data taking so as to concentrate on the detection of stellar collapse and to continue searching for nucleon decay into the Pati Salam type modes.

We thank J. Wilson, S. Bludman and A. Burrows for helpful discussions.

References

1. J. R. Wilson, et al., Ann. New York Acad. Sci. 262, 54 (1975).

2. V. C. Chechetkin, et al., Astronon. Ah. 54, N3 (1977).

3. G. V. Domogatsky, et al., Appl. Math of the USSR. Ac Sci N40 (1977).

4. V. S. Imshenik and D. K. Nadyozhin, Astron. Zh. Letter (1977).

5. S. Budman, private communication.

6. T. P. Khalchulzov, et al., Proceedings of the Bangalore Cosmic Ray Conference (1983).

7. A. Chudakov, et al., Proc. Int. Conf. Neutrino 77 1, 155, Nawha, Moscow (1978).

8. A. Burrows, private communication, J. Wilson, private communication.

9. M. Cherry, et al., Nucl. Phys. 8, 879 (1982).

10. D. Cline, Detecting Stellar Collapse in Our Galaxy with the HPW Detector Operating as an Antineutrino Telescope, HPW Internal Note, 1984.

Table 1

Number of Events Expected for "Normal" Stellar Collapse Process in HPW Detector ($E_{VIS \bar{\nu}_e}$ > 10 MeV)

Distance To Stellar Collapse (kp)	Number Expected ν_e	$\bar{\nu}^e$
1	492	37,200
5	19.5	1486
10	5	372
2°	1.2	93
30	.6	41
50	.2	14.9
100	.05	3.7

Table 2

Results of the Search

No Cuts - No Events with
N > 7

Rules out normal supernove up to 50 kp

Full Cuts - No Events with
N > 3

Rules out supernove with energy release into $\bar{\nu}_e$ decreased by 1/10 - up to 30 Kp
(Rotating stars??)

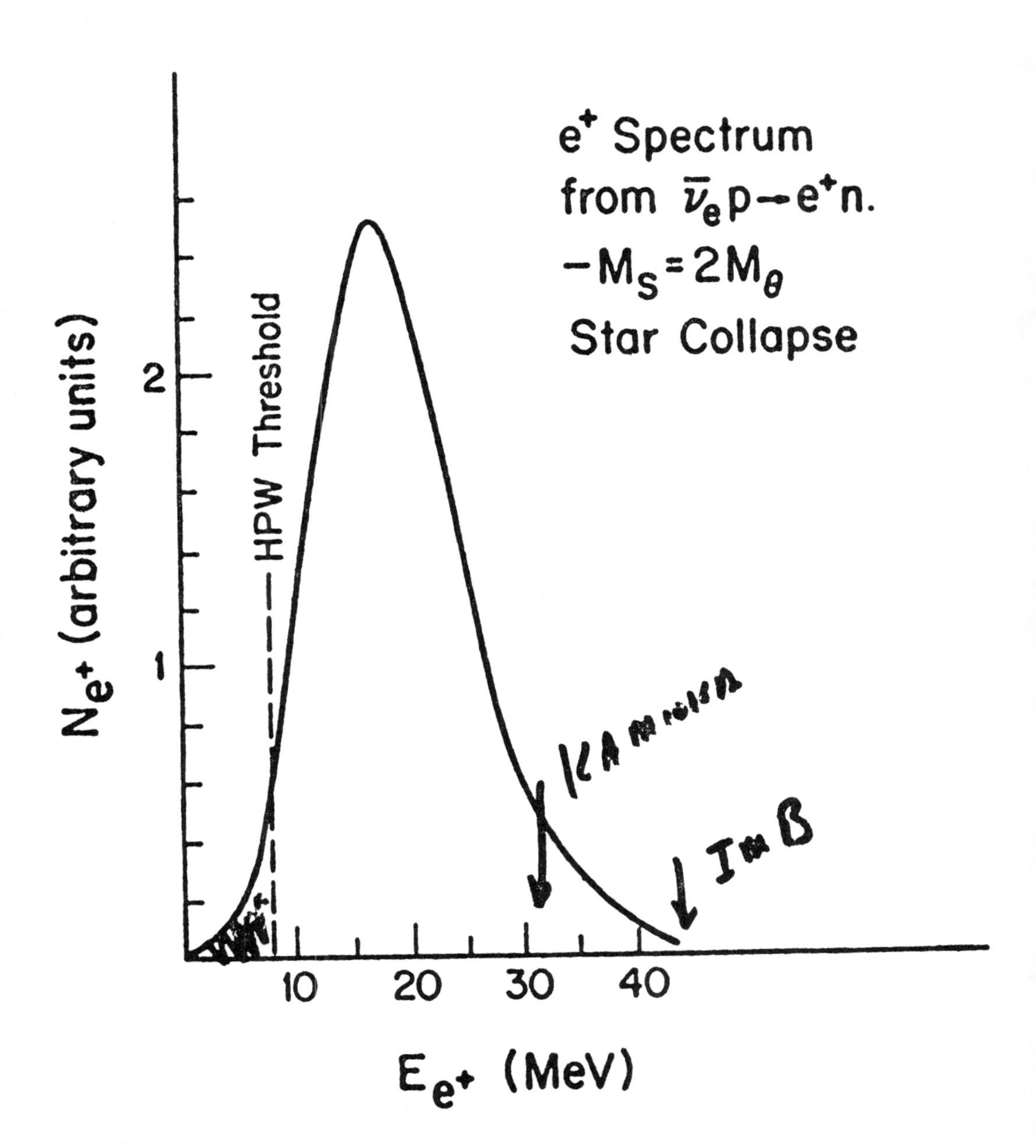

e^{+} Spectrum
from $\bar{\nu}_e p \rightarrow e^{+}n$.
$-M_S = 2M_\theta$
Star Collapse
HPW Threshold
$N_{e^{+}}$ (arbitrary units)
2
1
10
20
30
40
$E_{e^{+}}$ (MeV)

Fig. 1

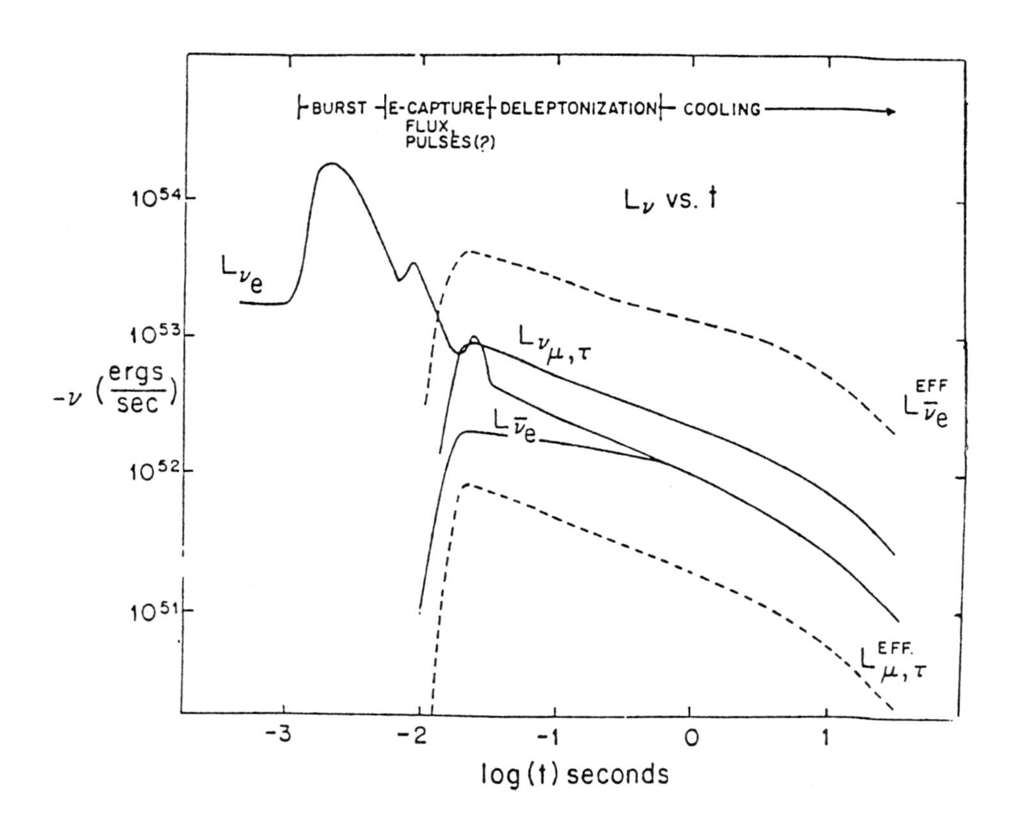
BURST
E-CAPTURE
DELEPTONIZATION
COOLING
FLUX PULSES(?)
L_ν vs. t
L_{ν_e}
$L_{\nu_{\mu,\tau}}$
$L_{\bar{\nu}_e}$
$L^{EFF}_{\bar{\nu}_e}$
$L^{EFF.}_{\mu,\tau}$
10^{54}
10^{53}
10^{52}
10^{51}
L_ν (ergs/sec)
-3
-2
-1
0
1
log(t) seconds

Fig. 2

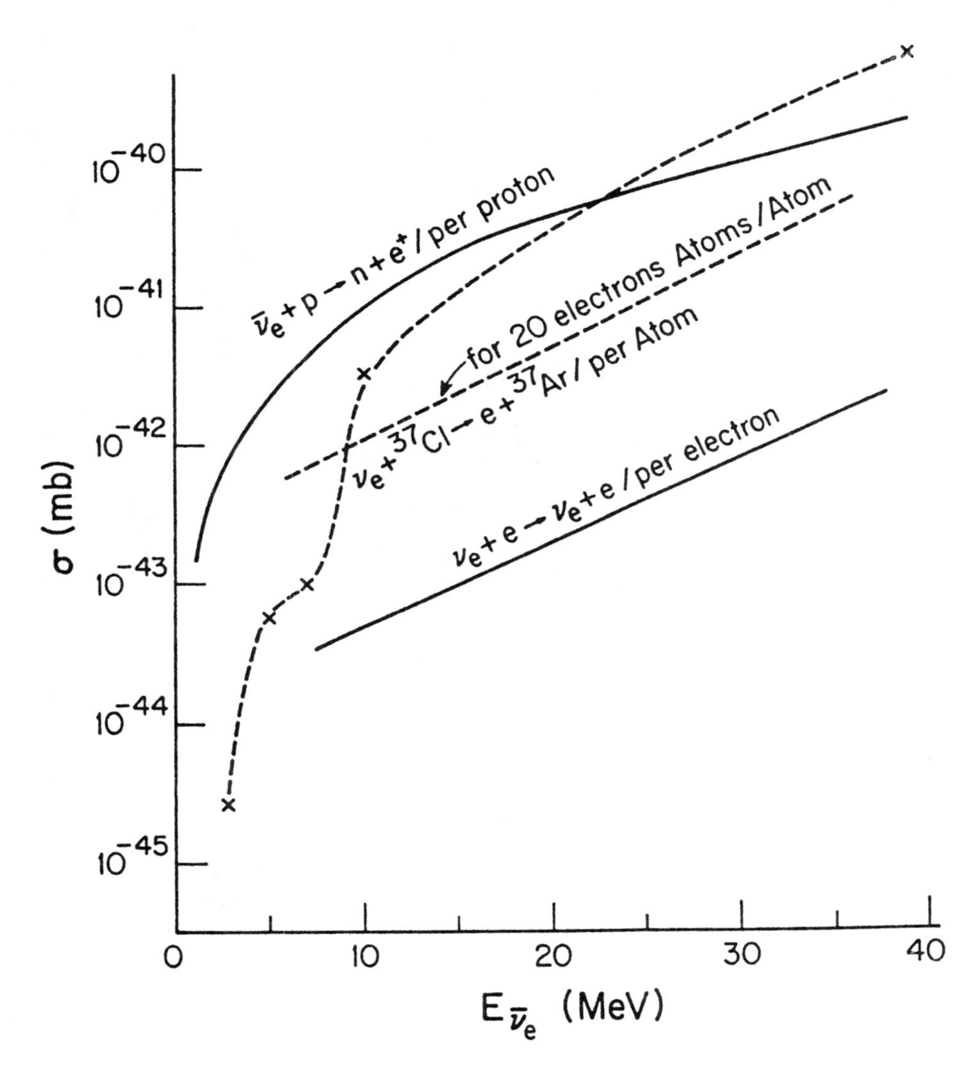

$\bar{\nu}_e + p \to n + e^+$ / per proton
for 20 electrons Atoms / Atom
$\nu_e + {}^{37}Cl \to e + {}^{37}Ar$ / per Atom
$\nu_e + e \to \nu_e + e$ / per electron
10^{-40}
10^{-41}
10^{-42}
10^{-43}
10^{-44}
10^{-45}
σ (mb)
0
10
20
30
40
$E_{\bar{\nu}_e}$ (MeV)

Fig. 3

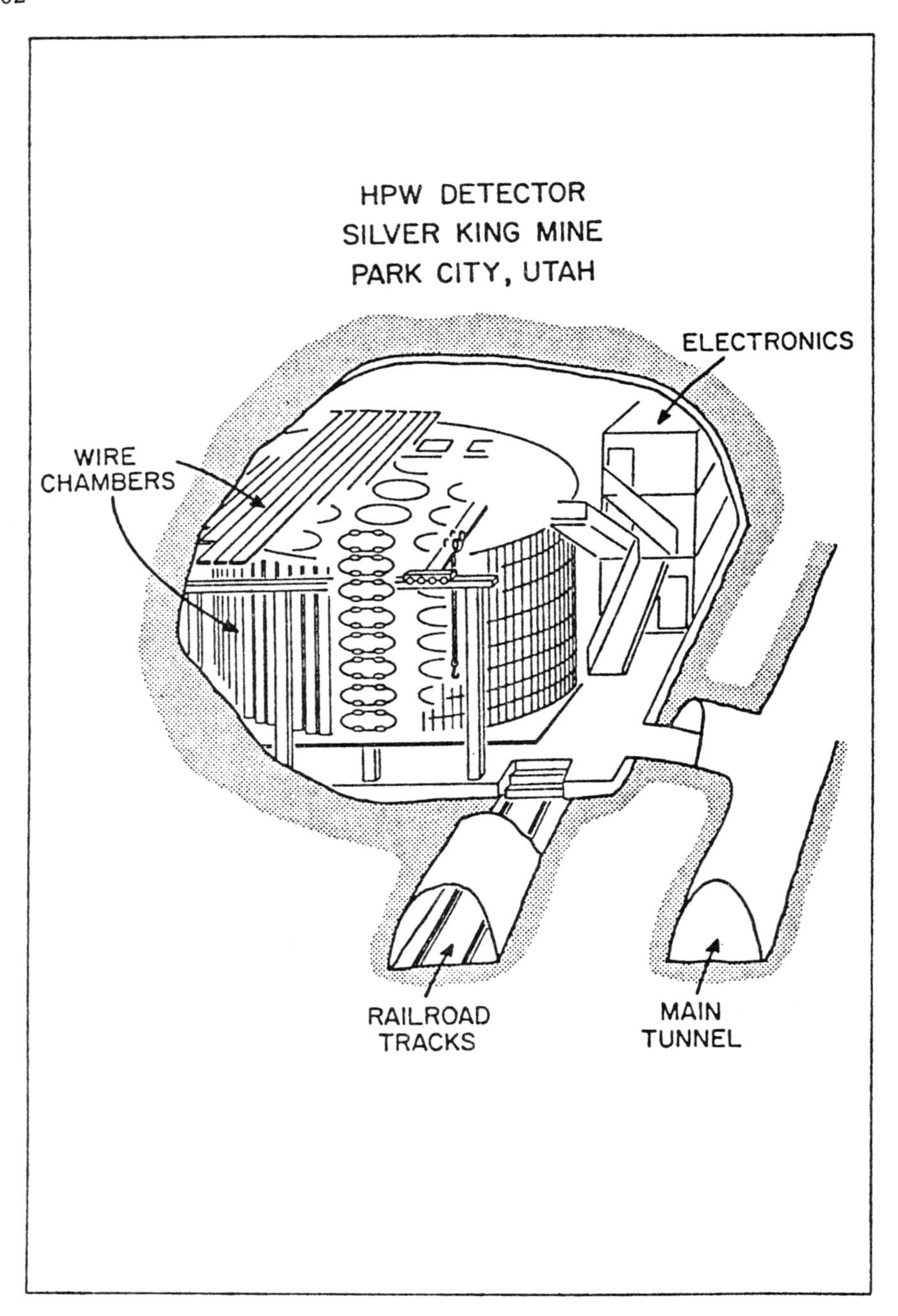
HPW DETECTOR
SILVER KING MINE
PARK CITY, UTAH
ELECTRONICS
WIRE
CHAMBERS
RAILROAD
TRACKS
MAIN
TUNNEL

Fig. 4

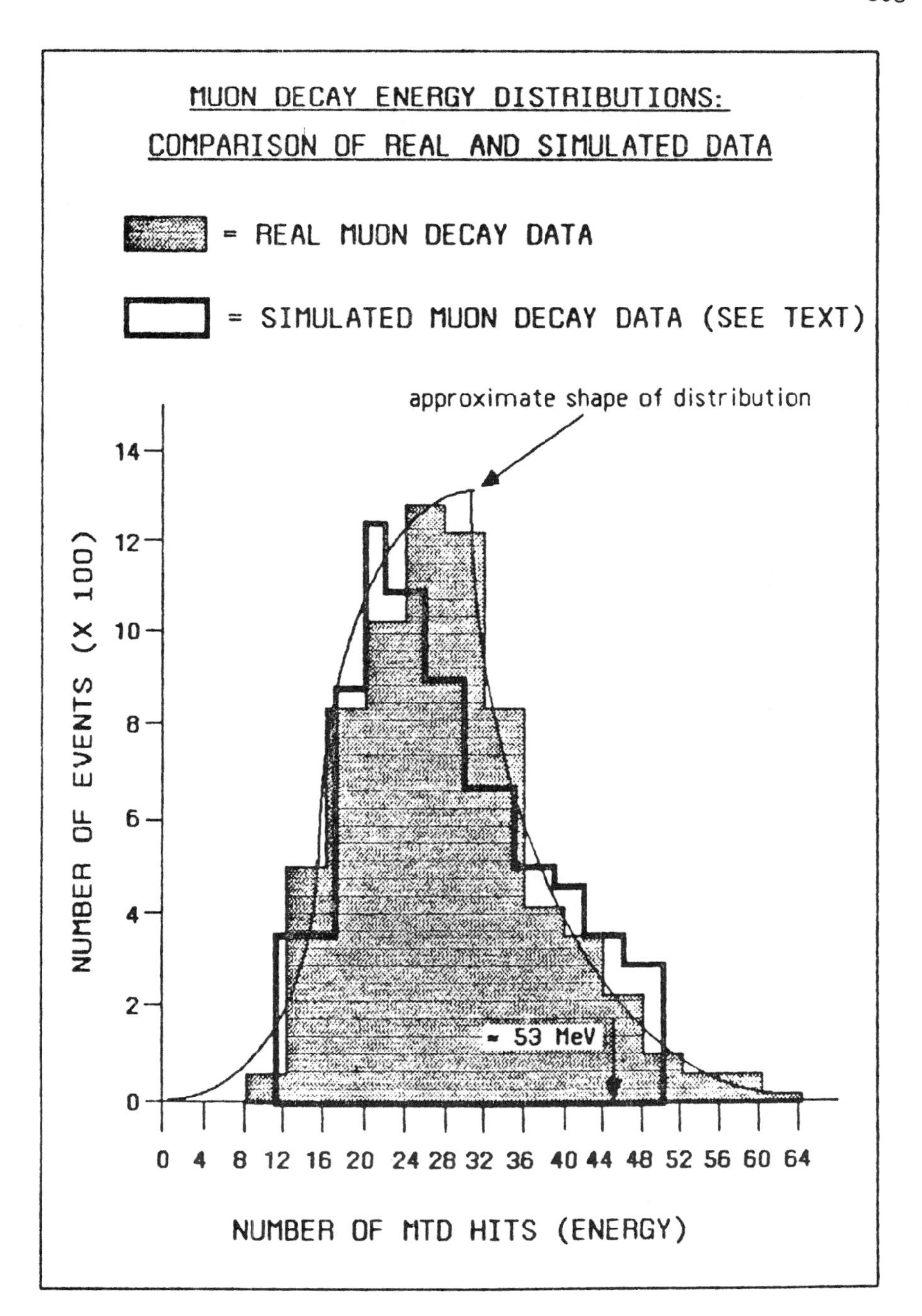
MUON DECAY ENERGY DISTRIBUTIONS:
COMPARISON OF REAL AND SIMULATED DATA
= REAL MUON DECAY DATA
= SIMULATED MUON DECAY DATA (SEE TEXT)
approximate shape of distribution
≈ 53 MeV
NUMBER OF EVENTS (X 100)
0
2
4
6
8
10
12
14
0 4 8 12 16 20 24 28 32 36 40 44 48 52 56 60 64
NUMBER OF MTD HITS (ENERGY)

Fig. 5

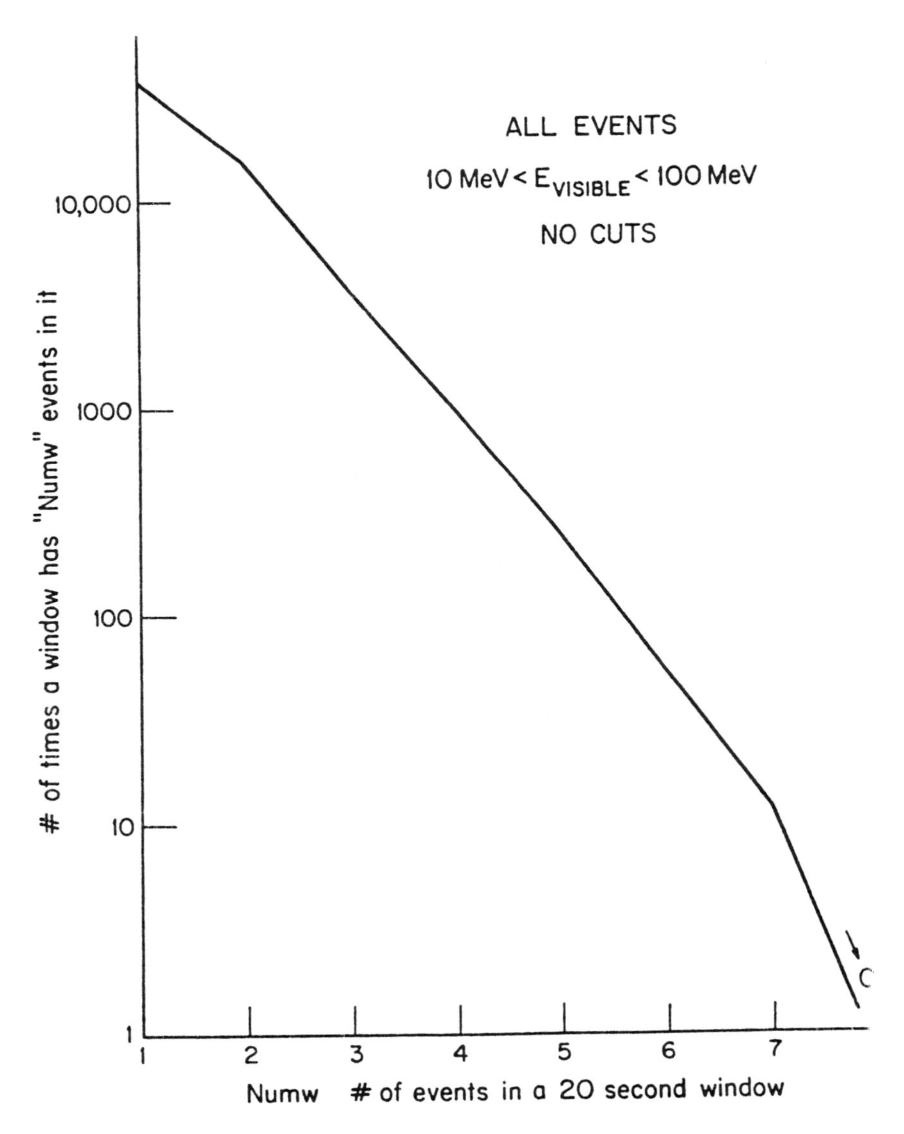

ALL EVENTS
10 MeV < $E_{VISIBLE}$ < 100 MeV
NO CUTS
10,000
1000
100
10
1
# of times a window has "Numw" events in it
1
2
3
4
5
6
7
Numw # of events in a 20 second window

Fig. 6

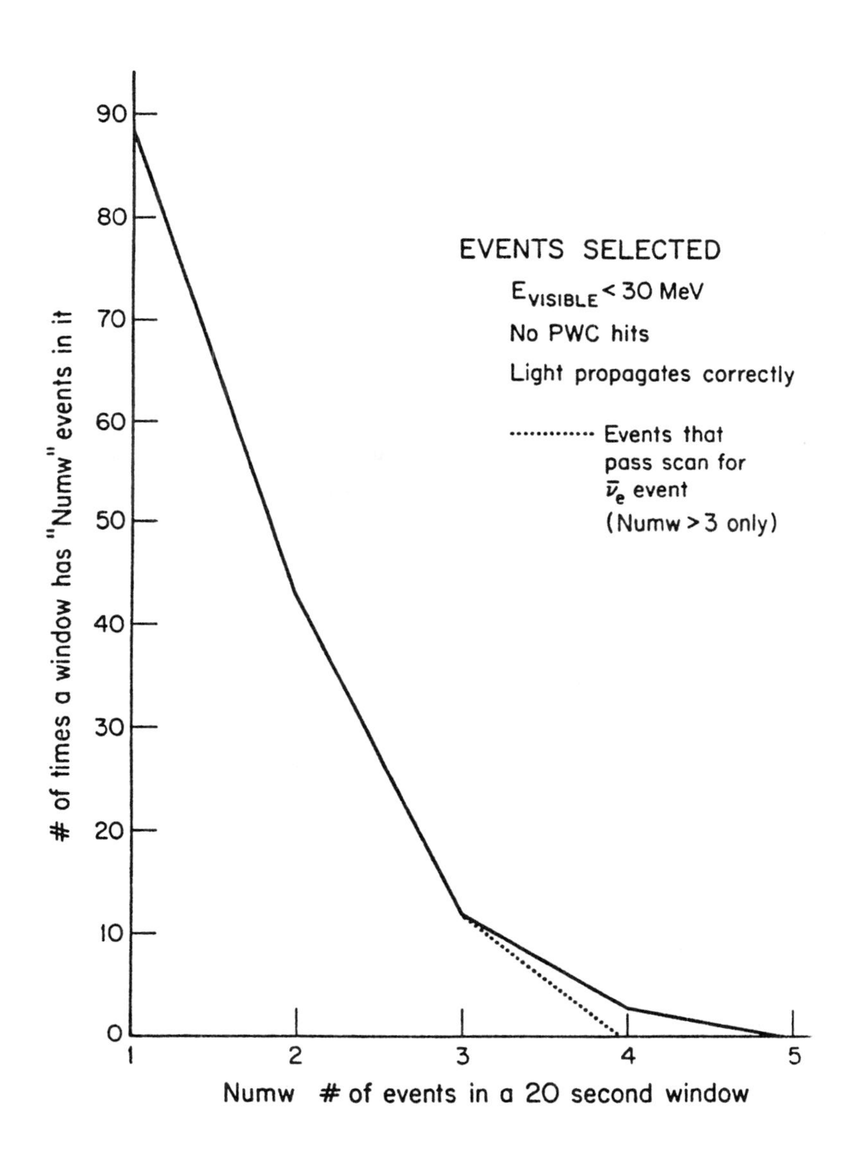

90
80
70
60
50
40
30
20
10
0
1
2
3
4
5
# of times a window has "Numw" events in it
Numw # of events in a 20 second window
EVENTS SELECTED
$E_{VISIBLE} < 30$ MeV
No PWC hits
Light propagates correctly
Events that pass scan for $\bar{\nu}_e$ event (Numw > 3 only)

Fig. 7

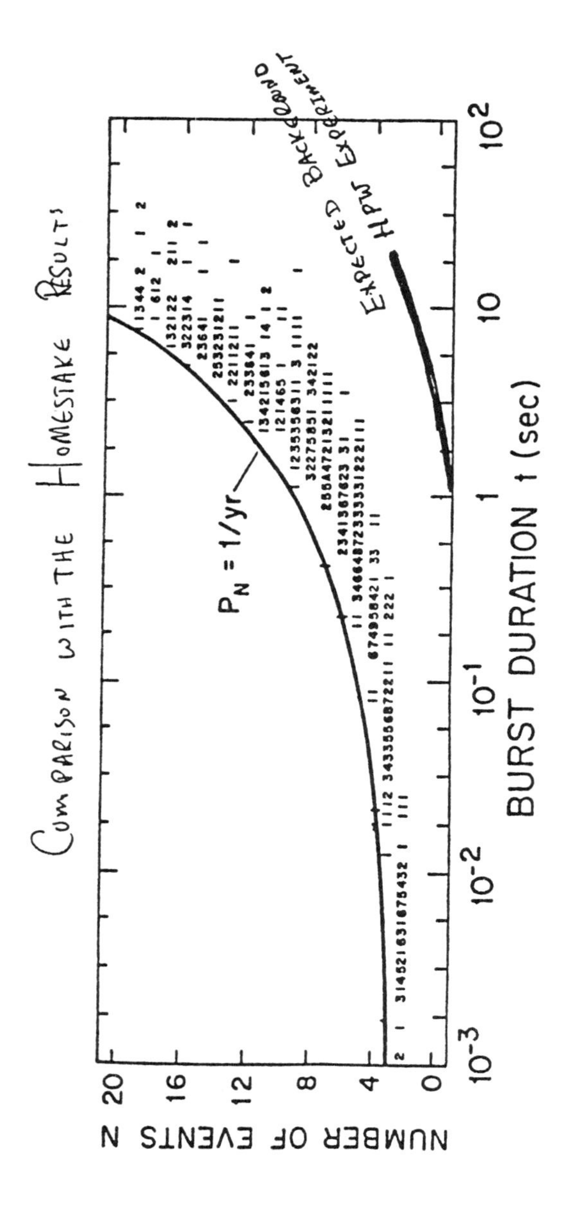

COMPARISON WITH THE HOMESTAKE RESULTS
NUMBER OF EVENTS N
BURST DURATION t (sec)
P_N = 1/yr
EXPECTED BACKGROUND
HPW EXPERIMENT

Fig. 8

LIMITS ON THE $>10^{17}$ eV COSMIC RAY ν_e FLUX

R.M. Baltrusaitis, G.L. Cassiday, J.W. Elbert, P.R. Gerhardy,
E.C. Loh., Y. Mizumoto, P. Sokolsky, and D. Steck.
Physics Department, University of Utah, Salt Lake City, UT 84112

ABSTRACT

We report on a search for upward-going extensive aire showers using the University of Utah Fly's Eye detector. No events have been found in 6×10^6s of running time. The resultant ν_e flux limit for $\sigma_\nu = 10^{-33}cm^2$ varies from 7.2×10^{-14} to 5.0×10^{-17} ν/cm^2-sec-ster for E_ν between 10^{18} and 10^{21} eV. We also present limits for larger σ_ν using near-horizontal events originating in the atmosphere.

I. INTRODUCTION

We have recently reported limits on astrophysical neutrino fluxes at energies $>10^{19}$ eV[1]. Here we update these flux limits, extend them to lower energies, and describe the calculations in more detail.

The most intense expected source of $>10^{17}$ eV neutrinos comes from the interaction of the primary cosmic ray flux with the 2.7° black body radiation. At proton energies of $>10^{19}$ eV, the reaction $\gamma + p \rightarrow N^*(1238) \rightarrow N\pi \rightarrow N\mu\nu_\mu \rightarrow Ne\nu_e\nu_\mu\nu_\mu$ is above threshold for a significant fraction of 2.7°K photons. The onset of such a threshold imples a reduction in the mean-free-path for protons of energy $>10^{19}$ eV and leads to the well-known prediction of a cut-off in the cosmic ray flux at 5×10^{19} eV[2]. A direct consequence is the existence of a flux of ν_μ and ν_e's at $\geq 10^{17}$ eV in a ratio of 2 to 1. The contribution of neutrinos from atmospheric EAS and other sources at these energies is expected to be many orders of magnitude below the contribution from this source. There have been several calculations of this effect with respect to both the primary spectrum cut-off and the consequent neutrino flux[3]. The flux expectations from various authors range from 10^{-17} to $10^{-18}\nu/cm^2$-sec-ster at $E_\nu = 10^{19}$ eV>.

The theoretical assumptions leading to this flux include:

a. The universality and black-body spectral shape of the 2.7°K radiation.
b. The universality, high energy shape and extent of source distribution and evolution of the primary cosmic ray spectrum.
c. The 300 MeV/c photoproduction cross-section and π and μ decay kinematics.

Since c. is well know, observation or non-observation of such neutrinos test issues a and b.

Contributions to the UHE neutrino flux from point sources is possible but completely speculative. Weiler[4] has pointed out that if a substantial flux exists above 10^{21} eV, the reaction $\nu\nu_r \rightarrow Z°$ can be used to search

0094-243X/85/1260307-14 $3.00

for the remnant neutrino flux ν_r. Observation of the neutrino flux from a highly red-shifted source would show an absorption dip at an energy $>10^{20}$ eV which depends on the neutrino mass. This appears to be the only hope of directly measuring the relic neutrino background. We search for the neutrino flux at energies $>10^{20}$ eV to examine the feasibility of this proposal as well as for the intrinsic interest in observing such high energy sources.

II. THE DETECTOR

The Fly's Eye detector[5] is an array of 67, 1.5m diameter mirrors each with twelve or fourteen phototubes located at the focal plane. Extensive air showers (EAS) with $E > 10^{17}$ eV passing through the atmosphere near the detector generate sufficient Nitrogen scintillation light to allow imaging of the shower by the phototubes. Phototubes whose direction vectors intercept the EAS axis receive scintillation light. A combination of tube hit geometry and timing of the relative delay between hit tubes allows the complete geometrical reconstruction of the shower. The variables chosen to describe the geometry are θ, ϕ and R_p, the zenith angle, azimuthal angle and impact parameter, respectively. A typical shower geometry is shown in Fig. 1. The reconstruction accuracy depends on the total track length projected on the celestial sphere. For tracks with track length >50° the errors in θ and R_p are typically $\Delta\theta=\pm2°$ and $\Delta R_p/R_p \sim .1$. The reconstruction algorithm and error estimation had been checked by examining a subset of the data where the EAS is visible in a second, smaller Fly's Eye composed of eight mirrors. The stereo reconstruction available for such events give additional constraints and confirms the adequacy of monocular reconstruction and error estimation.

The total energy of the incoming shower is determined from the analysis of the pulse height of hit tubes in a calorimetric way. Since the total scintillation light observed by any tube is proportional to the total number of ionizing particles traversing the field of view of the tube, the pulse height distribution can be converted to a size distribution. The total energy of the shower is then derived from the size of the shower at maximum or by integrating the size curve. Fig. 2 shows a representative shower profile.

Since detection efficiency improves with increasing light output, very energetic EAS are visible over a larger fiducial volume. Fig. 3 shows the distribution of EAS energy versus R_p for all tracks detected. The maximum detectable R_p increases as a function of energy. At $E=10^{19}$ eV showers with R_p = 20 km are detectable while for 10^{18} eV, the maximum R_p is 5 km. Note that the R_p cutoff is due both to the $\Delta\theta$ cut and to atmospheric attenuation length, which is 15 km at nitrogen scintillation wavelengths. The approximate fiducial volume in which EAS are detected with good efficiency is then a cylinder of radius $R=R_p^{max}$ and height of 15 km, centered on the Fly's Eye.

A byproduct of the determination of shower profile is measurement of X_m, the depth of shower maximum in the atmosphere in gm/cm^2, and Xo, the depth of first shower observation. This measurement allows us to estimate how many interaction lengths of atmosphere were traversed by the initial particle before interacting. The distribution of Xo is shown in Fig. 4. Note that the Xo distribution extends beyond the expected distribution of the actual point of first interaction since Xo is always an upper limit on the actual interaction depth.

III. NEUTRINO SIGNATURES

In the standard model for the reaction $\nu N \rightarrow$ Lepton + X, when $E_\nu >> M_W c^2$, propagator effects distort the y distribution and $\langle y \rangle \rightarrow o$. Hence, essentially all the energy in the ν interaction is expected to be transferred to the final state lepton. We assume this to be the case at our energies. If we consider ν_e charged current interactions only, the expected neutrino signature will thus be an EAS produced by a $>10^{17}$ eV electron. However, electron and hadron showers of these energies are essentially indistinguishable by their profile in the atmosphere. The response and efficiency of the detector to such neutrino interactions is thus essentially identical to its response to hadronic cosmic rays. Hence our understanding of detector acceptance and efficiency based on our study of the hadronic cosmic ray spectrum can be applied here.

Table I gives the flux limits based on no observed downward events with interaction point deeper than 2500gm/cm^2. Note that the limits improve with increasing E_ν because the fiducial volume increases. Limits approaching flux levels calculated by Hill and Schramm[3] can only be achieved if σ_ν is orders of magnitude larger than predicted by the standard model.

We can also set flux times cross-section limits on hypothetical, weakly interacting components of the primary cosmic ray flux, such as photinos. The only requirements on such particles are that $\sigma_{int} < 10^{-29}$ cm^2 and that the final state particles in the interaction carrying most of the energy be electrons, photons and/or hadrons. Our data then implies a flux times cross-section for such particles in units of (sec-ster)$^{-1}$ of 1×10^{-45} at 10^{17} eV to 3.9×10^{-47} at 10^{20} eV.

IV. LIMIT BASED ON UPWARD EVENTS.

Since particles producing upward events ($\theta_Z > 90°$) must travel thru large numbers of interaction lengths of earth, there is no background from hadronic sources. Charged leptons will also be severely attenuated due to radiative energy losses, so that observation of upward events can be uniquely interpreted as observation of the UHE neutrino flux. The sensitivity to the neutrino flux is determined by two factors: a. the attenuation of neutrinos by the Earth as a function of zenith angle, and b. the depth into the Earth that a neutrino interaction can occur and still produce an atmospheric EAS sufficiently energetic to trigger the detector.

The attenuation length of the Earth is a function of the zenith angle since the average density of matter traversed by the neutrino depends on the depth of Earth's material sampled. We assume a density distribution for the Earth based on the Dziewenski-Anderson preliminary reference Earth model[6]. The resultant attenuation as a function of θ_z is shown in Fig. 6 for $\sigma_\nu = 10^{-33}$ cm^2. It is clear that the resultant angular distribution of events will peak near the horizontal direction.

Since we limit ourselves to the detection of the ν_e flux and most of the final state energy is carried by the electron, the visible depth into the Earth is determined by the radiation length of earth.

For electron energies $>10^{18}$ eV, the Landau-Pomeranchuk-Migdal[7] effect becomes important and the pair-production and bremsstrahlung cross-section are suppressed relative to Bethe-Heitler in dense materials The effect is much less pronounced in the atmosphere[8]. We have calculated, in a Monte Carlo program, the shower profiles of electrons produced in the earth's crust ($\rho_{crust} = 2.6$ gm/cm^3) and entering the atmosphere ($\rho_{atm} \sim 10^{-3}$ gm/cm^3). The net result is an elongation of the shower development while the shower is in the crust and a subsequent speed-up in development in the atmosphere (see Fig. 7). Since the Fly's Eye detection efficiency depends on the size of the atmospheric shower at maximum, we can estimate the detector's response to showers produced at different depths. Fig. 8 shows the shower size at maximum as a function of E_ν and depth into the crust. Showers are detected with good efficiency to depths of 40m for $E_\nu = 10^{18}$, 100m for $E_\nu = 10^{19}$, 300m for $E_\nu = 10^{20}$ and 1200 m for $E_\nu = 10^{21}$ eV.

We have observed no upward event candidates (see Fig. 5). We quote flux limits (see Table II) for upward events as a function of E_ν and σ_ν in the range 10^{-33} to 10^{-32} cm^2. The standard model with $M_W = 80$ GeV/c^2 predicts $\sigma_\nu \sim 10^{-33}$ cm^2. However, QCD effects may make this somewhat larger[15].

IV CONCLUSIONS

The present Fly's Eye has sensitivity to upward going neutrino events approaching some recent predictions of the flux from cosmic ray interactions with the 2.7°K black body radiation. Although these limits are not yet very restrictive, they are the first limits for such processes at these energies. We expect, as the sensitivity of the Fly's Eye improves and running time increases, that these limits will improve by an order of magnitude.

The limit on the neutrino flux above energies of 10^{21} eV of $<5\times10^{-17}$/cm^2-sec-ster makes the search for relic neutrinos by the Weiler method extremely difficult, since a sizable number of events would have to be collected to search for an absorption dip at some energy.

V ACKNOWLEDGMENTS

We gratefully acknowledge the United States National Science Foundation for providing the funds for the research.

XI REFERENCES

[1]R. M. Baltrusaitis et al., Astro. J. 281, L9 (1984).

[2]K. Greisen, Phys. Rev. Lett. 16, 748 (1966); V. A. Kuzmin, and G. T. Zatsepin, Zh. Eksp. Teor. Fiz. 4, 778 (1966).

[3]F. W. Stecker, Phys. Rev. Lett. 21, 101 (1968); V. S. Berezinsky and G. T. Zatsepin, Soviet J. Nucl. Phys. 11, 111 (1970); S. H. Margolis, D. N. Schramm and R. Silverberg, Ap. J. 221 990 (1978); F. W. Stecker, NASA Tech. Mem, 79609, 1979 (unpublished); C. T. Hill and D. N. Schramm, Phys. Lett. 131B, 247(1983); C. T. Hill and D. N. Schramm, Phys. Rev. D. (to be published).

[4]T. Weiler, Phys. Rev. Lett. 49, 234 (1982).

[5]R. M. Baltrusaitis et al., Nuclear Inst. and Meth. (to be published) (1984); R. Cady et al., in Proceedings of the 18th International Cosmic Ray Conference, edited by P. V. Ramana Murthy (Tata Institute of Fundamental Research, Bombay, India,1983) Vol. 9, p. 351.

[6]T. L. Wilson, Nature 309, 38 (1984); A. M. Dziewonski and D. L. Anderson, Phys. Earth Planet. Inter. 25, 297 (1981).

[7]L. D. Landau and I. Ya. Pomeranchuk, Dokl. Akad. Nauk SSSR 92, 535 (1953); A. B. Migdal, Soviet Phys. - JETP 5, 527 (1957).

[8]E. Konishi, A. Misaki and N. Fujimaki, N. C. 44A, 509 (1978).

[9]Yu. Andreyev, V. S. Berezinsky and A. Yu. Smirnov, Phys. Lett. 84B, 24 (1979).

TABLE I

Limits of ν Flux based on downward events (ν/cm^2-sec-ster)

$\sigma_\nu (cm^2)$ / $E_\nu (eV)$	10^{17}	10^{18}	10^{19}	10^{20}
1×10^{-31}	1.0×10^{-14}	3.8×10^{-15}	1.0×10^{-16}	3.8×10^{-16}
1×10^{-30}	1.0×10^{-15}	3.8×10^{-16}	1.0×10^{-16}	3.8×10^{-17}
1×10^{-29}	1.0×10^{-16}	3.8×10^{-17}	1.0×10^{-17}	3.8×10^{-18}

TABLE II

Limits on ν Flux based on upward events (ν/cm^2-sec-ster)

$\sigma_\nu (cm^2)$ / $E_\nu (eV)$	10^{18}	10^{19}	10^{20}	10^{21}
10^{-33}	7.2×10^{-14}	9×10^{-15}	3.8×10^{-16}	5.0×10^{-17}
3×10^{-33}	1.4×10^{-13}	7.8×10^{-15}	7.2×10^{-16}	1.1×10^{-16}
5×10^{-33}	4.1×10^{-13}	2.3×10^{-14}	2.2×10^{-15}	3.3×10^{-16}
1×10^{-32}	3.7×10^{-12}	2.1×10^{-13}	2.0×10^{-14}	3.0×10^{-15}

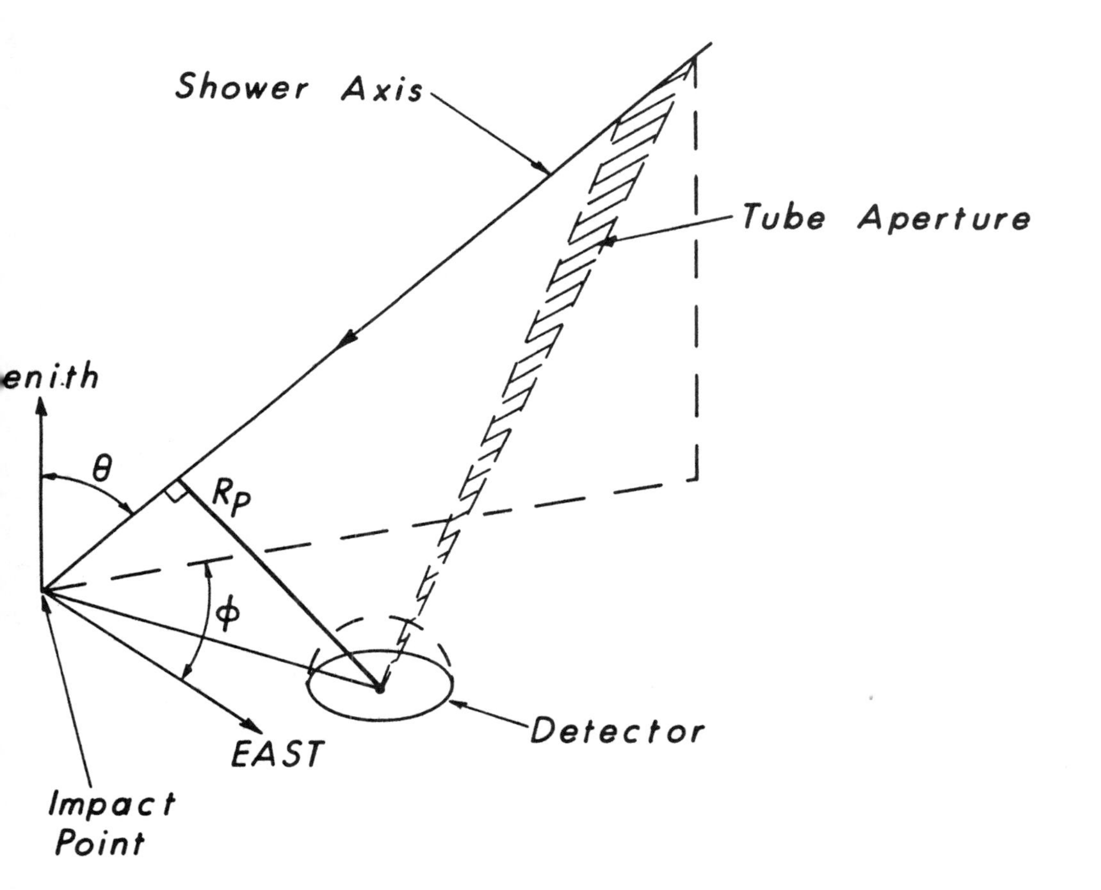

FIG. 1 Typical shower geometry

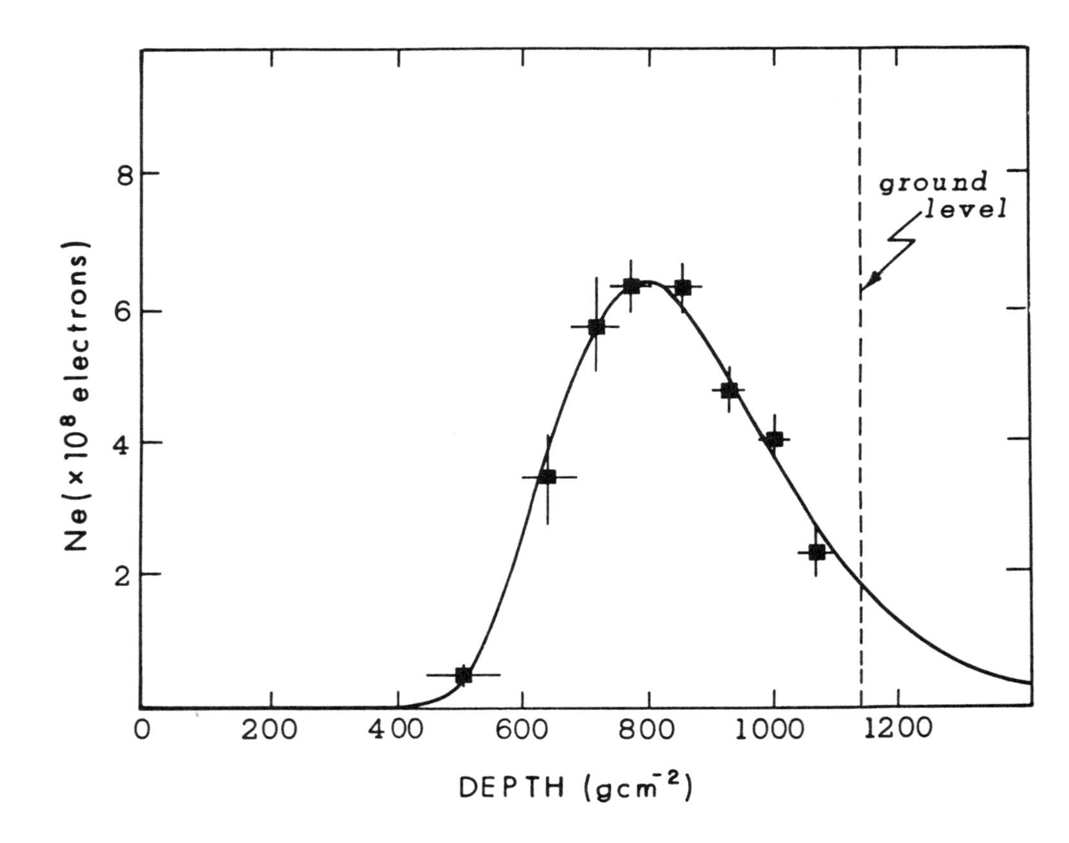

FIG. 2. Representative shower profile. The solid curve represents a fit using a Gaiser-Hillas [T. K. Gaisser and A. M. Hillas, in Proceedings of the 15th International Cosmic Ray Conference (Bulgarian Academy of Sciences, Plovdiv, Bulgaria, 1977)] parametrization for EAS development.

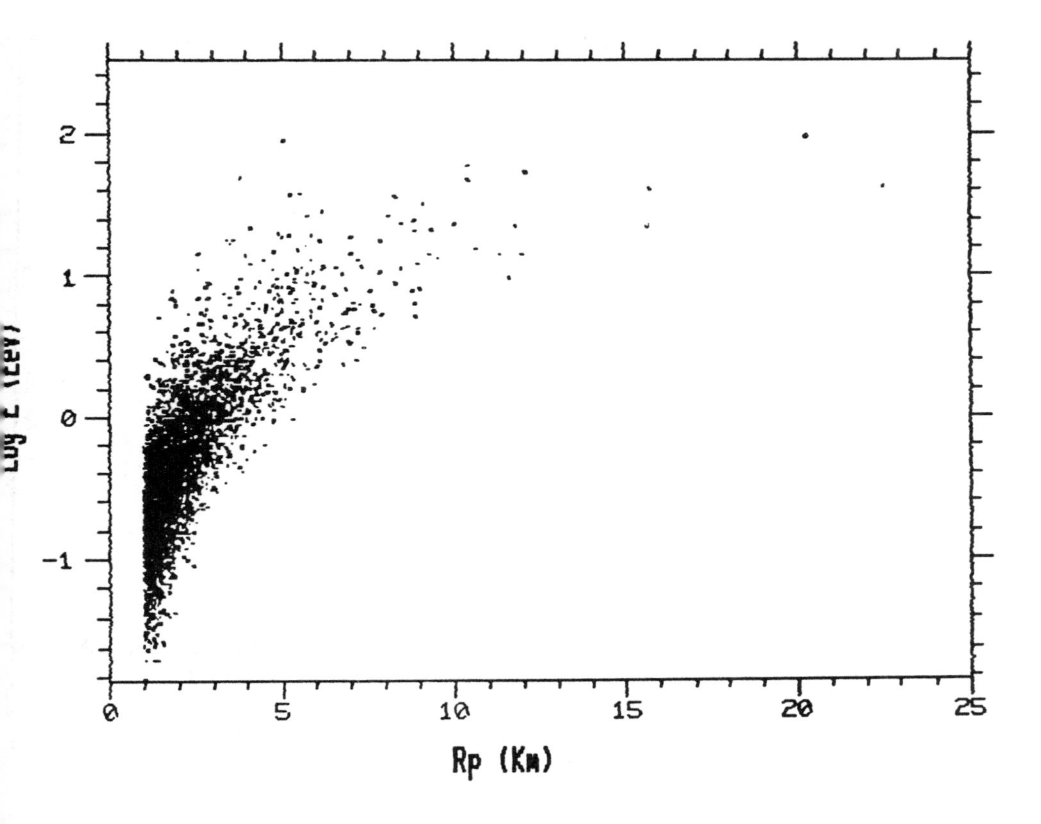

FIG. 3. Distribution of EAS energy versue R_p. One EeV is 10^{18} eV.

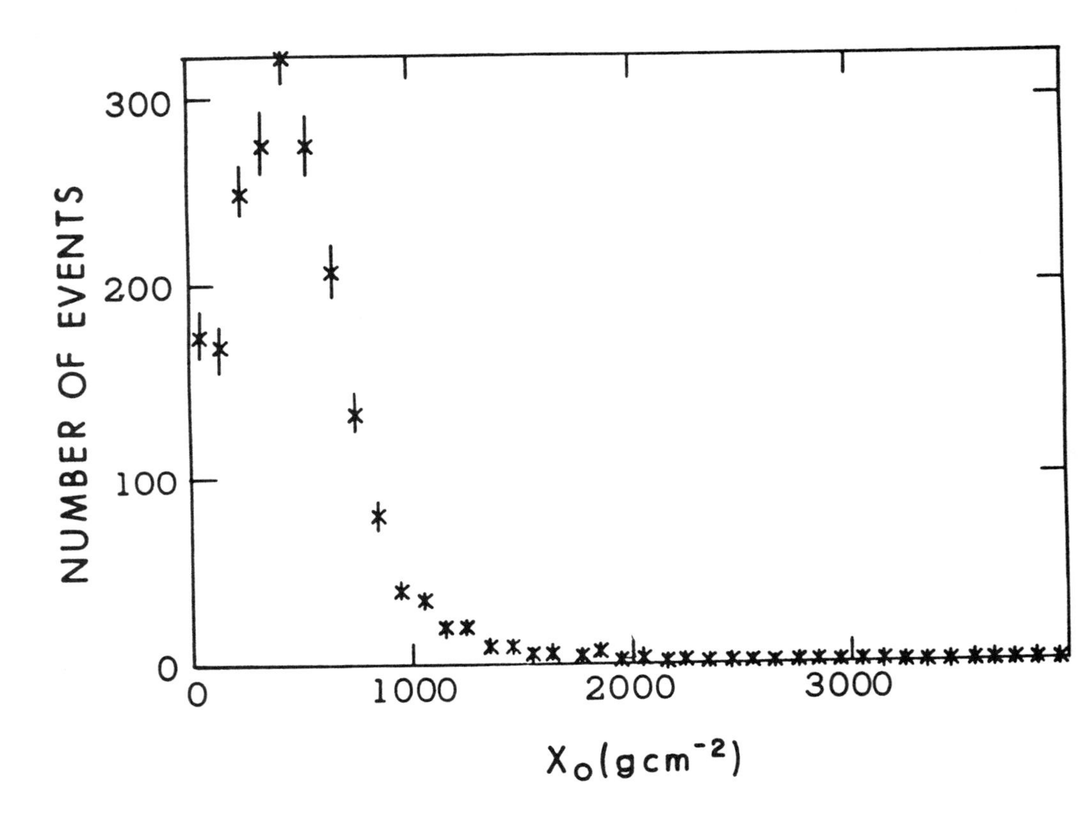

FIG. 4. Distribution of X_0, the depth of first observed interaction of EAS.

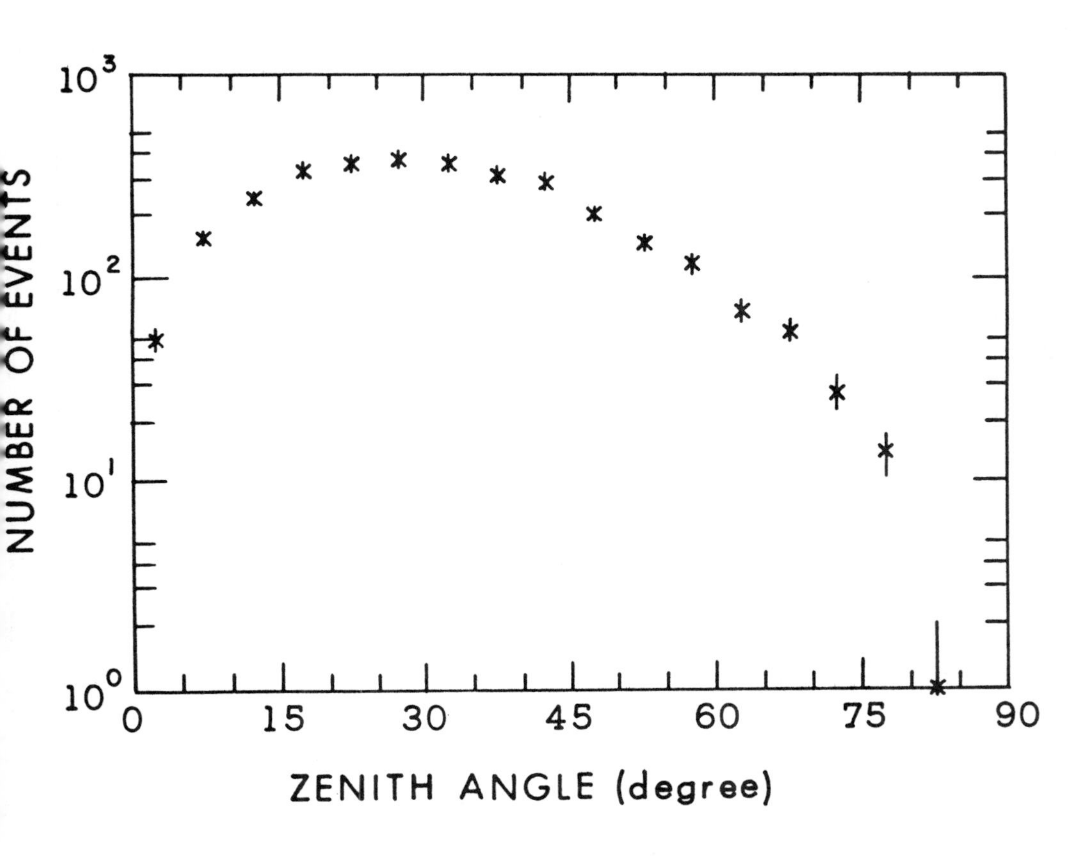

FIG. 5. Zenith angle distribution of events.

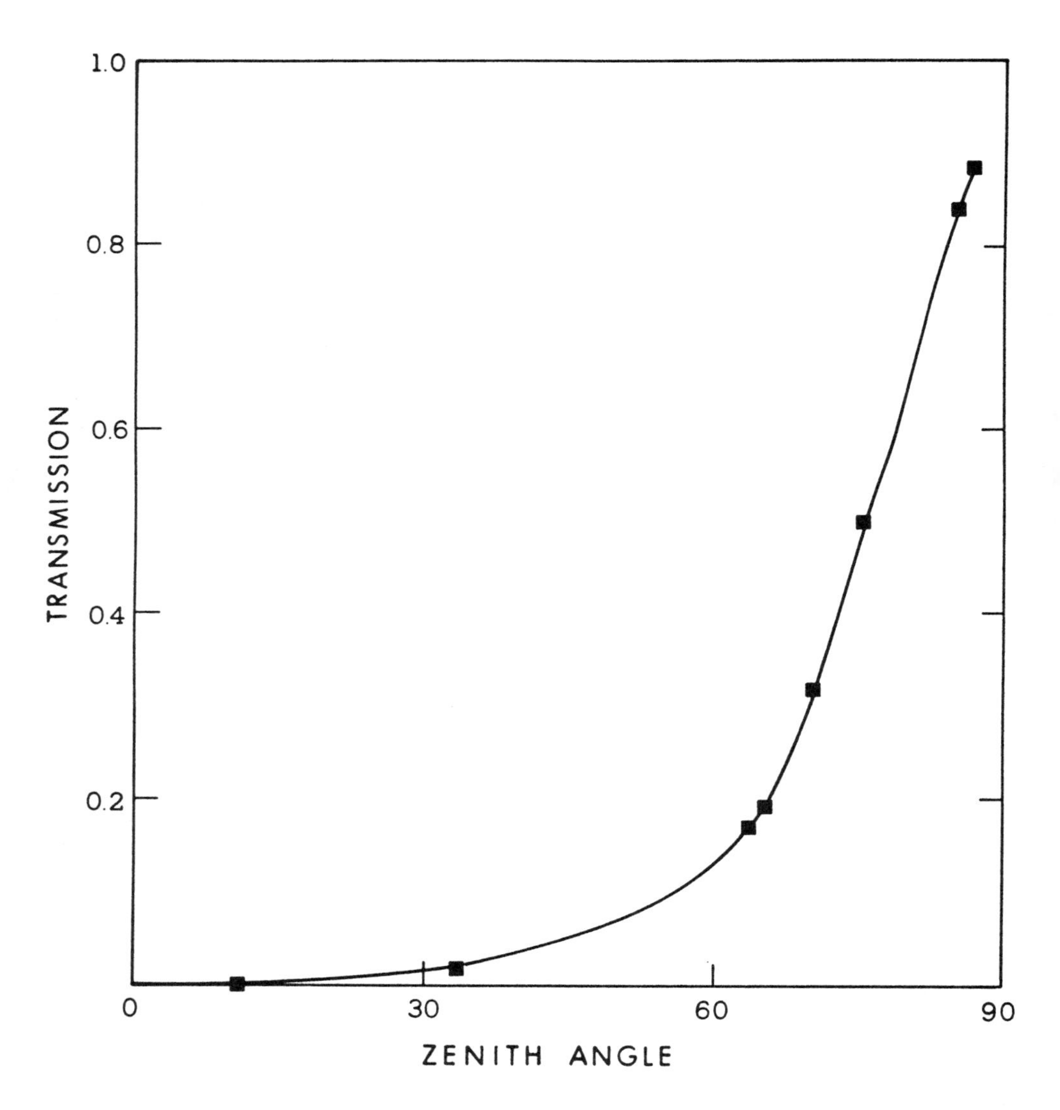

FIG. 6. Transmission of neutrino flux thru the Earth as a function of zenith angle for $\sigma_\nu = 10^{-33}\text{cm}^2$.

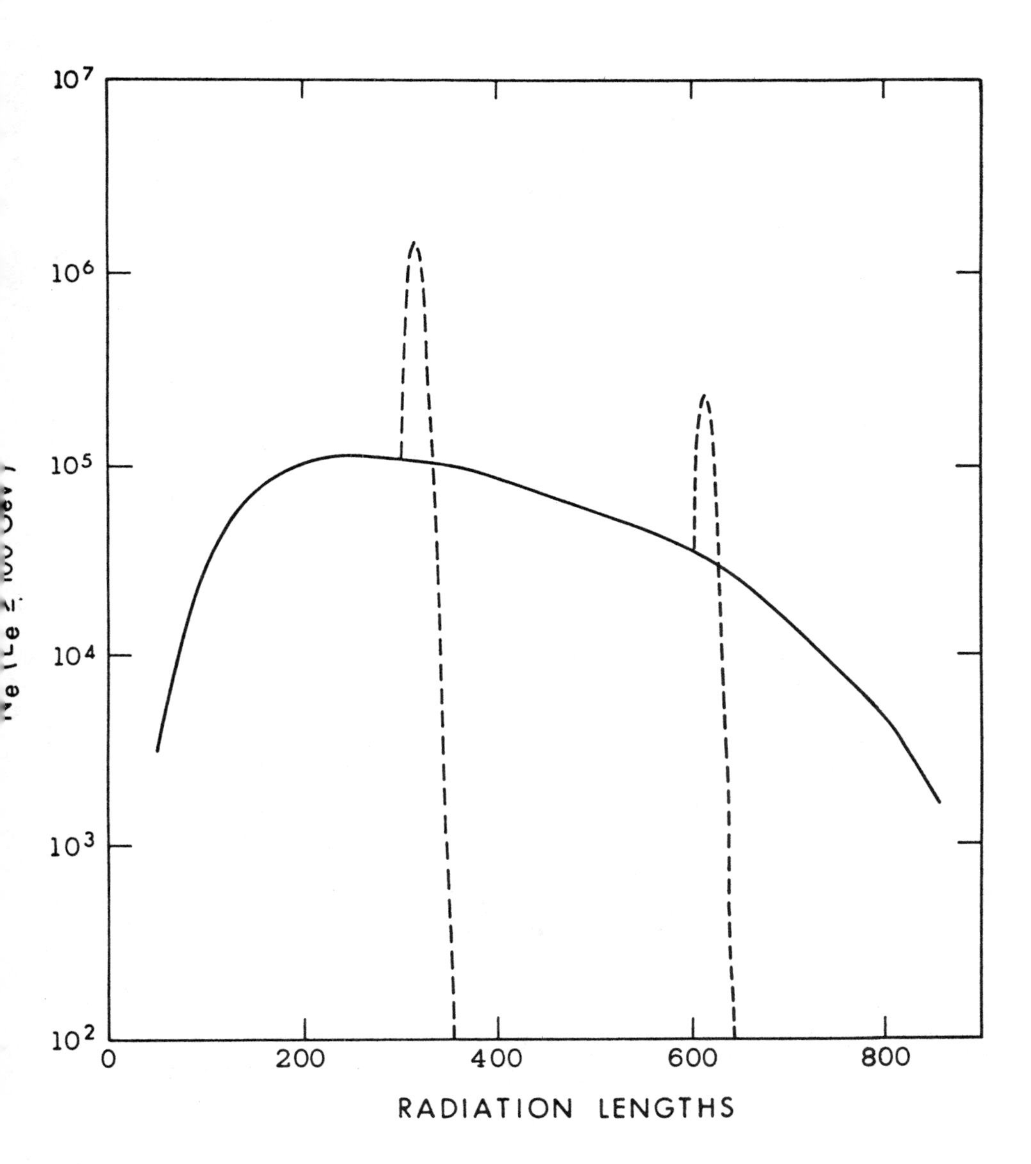

FIG. 7. Typical 10^{18} eV crust-air shower development. The solid line represents the shower profile in the Earth's crust. The dashed lines indicates the profile in the atmosphere for shower originating 300 and 600 rad. lengths into the crust. Electrons in the shower are followed to Ee > 100 GeV.

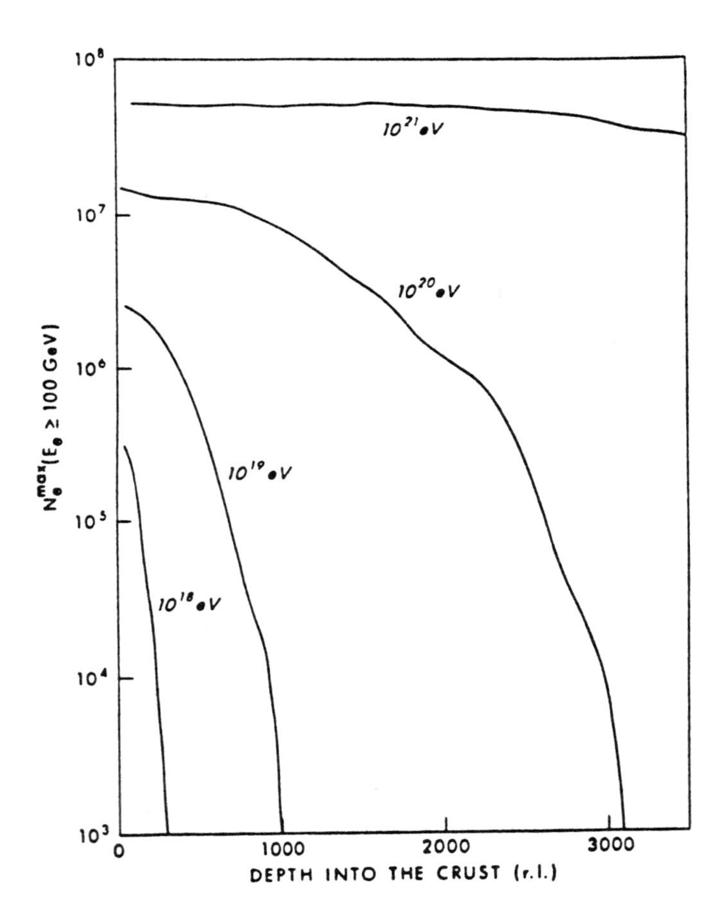

FIG. 8. Calculated shower size at shower maximum as a function of electron energy and depth of interaction in the crust.

AUTHOR INDEX

AIP Conference Proceedings

No. 29	Magnetism and Magnetic Materials - 1975 (21st Annual Conference, Philadelphia)	76-10931	0-88318-128-2
No. 30	Particle Searches and Discoveries - 1976 (Vanderbilt Conference)	76-19949	0-88318-129-0
No. 31	Structure and Excitations of Amorphous Solids (Williamsburg, VA., 1976)	76-22279	0-88318-130-4
No. 32	Materials Technology - 1976 (APS New York Meeting)	76-27967	0-88318-131-2
No. 33	Meson-Nuclear Physics - 1976 (Carnegie-Mellon Conference)	76-26811	0-88318-132-0
No. 34	Magnetism and Magnetic Materials - 1976 (Joint MMM-Intermag Conference, Pittsburgh)	76-47106	0-88318-133-9
No. 35	High Energy Physics with Polarized Beams and Targets (Argonne, 1976)	76-50181	0-88318-134-7
No. 36	Momentum Wave Functions - 1976 (Indiana University)	77-82145	0-88318-135-5
No. 37	Weak Interaction Physics - 1977 (Indiana University)	77-83344	0-88318-136-3
No. 38	Workshop on New Directions in Mossbauer Spectroscopy (Argonne, 1977)	77-90635	0-88318-137-1
No. 39	Physics Careers, Employment and Education (Penn State, 1977)	77-94053	0-88318-138-X
No. 40	Electrical Transport and Optical Properties of Inhomogeneous Media (Ohio State University, 1977)	78-54319	0-88318-139-8
No. 41	Nucleon-Nucleon Interactions - 1977 (Vancouver)	78-54249	0-88318-140-1
No. 42	Higher Energy Polarized Proton Beams (Ann Arbor, 1977)	78-55682	0-88318-141-X
No. 43	Particles and Fields - 1977 (APS/DPF, Argonne)	78-55683	0-88318-142-8
No. 44	Future Trends in Superconductive Electronics (Charlottesville, 1978)	77-9240	0-88318-143-6
No. 45	New Results in High Energy Physics - 1978 (Vanderbilt Conference)	78-67196	0-88318-144-4
No. 46	Topics in Nonlinear Dynamics (La Jolla Institute)	78-057870	0-88318-145-2
No. 47	Clustering Aspects of Nuclear Structure and Nuclear Reactions (Winnepeg, 1978)	78-64942	0-88318-146-0
No. 48	Current Trends in the Theory of Fields (Tallahassee, 1978)	78-72948	0-88318-147-9
No. 49	Cosmic Rays and Particle Physics - 1978 (Bartol Conference)	79-50489	0-88318-148-7
No. 50	Laser-Solid Interactions and Laser Processing - 1978 (Boston)	79-51564	0-88318-149-5
No. 51	High Energy Physics with Polarized Beams and Polarized Targets (Argonne, 1978)	79-64565	0-88318-150-9
No. 52	Long-Distance Neutrino Detection - 1978 (C.L. Cowan Memorial Symposium)	79-52078	0-88318-151-7
No. 53	Modulated Structures - 1979 (Kailua Kona, Hawaii)	79-53846	0-88318-152-5

AIP Conference Proceedings

No. 54	Meson-Nuclear Physics - 1979 (Houston)	79-53978	0-88318-153-3
No. 55	Quantum Chromodynamics (La Jolla, 1978)	79-54969	0-88318-154-1
No. 56	Particle Acceleration Mechanisms in Astrophysics (La Jolla, 1979)	79-55844	0-88318-155-X
No. 57	Nonlinear Dynamics and the Beam-Beam Interaction (Brookhaven, 1979)	79-57341	0-88318-156-8
No. 58	Inhomogeneous Superconductors - 1979 (Berkeley Springs, W.V.)	79-57620	0-88318-157-6
No. 59	Particles and Fields - 1979 (APS/DPF Montreal)	80-66631	0-88318-158-4
No. 60	History of the ZGS (Argonne, 1979)	80-67694	0-88318-159-2
No. 61	Aspects of the Kinetics and Dynamics of Surface Reactions (La Jolla Institute, 1979)	80-68004	0-88318-160-6
No. 62	High Energy e^+e^- Interactions (Vanderbilt, 1980)	80-53377	0-88318-161-4
No. 63	Supernovae Spectra (La Jolla, 1980)	80-70019	0-88318-162-2
No. 64	Laboratory EXAFS Facilities - 1980 (Univ. of Washington)	80-70579	0-88318-163-0
No. 65	Optics in Four Dimensions - 1980 (ICO, Ensenada)	80-70771	0-88318-164-9
No. 66	Physics in the Automotive Industry - 1980 (APS/AAPT Topical Conference)	80-70987	0-88318-165-7
No. 67	Experimental Meson Spectroscopy - 1980 (Sixth International Conference, Brookhaven)	80-71123	0-88318-166-5
No. 68	High Energy Physics - 1980 (XX International Conference, Madison)	81-65032	0-88318-167-3
No. 69	Polarization Phenomena in Nuclear Physics - 1980 (Fifth International Symposium, Santa Fe)	81-65107	0-88318-168-1
No. 70	Chemistry and Physics of Coal Utilization - 1980 (APS, Morgantown)	81-65106	0-88318-169-X
No. 71	Group Theory and its Applications in Physics - 1980 (Latin American School of Physics, Mexico City)	81-66132	0-88318-170-3
No. 72	Weak Interactions as a Probe of Unification (Virginia Polytechnic Institute - 1980)	81-67184	0-88318-171-1
No. 73	Tetrahedrally Bonded Amorphous Semiconductors (Carefree, Arizona, 1981)	81-67419	0-88318-172-X
No. 74	Perturbative Quantum Chromodynamics (Tallahassee, 1981)	81-70372	0-88318-173-8
No. 75	Low Energy X-ray Diagnostics - 1981 (Monterey)	81-69841	0-88318-174-6
No. 76	Nonlinear Properties of Internal Waves (La Jolla Institute, 1981)	81-71062	0-88318-175-4
No. 77	Gamma Ray Transients and Related Astrophysical Phenomena (La Jolla Institute, 1981)	81-71543	0-88318-176-2
No. 78	Shock Waves in Condensed Matter - 1981 (Menlo Park)	82-70014	0-88318-177-0
No. 79	Pion Production and Absorption in Nuclei - 1981 (Indiana University Cyclotron Facility)	82-70678	0-88318-178-9
No. 80	Polarized Proton Ion Sources (Ann Arbor, 1981)	82-71025	0-88318-179-7
No. 81	Particles and Fields - 1981: Testing the Standard Model (APS/DPF, Santa Cruz)	82-71156	0-88318-180-0

AIP Conference Proceedings

AIP Conference Proceedings

No.	Title	LC Number	ISBN
No. 106	Predictability of Fluid Motions (La Jolla Institute, 1983)	83-73641	0-88318-305-6
No. 107	Physics and Chemistry of Porous Media (Schlumberger-Doll Research, 1983)	83-73640	0-88318-306-4
No. 108	The Time Projection Chamber (TRIUMF, Vancouver, 1983)	83-83445	0-88318-307-2
No. 109	Random Walks and Their Applications in the Physical and Biological Sciences (NBS/La Jolla Institute, 1982)	84-70208	0-88318-308-0
No. 110	Hadron Substructure in Nuclear Physics (Indiana University, 1983)	84-70165	0-88318-309-9
No. 111	Production and Neutralization of Negative Ions and Beams (3rd Int'l Symposium, Brookhaven, 1983)	84-70379	0-88318-310-2
No. 112	Particles and Fields - 1983 (APS/DPF, Blacksburg, VA)	84-70378	0-88318-311-0
No. 113	Experimental Meson Spectroscopy - 1983 (Seventh International Conference, Brookhaven)	84-70910	0-88318-312-9
No. 114	Low Energy Tests of Conservation Laws in Particle Physics (Blacksburg, VA, 1983)	84-71157	0-88318-313-7
No. 115	High Energy Transients in Astrophysics (Santa Cruz, CA, 1983)	84-71205	0-88318-314-5
No. 116	Problems in Unification and Supergravity (La Jolla Institute, 1983)	84-71246	0-88318-315-3
No. 117	Polarized Proton Ion Sources (TRIUMF, Vancouver, 1983)	84-71235	0-88318-316-1
No. 118	Free Electron Generation of Extreme Ultraviolet Coherent Radiation (Brookhaven/OSA, 1983)	84-71539	0-88318-317-X
No. 119	Laser Techniques in the Extreme Ultraviolet (OSA, Boulder, Colorado, 1984)	84-72128	0-88318-318-8
No. 120	Optical Effects in Amorphous Semiconductors (Snowbird, Utah, 1984)	84-72419	0-88318-319-6
No. 121	High Energy e^+e^- Interactions (Vanderbilt, 1984)	84-72632	0-88318-320-X
No. 122	The Physics of VLSI (Xerox, Palo Alto, 1984)	84-72729	0-88318-321-8
No. 123	Intersections Between Particle and Nuclear Physics (Steamboat Springs, 1984)	84-72790	0-88318-322-6
No. 124	Neutron-Nucleus Collisions - A Probe of Nuclear Structure (Burr Oak State Park-1984)	84-73216	0-88318-323-4
No. 125	Capture Gamma-Ray Spectroscopy and Related Topics-1984 (Internat. Symposium, Knoxville)	84-73303	0-88318-324-2
No. 126	Solar Neutrinos and Neutrino Astronomy (Homestake, 1984)	84-63143	0-88318-325-0
NO. 127	Physics of High Energy Particle Accelerators (BNL/SUNY Summer School, 1983)		0-88318-326-9